W0111130

Studies in Energy, Resource and Environmental Economics

Series Editors

Georg Erdmann, Former Chair of Energy Systems
Technical University of Berlin
Oberrieden, Zürich, Switzerland

Anne Neumann, Norwegian University of Science and Tech
Trondheim, Norway

Andreas Loeschel, Ruhr-Universität Bochum
Bochum, Nordrhein-Westfalen, Germany

This book series offers an outlet for cutting-edge research on all areas of energy, environmental and resource economics. The series welcomes theoretically sound and empirically robust monographs and edited volumes, as well as textbooks and handbooks from various disciplines and approaches on topics such as energy and resource markets, the economics of climate change, environmental evaluation, policy issues, and related fields. All titles in the series are peer-reviewed.

Fateh Belaïd • Anvita Arora
Editors

Smart Cities

Social and Environmental Challenges
and Opportunities for Local Authorities

 Springer

Editors
Fateh Belaïd
King Abdullah Petroleum Studies
and Research Center
Riyadh, Saudi Arabia

Anvita Arora
King Abdullah Petroleum Studies
and Research Center
Riyadh, Saudi Arabia

ISSN 2731-3409   ISSN 2731-3417   (electronic)
Studies in Energy, Resource and Environmental Economics
ISBN 978-3-031-35663-6   ISBN 978-3-031-35664-3   (eBook)
https://doi.org/10.1007/978-3-031-35664-3

© King Abdullah Petroleum Studies and Research Center 2024, Corrected Publication 2024. This book
is an open access publication.
Open Access   This book is licensed under the terms of the Creative Commons Attribution 4.0
International License (http://creativecommons.org/licenses/by/4.0/), which permits use, sharing,
adaptation, distribution and reproduction in any medium or format, as long as you give appropriate credit
to the original author(s) and the source, provide a link to the Creative Commons license and indicate if
changes were made.
The images or other third party material in this book are included in the book's Creative Commons
license, unless indicated otherwise in a credit line to the material. If material is not included in the book's
Creative Commons license and your intended use is not permitted by statutory regulation or exceeds the
permitted use, you will need to obtain permission directly from the copyright holder.
The use of general descriptive names, registered names, trademarks, service marks, etc. in this publication
does not imply, even in the absence of a specific statement, that such names are exempt from the relevant
protective laws and regulations and therefore free for general use.
The publisher, the authors, and the editors are safe to assume that the advice and information in this book
are believed to be true and accurate at the date of publication. Neither the publisher nor the authors or the
editors give a warranty, expressed or implied, with respect to the material contained herein or for any
errors or omissions that may have been made. The publisher remains neutral with regard to jurisdictional
claims in published maps and institutional affiliations.

This Springer imprint is published by the registered company Springer Nature Switzerland AG
The registered company address is: Gewerbestrasse 11, 6330 Cham, Switzerland

Paper in this product is recyclable.

Foreword

Many articles and books have been devoted to the rise of the smart city. However, the collection of essays gathered in this volume arrives at a moment in the development of this set of ideals, projects, and realizations, which makes necessary a new assessment of where things stand. Indeed, the smart city is no longer a remote perspective made believable by a series of pioneering experiments. It has become a pervasive urban reality, a fully fledged domain of urban planning and policies in developing as well as in developed economies. A growing body of academic and professional literature reflects this spread. In such a context, the aim of this volume is twofold. First, it proposes an overview of some of the key directions in which the smart city is continuing to develop, from data management to transportation. Second, it intersects these trends with the specificity of a series of regions and cities, from Western Europe to North Africa and the Middle East.

To fully grasp the difference between the type of analysis developed by the authors of this volume, it is worth returning a moment to the origins of the smart city. Initially, the concept was born approximately in 2008 from the desire of large digital companies, CISCO and IBM in particular, to create new markets for themselves by proposing to cities to improve the management of their infrastructures through applications adapted to their needs. The use of the term "smart" to label cities but also territories operated with the help of software suites informed by sensors was introduced by IBM. The digital giants' initiative quickly met with interest from mayors faced with growing management difficulties. In France, cities such as Montpellier and Nice, for example, conducted pioneering experiments with CISCO and IBM.

The realization of the strategic nature of the massive amounts of data produced by cities quickly reinforced the craze for the smart city. With it came the arrival of other players, platforms offering services often based on algorithmic matching, such as Airbnb and Uber. Their rise to power was inextricably linked to the widespread use of information and communication technologies, particularly to the diffusion of smartphones.

By the mid-2010s, smart city dynamics were well underway, with recurring questions such as whether to give precedence to a top-down, neo-cybernetic approach to urban problems or, to the contrary, promote bottom-up initiatives. The control room, the digital equivalent of a panopticon, was emblematic of the first approach, while the second was epitomized by collaborative platforms such as OpenStreetMap. Although they are still looming in the background of many contemporary smart city policies, this type of issue has been superseded by more concrete questions, which the following essays illustrate.

Transportation ranks among the issues approached by the authors of this volume, from the creation of mobility hubs that bring together several modes of transport to the promotion of sustainable mobility. Housing policies and questions related to health are also discussed, along with the financial tools necessary to achieve smartness in cities. Compared to the debates and experiments that had marked the rise of the smart city, the importance of environmental concerns is especially noticeable. Climate change has fostered a "green turn" epitomized by a series of contributions. While the smart city movement was spreading, concerns for social justice have also been ramping up. Faced with the perspective of a planetary environmental crisis and the dramatic increase in social inequalities, smart technologies appear to be a possible way to mitigate their effects.

The authors of this volume also discuss in detail the new tools that enable the development of better smart cities, such as big data, digital twins, artificial intelligence, and machine learning, which will enable the exploration of scenarios of development much more efficiently than in the past. The maturity of the smart city movement is also epitomized by the development of financial instruments as well as by the multiplication of key performance indicators that enable us to compare the approaches and results achieved, a comparison all the more important that cities both compete and collaborate beyond national boundaries today. Indeed, in addition to nation-states, international networks and leagues of major cities represent a rising political force at the scale of the globe that must be reckoned with.

Some issues still need discussion. Among the interrogations that this volume raises, one finds the question of situated smartness. Should the path to the smart city be more forcefully diversified according to local circumstances and culture, or should it remain based on principles relatively indifferent to these local factors? Should one, for instance, imagine a specific Middle Eastern trajectory leading to the smart city different from the Northern European or North American ones? This could very well represent a new challenge awaiting us in the years to come. Another challenge regards the way smartness will lead to the redefinition of the relations between cities and regions. Are smart regions to be composed of loose aggregates of smart cities or should they develop specific approaches and tools with an integrating ambition? The rapid urbanization of the planet at a scale never seen before makes this question almost unavoidable. As an essential contribution to the study of the present state of smart technologies and cities, this book provides an excellent platform to start dealing with these new challenges.

Antoine Picon is the G. Ware Travelstead Professor of the History of Architecture and Technology at Harvard University – Graduate School of Design. He teaches courses in the history and theory of architecture and technology. Trained as an engineer, architect, and historian, Picon works on the history of architectural and urban technologies from the eighteenth century to the present. Picon is the author of numerous books and articles on the relationship between science and technology on the one hand and architecture and technology on the other hand. Picon's most recent books offer a comprehensive overview of the changes brought by the computer and digital culture to the theory and practice of architecture as well as to the planning and experience of the city. He has published, in particular, *Digital Culture in Architecture: An Introduction for the Design Profession* (2010), *Ornament: The Politics of Architecture and Subjectivity* (2013), *Smart Cities: Théorie et Critique d'un Idéal Autoréalisateur* (2013), and *Smart Cities: A Spatialised Intelligence* (2015). Picon has received a number of awards for his writings, including the Médaille de la Ville de Paris and twice the Prix du Livre d'Architecture de la Ville de Briey, as well as the Georges Sarton Medal of the University of Gand. In 2010, he was elected a member of the French Académie des Technologies. He has been Chevalier des Arts et Lettres since 2014. He is also Chairman of the Fondation Le Corbusier.

Antoine Picon
Harvard University – Graduate School of Design
Cambridge, MA, USA

Introduction

Cities are the origin of 75% of all carbon dioxide emissions from energy use. The urbanization of the world's population is accelerating steadily. The global population is projected to increase to approximately ten billion, and more than two-thirds of the world's population will live in urban areas by 2050. The construction of infrastructure for rapidly growing cities in emerging economies could generate 226 gigatons of CO_2 by 2050, which is over four times the amount generated by constructing existing infrastructure in developed economies.[1] Cities are particularly well positioned to lead in mitigating climate change impacts and fostering the transition to a more sustainable world.

The looming intensive urbanization will result in complex challenges, from environmental degradation to social inequalities, unsustainable energy use, and natural and human-made disasters. Mitigating these challenges and making the urban space more resilient will require multiple strategies, including recognizing the global sustainability context, understanding climate interactions, and combining technological transformation and behavioral change. By harnessing disruptive technologies, smart cities promise to address these challenges and make cities more sustainable, resilient, eco-friendly, and livable.

As many countries strive to go carbon neutral by 2050, cities are central to achieving the Paris agreement agenda. Smart cities are pivotal to supporting the dynamic growth of the population and easing the tension between economic development and sustainability.

Building on this conjecture, this volume aims to inform the policy-making process within the framework of the carbon neutrality objectives and the ongoing cities' transformation. It also aspires to contribute to the literature on the global governance of energy and ecological transition. It federates a large international research community and stakeholders, including players in energy policy, urban and construction sectors, cities, and policymakers.

---

[1] Bai, Xuemei, Richard J. Dawson, Diana Ürge-Vorsatz, Gian C. Delgado, Aliyu Salisu Barau, Shobhakar Dhakal, David Dodman et al. 2018. "Six research priorities for cities and climate change.": 23–25.

Although the concept of smart cities is widespread, research in this area is still in its infancy. The notion is still ambiguous, with limited conceptualizations and practical frameworks that could assist policymakers in realizing their smart city initiatives. As many organizations and policymakers are under constant pressure to collect, process, and disclose detailed and accurate information on the considerable challenges posed by increased energy demand and urbanization, a systematic understanding of the complex nature of smart and sustainable cities becomes paramount. Especially in light of recent challenges facing urban developments (e.g., energy transition and consumption, improving air quality, adapting to climate change, improving interaction/integration between transportation and buildings, biodiversity preservation, etc.), aggressive urban agenda development becomes necessary to share information in real time, identify problems, anticipate risks, and design solutions that enhance cooperation among stakeholders to improve growth, quality of life, and innovation in cities and resolve societal challenges.

The volume collects a wide range of new high-quality theoretical, empirical, and case studies at the nexus of urban, socioeconomic, energy, governmental, and ecological transformation. More precisely, in addition to exploring smart cities' concepts and benefits, the volume contains state-of-the-art empirical and methodological advances to explore and document the economic and environmental implications of the urban transformation process. The different chapters focus on the following major themes: (i) smart city concepts and benefits; (ii) the merging concept of smart energy cities; (iii) the role of urban modeling and digital twins in improving cities' sustainability; (iv) smart mobility in cities; and (v) case studies and city initiatives in emerging economies and Saudi Arabia.

The distinctiveness of this volume makes the impacts of the proposed research framework for city leaders and ecosystems compelling for a number of reasons. Specifically, this volume provides several contributions seeking to inform policymakers and inspire further research in this dynamic area.

First, the book thoroughly examines cities' issues and challenges and how they are addressed by identifying, diagnosing, and tackling them.

Second, the results of the different chapters will help policymakers and practitioners navigate the global smart city research agenda, provide a map of existing works and case studies, recommend promising avenues for future research, and inspire future collaboration.

Third, the volume captures the most relevant thinking on smart and sustainable cities from international experts from various disciplines and explores some salient factors that can assist in creating resilient and sustainable cities.

With energy powering the most attractive aspects of modern society's urban environment, smart cities are ideally situated to shape carbon-neutral built environments. Low-carbon cities will rely extensively on a combination of green technological innovation – improving new technologies and system performance – and sustainable behaviors such as energy sobriety and waste reduction.

Contents

Contributors

Razan Amine Cambridge University, The Old Schools, Trinity Ln, Cambridge, UK
King Abdullah Petroleum Studies and Research Center, Riyadh, Saudi Arabia

Alberto Barbaresi Department of Agricultural and Food Sciences, University of Bologna, Bologna, Italy

Fateh Belaïd King Abdullah Petroleum Studies and Research Center, Riyadh, Kingdom of Saudi Arabia

Souhir Bennaya LEM Lille Economie Management, UMR 9221, Lille Cedex, France

Yagyavalk Bhatt King Abdullah Petroleum Studies and Research Center, Riyadh, Kingdom of Saudi Arabia

Curtis B. Charles Academic Affairs, The UWI Five Islands Campus, Five Islands Village, Antigua and Barbuda

Mohammad I. Elian Department of Economics and Finance, Gulf University for Science and Technology, Hawally, Kuwait

Jamila El-Mir MENA Environmental and Sustainability Policy and Strategy Expert, Government of Dubai, Dubai, United Arab Emirates

Nathalie Gaussier Université de Bordeaux, CNRS, BSE, UMR 6060, Pessac, France

Mansoureh Gholami Department of Agricultural and Food Sciences, University of Bologna, Bologna, Italy

Wassim Hached LVMT, Univ Gustave Eiffel, IFSTTAR, Ecole des Ponts, Marne-la-Vallée, France

Peter Kawalek Centre for Information Management, School of Business & Economics, Loughborough University, Loughborough, UK

Moez Kilani University of Littoral, Opal Coast, Dunkerque, France

Khalid M. Kisswani Department of Economics and Finance, Gulf University for Science and Technology, Hawally, Kuwait

Center for Sustainable Energy and Economic Development (SEED), Gulf University for Science and Technology, Hawally, Kuwait

Nathalie Lazaric GREDEG, University of Côte d'Azur, Valbonne, France

University of Gothenburg, Göteborg, Sweden

Alain L'Hostis LVMT, Univ Gustave Eiffel, IFSTTAR, Ecole des Ponts, Marne-la-Vallée, France

Camille Massie King Abdullah Petroleum Studies and Research Center, Riyadh, Saudi Arabia

A. H. M. Mehbub Anwar Transport and Infrastructure Department, King Abdullah Petroleum Studies and Research Center (KAPSARC), Riyadh, Saudi Arabia

Dylan Moinse LVMT, Université Gustave Eiffel, IFSTTAR, École des Ponts, Marne-la-Vallée, Paris, France

Abu Toasin Oakil Transport and Infrastructure Department, King Abdullah Petroleum Studies and Research Center (KAPSARC), Riyadh, Saudi Arabia

Anna Laura Petrucci Universita' La Sapienza, Rome, Italy

King Saud University, Riyadh, Saudi Arabia

Eliane Propeck-Zimmermann Laboratoire Image Ville Environnement (LIVE), UMR 7362 CNRS, University of Strasbourg, Strasbourg, France

Boumediene Ramdani Centre for Entrepreneurship and Organizational Excellence, College of Business & Economics, Qatar University, Doha, Qatar

Jitendra Roychoudhury King Abdullah Petroleum Studies and Research Center, Riyadh, Kingdom of Saudi Arabia

Isam Shahrour Laboratoire de Génie Civil et géo-Environnement, Université de Lille, Lille, France

Patrizia Tassinari Department of Agricultural and Food Sciences, University of Bologna, Bologna, Italy

Daniele Torreggiani Department of Agricultural and Food Sciences, University of Bologna, Bologna, Italy

Mira Toumi GREDEG, University of Côte d'Azur, Valbonne, France

Densil A. Williams The University of West Indies, Five Islands Campus, Five Islands Village, Antigua and Barbuda

Seghir Zerguini Université de Bordeaux, CNRS, BSE, UMR 6060, Pessac, France

About the Editors

Fateh Belaïd is a Fellow at KAPSARC. Before joining KAPSARC, he was a full professor of Economics at Lille Catholic University and director of the Smart & Sustainable Cities research unit. Fateh has also held various positions at the French Scientific and Technical Center for Building and led multiple collaborative projects for the French Ministry of Ecological Transition and the European Commission. He is an energy and environmental economist drawing from the fields of applied micro-economics, energy modeling, and econometrics. He has published widely on household energy consumption, energy-saving behaviors, individual preference and investment in energy efficiency, energy poverty, renewables, and energy policy. He received a habilitation for supervising doctoral research from Orléans University, a Ph.D. in Economics, an M.S. in Applied Economics & Decision Theory from Littoral University, and an engineering degree in statistics. His work has been published in journals including *Ecological Economics*, *The Energy Journal*, *Energy Economics*, *Economic Surveys*, *Energy Policy*, and *Environmental Management.*

Anvita Arora is an architect and transport planner whose current areas of research at KAPSARC (Saudi Arabia) include smart cities, electric vehicles, and freight mobility. Before joining the Center in February 2018, she was the managing director and CEO of Innovative Transport Solutions (iTrans), an incubator company of IIT Delhi, where she led over 40 applied research and planning projects for 10 years for clients ranging from city-level and country-level authorities to funding agencies including the UNEP, World Bank, Asian Development Bank, and DFID. Anvita has been teaching transport planning in the Urban Design Department of the School of Planning and Architecture in Delhi for the past 12 years and was visiting faculty at the TERI University, Delhi. She was also associated with the Transportation Research and Injury Prevention Program (TRIPP) at IIT Delhi, a Volvo Research and Educational Foundations (VREF) Centre of Excellence, for nearly 12 years.

Innovation and Smart Cities Research: A Review and Future Directions

Boumediene Ramdani and Peter Kawalek

Abstract This chapter aims to review existing evidence and map research on innovation in smart cities. Based on data from 822 articles and chapters, bibliometric analyses were performed to capture descriptive statistics and key themes of this field of research. The results of our descriptive analysis show that interest in this field of research is increasing, and substantial contributions have been made in the past 12 years. Moreover, the results from co-citation analysis show that innovation in smart city research is grounded in four clusters: open, urban, sustainable, and digital innovation. Key contributions within each theme will be discussed, and future research opportunities will be highlighted.

Keywords Smart cities · Innovation · Review · Bibliometric analysis

1 Introduction

Smart cities are defined as "initiatives or approaches that effectively leverage digitalisation to boost citizen well-being and deliver more efficient, sustainable and inclusive urban services and environments as part of a collaborative, multistakeholder process" (OECD, 2018). With the proliferation of smart city initiatives around the world, greater attention has been given to innovation as a novel way to build new smarter cities or regenerate older ones. Innovation can be embedded in every stage of development, from planning to construction to management and operations,

B. Ramdani (✉)
Centre for Entrepreneurship and Organizational Excellence, College of Business & Economics, Qatar University, Doha, Qatar
e-mail: b.ramdani@qu.edu.qa

P. Kawalek
Centre for Information Management, School of Business & Economics, Loughborough University, Loughborough, UK
e-mail: P.Kawalek@lboro.ac.uk

© The Author(s) 2024

F. Belaïd, A. Arora (eds.), *Smart Cities*, Studies in Energy, Resource and Environmental Economics, https://doi.org/10.1007/978-3-031-35664-3_1

1

implementation, and support. A recent UNDP report (2021) claims that smart inno-vations are shaping urban cities across the globe by addressing citizen priorities.

Innovation in smart city research is divided between a technocentric perspective originating from the American business community and a holistic perspective instigated from European institutions (Mora et al., 2017). Advocates of the technocentric perspective suggest that digital innovation should be at the heart of successful smart city initiatives (Cardullo & Kitchin, 2019; Yang et al., 2017), whereas supporters of the holistic perspective claim that other urban innovations should be considered in addition to digital innovation to build successful smart cities (Yigitcanlar et al., 2018; Batty et al., 2012a, b). Building smart cities involves not only technological changes but also changes in regulations, infrastructure, industrial networks, practices, and culture (Geels, 2002). Therefore, other types of innovation may be necessary for these changes to take place. However, it is unclear what these innovations are and what evidence exists to support them.

This chapter aims to review existing evidence on innovation in smart city research. This is timely for several reasons. First, smart city research is interdisciplinary, and existing reviews do not tackle innovation head-on. Second, this review will bring together key contributions in this field of research to shed light on how urban problems have been resolved. Third, this review is timely to show where future research efforts should focus to move this area of research forward. Using the Clarivate Analytics Web of Science (WoS) database, we will identify the key publications and cluster them to show existing evidence.

The remainder of this chapter proceeds as follows. The next section covers the research method (Sect. 2). Section 3 presents the results of our analyses. The following section (Sect. 4) discusses the main findings. Finally, Section 5 concludes the chapter, and Sect. 6 discusses future research on smart city innovation.

2   Research Method

Following the standard workflow of science mapping (Zupic & Čater, 2015), we performed bibliometric analyses using the five-stage workflow recommended by Aria and Cuccurullo (2017). First, this review aims to map the existing evidence on innovation and smart city research using the ISI Web of Science (WoS) database. Second, data were collected from the WoS database using search strings ("innovation" and "smart cit*"). Our search returned 1213 documents. By limiting our search criteria to only articles, reviews, and book chapters in English, we ended up with 822 documents (770 articles, 48 reviews, and 4 book chapters). Third, data were analyzed using the *bibliometrics* R package to retrieve statistics on journals, authors, countries, affiliations, and co-citations. Fourth, network analysis was performed to visualize the data using *VOSViewer*. Fifth, we use both *bibliometrics and VOSViewer* to interpret the results through topical analysis.

3 Results

The evolution of this field is captured with descriptive statistics on publications over time, most cited authors, key journals, most cited sources, corresponding author's country and their affiliations. Moreover, co-citation analysis will be used to identify the clusters in this field of research.

3.1 Descriptive Analysis

Over the past 12 years, 822 articles and book chapters were published in innovation and smart city research. Interest in this field started in 2010 with the first publication in this field on learning cities and regions (Longworth & Osborne, 2010). Since 2015, interest in this field started a momentum that continuously accelerated with outputs that peaked in 2021 with 187 publications. Up to the end of March, researchers have published 60 articles and chapters in 2022 (Fig. 1).

Innovation and smart cities research had contributions from 2254 authors. Most of these authors had only one publication, and only 229 authors had more than one output. Table 1 lists the most cited authors in this field in descending order of total citations (TS), number of publications (NB), and h-index. Michael Batty leads with number of citations, whereas Tan Yigitcanlar leads with number of publications. Alberto Ferraris comes second with 606 citations and 9 publications, followed by Mark Deakin with 561 citations and 8 publications.

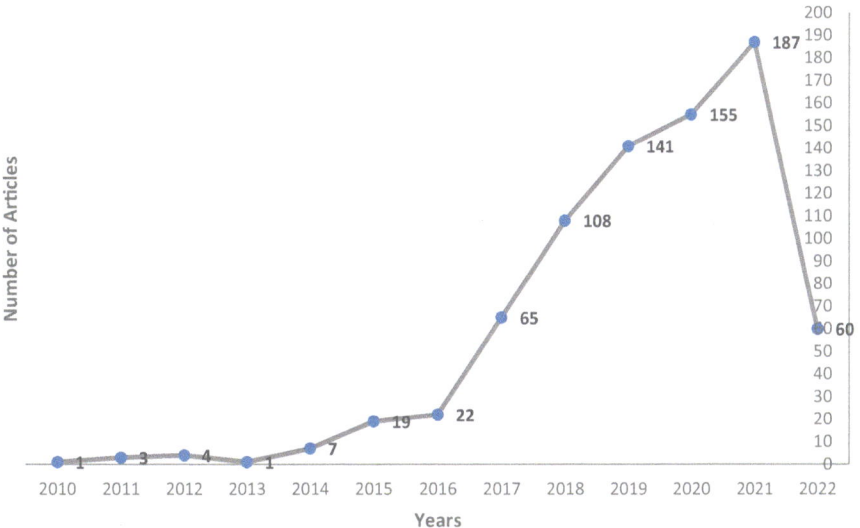

Fig. 1 Publications over time (2010–2022) in WoS

Table 1 Most cited authors

#	Authors	NP	TC	h-index
1	Michael Batty	4	863	3
2	Alberto Ferraris	9	606	9
3	Mark Deakin	8	561	7
4	Margarita Angelidou	5	560	5
5	Stefano Bresciani	6	521	6
6	Tan Yigitcanlar	13	478	9
7	Luca Mora	9	380	7
8	Rob Kitchin	6	359	6
9	Jamile Sabatini-Marques	5	254	4
10	Alasdair Reid	6	209	5
11	Igor Calzada	6	178	5
12	Anastasia Panori	6	146	5
13	Luis Carvalho	4	125	3
14	Christina Kakderi	6	72	5
15	Yuanping Wang	5	65	2
16	Yajing Zhang	8	64	3
17	Nathalie Crutzen	4	52	3
18	Stan Geertman	4	51	3
19	Mário Franco	8	39	3
20	Margarida Rodrigues	8	39	3

NP: total number of publications; TC: total citations; h-Index: calculated based on author contributions to this area. Authors listed in descending order of TC

Articles have been published in several journal outlets. Journals with the most publications are *Sustainability* (88 articles), *Technological Forecasting and Social Change* (37 articles), and *Cities* (27 articles). The top 20 journals are listed in Table 2 with the total number of articles (NP), total citations (TC), and average citation scores (TC/NP). *Technological Forecasting and Social Change* has the highest score (46.81), followed by *Cities* (44.52) and *Government Information Quarterly* (40.00).

Total number of citations (TC) can be used as a measure of the impact of an article. Moreover, the yearly number of citations (TC/Y) score shows the yearly relevance of an article since it was published. Table 3 lists the 10 most cited sources in this field of research. The most cited article is Batty et al. (2012a, b), with 826 citations and a yearly citation score of 75.09. In this key contribution, Batty et al. defined smart cities, outlined research challenges, and outlined the paradigm shift from older to smarter cities. The second most cited article is Lombardi et al. (2012), with 381 citations. Lombardi et al. modeled smart city performance using the triple helix model and analytic network process. Their results indicate four categories of smart cities: entrepreneurial, pioneer, liveable, and connected cities. Lee et al. (2014) is the third most cited article with 333 citations. They developed a framework for smart city analysis with dimensions and subdimensions to help implement new smart cities and learn from Seoul and San Francisco. The fourth most cited article is

Table 2 List of journals

#	Journals	NP	%	TC	TC/NP
1	*Technological Forecasting and Social Change*	37	4.5	1732	46.81
2	*Cities*	27	3.28	1202	44.52
3	*Government Information Quarterly*	9	1.09	360	40.00
4	*Journal of Urban Technology*	22	2.68	843	38.32
5	*Journal of Cleaner Production*	16	1.95	513	32.06
6	*Urban Studies*	7	0.85	159	22.71
7	*Journal of Science and Technology Policy Management*	6	0.73	125	20.83
8	*European Planning Studies*	7	0.85	129	18.43
9	*Sustainable Cities and Society*	18	2.19	314	17.44
10	*Sensors*	16	1.95	205	12.81
11	*Sustainability*	88	10.71	907	10.31
12	*IEEE Access*	16	1.95	146	9.13
13	*Energies*	17	2.07	136	8.00
14	*International Entrepreneurship and Management Journal*	6	0.73	40	6.67
15	*Technology Innovation Management Review*	7	0.85	36	5.14
16	*Smart Cities*	16	1.95	75	4.69
17	*International Journal of E-Planning Research*	7	0.85	27	3.86
18	*IEEE Transactions on Engineering Management*	5	0.61	18	3.60
19	*Regional Studies*	6	0.73	19	3.17
20	*Wireless Personal Communications*	6	0.73	5	0.83

NP: total number of publications; %: percentage of publications in the dataset of 822 publications; TC: the total citations of a journal; TC/NP: average number of overall citations per article of a journal

Angelidou (2015), with 310 citations. Angelidou outlined the four forces shaping smart cities, including urban futures, knowledge and innovation economy, technology push, and application pull. Although Yang et al. (2017) is the fifth most cited article, it has one of the second highest yearly number of citations score (47.50). Yang et al. (2017) explored the technologies and solutions addressing big data challenges. Another article with a high yearly number of citations score is Cardullo and Kitchin (2019). They framed "citizen-centric" smart cities by rethinking "smart citizens" and "smart citizenship".

The ten most productive universities are listed in Table 4. Queensland University of Technology leads with 28 publications, followed by Edinburgh Napier University with 25 publications, then the University of Beira Interior with 24 publications. Aristotle University Thessaloniki and University Turin have 22 publications.

Authors from 76 countries published articles and chapters in this field of research. Table 5 lists the top 20 countries with the most publications. China leads with 94 publications, followed by the United Kingdom with 88 publications, then Italy with 83 publications. China's national development strategy and the 13th Five-Year Plan (2016–2020) include smart cities (Atha et al., 2020). Korea leads with an average citation score of 60.40 with only 15 publications, followed by Ireland (37.45) with only 11 publications and Greece (33.83) with 23 publications.

Table 3 The most cited sources

#	Authors	Year	Title	Journal	TC	TC/Y
1	Batty M, Axhausen KW, Giannotti F, Pozdnoukhov A, Bazzani A, Wachowicz M, Ouzounis G, and Portugali Y.	2012	Smart cities of the future	*The European Physical Journal Special Topics*	826	75.09
2	Lombardi P, Giordano S, Farouh H, and Yousef W.	2012	Modeling the smart city performance	*Innovation: The European Journal of Social Science Research*	381	34.64
3	Lee JH, Hancock MG, and Hu MC.	2014	Toward an effective framework for building smart cities: Lessons from Seoul and San Francisco	*Technological Forecasting and Social Change*	333	37.00
4	Angelidou M.	2015	Smart cities: A conjuncture of four forces	*Cities*	310	38.75
5	Yang C, Huang Q, Li Z, Liu K, and Hu F.	2017	Big Data and cloud computing: innovation opportunities and challenges	*International Journal of Digital Earth*	285	47.50
6	Gretzel U, Werthner H, Koo C, and Lamsfus C.	2015	Conceptual foundations for understanding smart tourism ecosystems	*Computers in Human Behavior*	252	31.50
7	Mora L, Bolici R, and Deakin M.	2017	The first two decades of smart-city research: A bibliometric analysis	*Journal of Urban Technology*	194	32.33
8	Cardullo P, and Kitchin R.	2019	Being a "citizen" in the smart city: Up and down the scaffold of smart citizen participation in Dublin, Ireland	*GeoJournal*	168	42.00
9	Leydesdorff L, and Deakin M.	2011	The triple-helix model of smart cities: A neo-evolutionary perspective	*Journal of Urban Technology*	166	13.83
10	Yigitcanlar T, Kamruzzaman M, Buys L, Ioppolo G, Sabatini-Marques J, da Costa EM, and Yun JJ.	2018	Understanding "smart cities": Intertwining development drivers with desired outcomes in a multidimensional framework	*Cities*	165	33.00

TC: the total citations per source; TC/Y: yearly number of citations

3.2   Co-citation Analysis

Co-citation analysis helps identify the most relevant and impactful sources based on their citations. In addition, it is useful in detecting schools of thought, as it maps articles cited by identified samples (Aria & Cuccurullo, 2017). A co-citation

Table 4 The most productive universities

#	Affiliations	Country	NP
1	Queensland University of Technology	Australia	28
2	Edinburgh Napier University	The United Kingdom	25
3	University of Beira Interior	Portugal	24
4	Aristotle University Thessaloniki	Greece	22
5	University Turin	Italy	22
6	Utrecht University	Netherlands	21
7	Open University	The United Kingdom	19
8	Oxford University	The United Kingdom	19
9	Delft University of Technology	Netherlands	18
10	Erasmus University	Netherlands	18

Table 5 Corresponding author country

#	Country	NP	%	TC	TC/NP
1	China	94	11.44	759	8.07
2	The United Kingdom	88	10.71	2401	27.28
3	Italy	83	10.10	2188	26.36
4	Spain	54	6.57	1008	18.67
5	The United States	52	6.33	1251	24.06
6	Netherlands	35	4.26	1032	29.49
7	Australia	29	3.53	729	25.14
8	Brazil	29	3.53	238	8.21
9	Portugal	25	3.04	146	5.84
10	Greece	23	2.80	778	33.83
11	Sweden	18	2.19	336	18.67
12	India	18	2.19	253	14.06
13	Finland	17	2.07	172	10.12
14	Korea	15	1.82	906	60.40
15	Canada	15	1.82	147	9.80
16	France	14	1.70	189	13.50
17	Norway	12	1.46	204	17.00
18	Ireland	11	1.34	412	37.45
19	Belgium	10	1.22	163	16.30
20	Denmark	9	1.09	141	15.67

NP: total number of publications; %: percentage of publications in the dataset of 822 publications; TC: total citations per country; TC/NP: average number of overall citations per country

analysis was performed using the Louvain clustering algorithm (Blondel et al., 2008) generating the co-citation network. This network can be visualized using VOSViewer, as illustrated in Fig. 2. Innovation in smart city research contains four clusters (Table 6): open innovation (in red), urban innovation (in green), sustainable innovation (in blue), and digital innovation (in yellow).

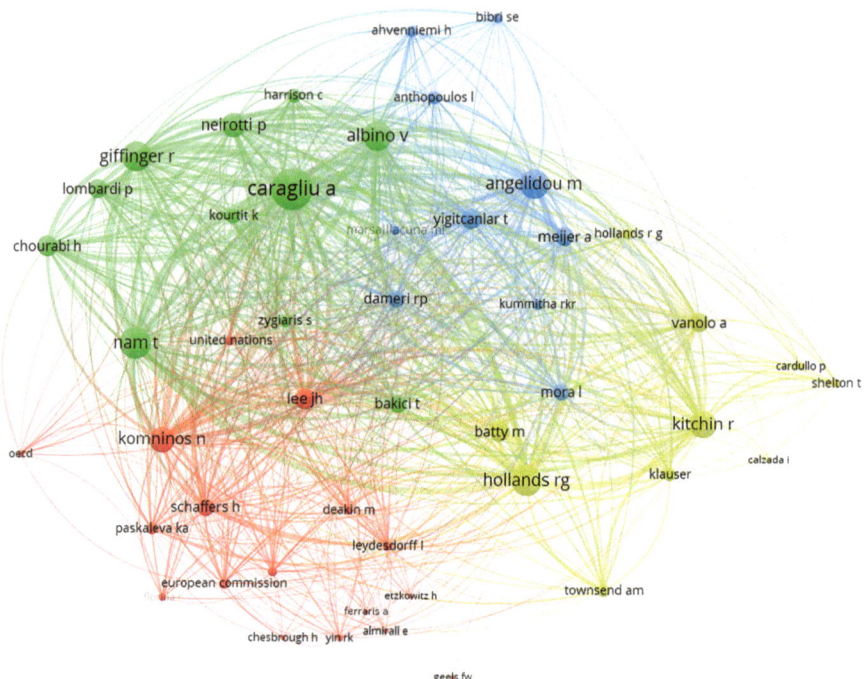

Fig. 2 Co-citation network among authors

Table 6 Major clusters in SSC research

Cluster	Research foci	Studies
Cluster 1	Open Innovation	Etzkowitz and Leydesdorff (2000); Geels (2002); Chesbrough (2003); Leydesdorff and Deakin (2011); Paskaleva (2011); Schaffers et al. (2011); Almirall et al. (2016); Cohen et al. (2016); Ferraris et al. (2020)
Cluster 2	Urban Innovation	Giffinger and Haindlmaier (2010); Harrison et al. (2010); Nam and Pardo (2011); Chourabi et al. (2012); Kourtit and Nijkamp (2012); Lombardi et al. (2012); Bakıcı et al. (2013); Zygiaris (2013); Neirotti et al. (2014); Albino et al. (2015); Caragliu and Del Bo (2019)
Cluster 3	Sustainble Innovation	Dameri (2013); Yigitcanlar and Lee (2014); Angelidou (2015); Marsal-Llacuna et al. (2015); Meijer and Bolívar (2016); Ahvenniemi et al. (2017); Anthopoulos (2017); Bibri and Krogstie (2017); Kummitha and Crutzen (2017); Mora et al. (2017); Yigitcanlar et al. (2019)
Cluster 4	Digital Innovation	Hollands (2008); Batty et al. (2012a, b); Townsend (2013); Kitchin (2014); Vanolo (2014); Calzada and Cobo (2015); Shelton et al. (2015); Cardullo and Kitchin (2019); Söderström et al. (2020)

4 Discussion

We performed two sets of analyses: descriptive and co-citation. The former indicates that interest in this field of research is increasing, and substantial contributions have been made in the past 12 years, with 822 articles and book chapters. The most cited authors include Michael Batty, Alberto Ferraris, and Mark Deakin. Tan Yigitcanlar leads with most published articles and book chapters. Research has been published in leading journals such as *Technological Forecasting and Social Change, Cities, Government Information Quarterly, Journal of Urban Technology,* and *Journal of Cleaner Production.* The most cited articles include Batty et al. (2012a, b), Lombardi et al. (2012), and Lee et al. (2014). In addition, the most productive universities in this field of research are Queensland University of Technology, Edinburgh Napier University, and University of Beira Interior. While China, the United Kingdom, and Italy lead in terms of the number of publications in this field, Korea, Ireland, and Greece lead with average citation scores. The latter analysis generated the co-citation network with four clusters: open, urban, sustainable, and digital innovation.

Cluster 1. Open Innovation
This cluster brings together research focusing on open innovation using the triple helix model (Etzkowitz & Leydesdorff, 2000), technological transitions (Geels, 2002), and open innovation (Chesbrough, 2003). First, advocates of the triple helix model suggest that instead of focusing on the national systems of innovation, institutional transformations can be achieved by rearranging university–industry–government relations (Etzkowitz & Leydesdorff, 2000). Using the triple helix model, Leydesdorff and Deakin (2011) argue that "cities can be considered as densities in networks among three relevant dynamics: the intellectual capital of universities, the wealth creation of industries, and the democratic government of civil society" (p. 53). They draw this argument from the experiences of cities such as Montreal and Edinburgh and demonstrate the transition of these cities to become "smarter" cities. Moreover, Ferraris et al. (2020) delved into the role of universities in smart city innovation, arguing that this can be achieved through multiple roles played by universities, such as a source of knowledge and financial mediator and an engager of different city stakeholders. Second, Geels (2002) argues that technological transitions not only involve technological changes but also affect regulations, infrastructure, industrial networks, practices, and culture. In city transformation, Almirall et al. (2016) discussed three tensions, namely, governance models—role as an orchestrator of ecosystems, as well as a collaborator, growth—maintaining as well as supporting new structures for innovation, and the sharing economy—resolving conflicts between two modes of production. Third, influenced by the work of Chesbrough (2003), several studies have emerged. Paskaleva (2011) probed European Union (EU) programs and found that an open innovation approach has emerged through linking urban territories, people, technologies, and other cities. She argues that this approach can be effective and sustainable as long as consistent frameworks, principles, and agendas are implemented. Schaffers et al. (2011)

looked at how the future of the Internet in smart cities can be explored although an open and user-driven innovation environment that enables experimentation in the domain of living labs. Finally, Cohen et al. (2017) explored the role of cities as a driver for open innovation and entrepreneurship. They argue that cities are becoming living labs for solving complex societal challenges through rapid prototyping and testing of innovations.

Cluster 2. Urban Innovation

Smart cities have become a cornerstone in urban planning (Kourtit & Nijkamp, 2012). Harrison et al. (2010) suggest that urban innovation can be achieved by using information technology (IT) to exploit existing data on traffic, energy, and citizen habits. They suggest that this could be achieved through instrumented, interconnected, and intelligent operations. While instrumental relates to real-time data, interconnected relates to the integration of data into one platform, and intelligent relates to modeling, optimization, and visualization of operations. To showcase urban innovation, studies have examined the performance of smart cities. Giffinger et al. (2010) have looked at the dimensions and subdimensions of ranking smart cities, including smart economy, smart people, smart governance, smart mobility, smart environment, and smart living. Using the triple helix model, Lombardi et al. (2012) looked at the different dimensions used to measure smart city performance in relation to university, civil society, and industry by listing subdimensions of smart governance, smart economy, smart human, smart living, and smart environment. By examining different performance measures of smart city initiatives, Albino et al. (2015) show the complexity of measuring smart city performance. Using the case of Barcelona, Zygiaris (2013) and Bakıcı et al. (2013) show how to build smart cities within an innovative ecosystem. Zygiaris (2013) introduced the Smart City Reference Model, which includes seven city planning layers: the city, green city, interconnection, instrumentation, open integration, application, and innovation layers. To implement innovation in smart cities, city planners need to comprehend the factors facilitating execution. These factors include technological, institutional, and human factors (Nam & Pardo, 2011) and IT infrastructure, security and privacy, and operational costs (Chourabi et al., 2012). The impact of smart city initiatives has been measured using total patent applications (Caragliu & Del Bo, 2019) and acceptance and use (Neirotti et al., 2014).

Cluster 3. Sustainable Innovation

A smart and sustainable city is "an innovative city that uses information and communication technologies (ICT) and other means to improve quality of life, efficiency of urban operation and services, and competitiveness, while ensuring that it meets the needs of present and future generations, with respect to economic, social and environmental aspects" (ITU, 2014). Before cities become smart, they need to be sustainable (Ahvenniemi et al., 2017; Yigitcanlar et al., 2019). Yigitcanlar et al. (2019) claim that three challenges face sustainable innovation in smart cities: technocentricity, practice complexities, and ad hoc notions of smart cities. Ahvenniemi et al. (2017) distinguished between the use of two terms "smart cities" and "sustainable cities" and suggested the use of a more accurate term "smart sustainable cities". The aim of smart and sustainable cities is "to maximize efficiency of energy and material resources, create a zero–waste system, support renewable energy production and consumption, promote carbon–neutrality and reduce pollution, decrease transport needs and encourage walking and cycling, provide efficient and sustainable transport, preserve ecosystems, emphasize design scalability and spatial proximity, and promote livability and sustainable community" (Bibri & Krogstie, 2017, p. 193). By examining ten smart city cases, Anthopoulos (2017) shows different aspects of sustainability for different smart cities. Smart and sustainable cities have been shown to create economic and public value (Dameri & Rosenthal-Sabroux, 2014).

Cluster 4. Digital Innovation
The quest for a new utopia for cities can be achieved through ubiquitous computing in urbanism (Townsend, 2013). On the one hand, and in their seminal piece, Batty et al. (2012a, b) worked on the *FutureICT* program that introduced an innovative approach to technological innovation. This approach advocates that technology is a social construction involving hardware, software, databases, and organizational technologies. They argue that the use of technological innovations can help city planners sense and measure, exchange in urban markets, and model. Moreover, digital innovation produces big data, enabling real-time city life analysis, novel approaches to city governance, and providing more efficient, productive, transparent, open, and sustainable cities (Kitchin, 2014). On the other hand, smart city agendas are driven by large IT corporations such as IBM (Söderström, 2014; Shelton et al., 2015). Calzada and Cobo (2015) criticize technological determinism and propose the ten dimensions of social innovation in smart cities. They argue that unplugging could be beneficial, and these benefits should not be disregarded because of the abundance of digital innovations. Moreover, Hollands (2008) argues that the smart city agenda assumes a positive impact of digital technologies with a hidden policy agenda of "high-tech urban entrepreneurialism". Other studies claim that smart cities enact hidden neoliberal agendas (Shelton et al., 2015; Cardullo & Kitchin, 2019). The techno-centric vision of smart cities in Europe comes from the availability of financial resources to reconstruct cities, involvement of large private corporations in digitization projects, creation of techno-centric solution-based rhetoric, and focus on sustainable smart cities to resolve economic crises (Vanolo, 2014).

5 Conclusions

The aim of this review chapter is to map existing evidence on innovation and smart city research. Using the ISI WOS database, we retrieved 822 articles and chapters. Bibliometric analyses were used to highlight descriptive statistics and key themes. The evolution of this field of research is captured through descriptive statistics on publications over time, most cited authors, key journals, most cited sources, corresponding author's country, and their affiliations. Co-citation analysis was used to identify the key themes in this field of research. Four clusters have been identified: open innovation, urban innovation, sustainable innovation, and digital innovation. Although existing evidence suggests that substantial research has been carried out to demonstrate innovation in smart cities, research gaps still exist, and we call for future research to document the innovation journey of smart cities.

6 Future Research

Reviewing existing work in innovation and smart city research, several promising avenues for future research have been highlighted in relation to the four clusters identified earlier. The key research questions are detailed in Table 7.

Although much work has focused on the four clusters, there remain many gaps that could be filled with future research. In the open innovation cluster, future research could exist, and future tensions and what open innovation mechanisms could be employed to resolve these tensions. Moving beyond the triple helix model, what theories can be used to enact open innovation within smart cities? We suggest using dynamic capabilities theory (Teece et al., 1997) to show the different capabilities achieved through open innovation in smart cities. In addition, more work is needed to show the experience of smart cities in using experimentation and rapid prototyping as open innovation methods. Urban innovation is the most

Table 7 Future research on smart city innovation

Cluster	Research questions
1. Open innovation	What are the existing and future tensions of innovation in smart cities? What open innovation methods and approaches will resolve these tensions? What theories can be used to move beyond the Triple Helix model in smart city innovation? How have cities used experimentation and rapid prototyping?
2. Urban innovation	What dimensions, layers, and types of urban innovation? What tools and frameworks can facilitate the implementation of urban innovation in smart cities? What internal and external forces influence urban innovation? What is the impact of urban innovation? How to measure it?
3. Sustainable innovation	What are the different aspects of sustainable innovation? How could city planners embed sustainable innovation in building smart cities? What policies are needed for implementing sustainable smart cities? How to measure the value of sustainable innovation in smart cities?
4. Digital innovation	What are the benefits and/or risks of digital innovations in smart cities? What methods and/or tools can help achieve the optimal balance between benefits and risks of digital innovations in smart cities? Why do digital innovations in smart cities fail and/or succeed? What policy agendas are driving and/or hindering digital innovations in smart cities?

promising research cluster. The UNDP (2021) suggests four types of innovations: community-organized, frugal, enterprise ventures, and institutional pioneers. More work is needed to unravel the dimensions, layers, and types of urban innovation. Furthermore, future research could look at the tools and frameworks that could facilitate the successful implementation of urban innovation in smart cities. Additionally, researchers need to examine the internal and external forces influencing urban innovation and measure the impact and outcomes of such innovations. In the sustainable innovation cluster, researchers can identify the different aspects of sustainable innovation. As cities cannot be smart without being sustainable, it is critical to explore the different ways of embedding sustainable innovation in smart city planning, construction, management and operations, and support. Another critical issue is demonstrating the value of sustainable innovation in smart cities, for which limited research exists. In the digital innovation cluster, researchers could assess the benefits and/or risks of digital innovations. Future research should also focus on methods and/or tools that can help achieve the optimal balance between the benefits and risks of digital innovations. Limited research exists on the success and failure of digital innovation in smart cities. Learning lessons are needed to replicate these innovations in different contexts. Moreover, research should examine existing policies driving and/or hindering digital innovations in smart cities. Finally, researchers need to identify other innovations in smart cities to move away from the holistic perspective of smart cities.

References

Ahvenniemi, H., Huovila, A., Pinto-Seppä, I., & Airaksinen, M. (2017). What are the differences between sustainable and smart cities? *Cities, 60*, 234–245.

Albino, V., Berardi, U., & Dangelico, R. M. (2015). Smart cities: Definitions, dimensions, performance, and initiatives. *Journal of Urban Technology, 22*(1), 3–21.

Almirall, E., Wareham, J., Ratti, C., Conesa, P., Bria, F., Gaviria, A., & Edmondson, A. (2016). Smart cities at the crossroads: New tensions in city transformation. *California Management Review, 59*(1), 141–152.

Angelidou, M. (2015). Smart cities: A conjuncture of four forces. *Cities, 47*, 95–106.

Anthopoulos, L. (2017). Smart utopia VS smart reality: Learning by experience from 10 smart city cases. *Cities, 63*, 128–148.

Aria, M., & Cuccurullo, C. (2017). bibliometrix: An R-tool for comprehensive science mapping analysis. *Journal of Informetrics, 11*(4), 959–975.

Atha, K., Callhan, J., Chen, J., Dru, J., Francis, E., Green, K., Lafferty, B., McReynolds, J., Mulvenon, J., Rosen, B., & Walz, E. (2020). *China's smart cities development*. Research report prepared on behalf of the U.S.–China Economic and Security Review Commission. SOS International LLC. https://www.uscc.gov/sites/default/files/2020-04/China_Smart_Cities_Development.pdf

Bakıcı, T., Almirall, E., & Wareham, J. (2013). A smart city initiative: The case of Barcelona. *Journal of the Knowledge Economy, 4*(2), 135–148.

Batty, M., Axhausen, K. W., Giannotti, F., Pozdnoukhov, A., Bazzani, A., Wachowicz, M., Ouzounis, G., & Portugali, Y. (2012a). Smart cities of the future. *The European Physical Journal Special Topics, 214*(1), 481–518.

Batty, M., Axhausen, K. W., Giannotti, F., Pozdnoukhov, A., Bazzani, A., Wachowicz, M., et al. (2012b). Smart cities of the future. *The European Physical Journal Special Topics, 214*(1), 481–518.

Bibri, S. E., & Krogstie, J. (2017). Smart sustainable cities of the future: An extensive interdisciplinary literature review. *Sustainable Cities and Society, 31*, 183–212.

Blondel, V. D., Guillaume, J. L., Lambiotte, R., & Lefebvre, E. (2008). Fast unfolding of communities in large networks. *Journal of statistical mechanics: theory and experiment, 2008*(10), P10008.

Calzada, I., & Cobo, C. (2015). Unplugging: Deconstructing the smart city. *Journal of Urban Technology, 22*(1), 23–43.

Caragliu, A., & Del Bo, C. F. (2019). Smart innovative cities: The impact of Smart City policies on urban innovation. *Technological Forecasting and Social Change, 142*, 373–383.

Cardullo, P., & Kitchin, R. (2019). Being a "citizen" in the smart city: Up and down the scaffold of smart citizen participation in Dublin, Ireland. *GeoJournal, 84*(1), 1–13.

Chesbrough, H. W. (2003). *Open innovation: The new imperative for creating and profiting from technology*. Harvard Business Press.

Chourabi, H., Nam, T., Walker, S., Gil-Garcia, J. R., Mellouli, S., Nahon, K., et al. (2012). Understanding smart cities: An integrative framework. In *2012 45th Hawaii international conference on system sciences* (pp. 2289–2297). IEEE.

Cohen, B., Almirall, E., & Chesbrough, H. (2016). The city as a lab: Open innovation meets the collaborative economy. *California Management Review, 59*(1), 5–13.

Cohen, B., Amorós, J. E., & Lundy, L. (2017). The generative potential of emerging technology to support startups and new ecosystems. *Business Horizons, 60*(6), 741–745.

Dameri, R. P. (2013). Searching for smart city definition: A comprehensive proposal. *International Journal of Computers & Technology, 11*(5), 2544–2551.

Dameri, R. P., & Rosenthal-Sabroux, C. (Eds.). (2014). *Smart city: How to create public and economic value with high technology in urban space*. Springer.

Etzkowitz, H., & Leydesdorff, L. (2000). The dynamics of innovation: From National Systems and "Mode 2" to a Triple Helix of university–industry–government relations. *Research Policy, 29*(2), 109–123.

Ferraris, A., Belyaeva, Z., & Bresciani, S. (2020). The role of universities in the Smart City innovation: Multistakeholder integration and engagement perspectives. *Journal of Business Research, 119*, 163–171.

Geels, F. W. (2002). Technological transitions as evolutionary reconfiguration processes: A multilevel perspective and a case-study. *Research Policy, 31*(8–9), 1257–1274.

Giffinger, R., & Haindlmaier, G. (2010). Smart cities ranking: An effective instrument for the positioning of the cities? *ACE: Architecture, City and Environment, 4*(12), 7–26.

Giffinger, R., Haindlmaier, G., & Kramar, H. (2010). The role of rankings in growing city competition. *Urban Research & Practice, 3*(3), 299–312.

Harrison, C., Eckman, B., Hamilton, R., Hartswick, P., Kalagnanam, J., Paraszczak, J., & Williams, P. (2010). Foundations for smarter cities. *IBM Journal of Research and Development, 54*(4), 1–16.

Hollands, R. G. (2008). Will the real smart city please stand up? Intelligent, progressive or entrepreneurial? *City, 12*(3), 303–320.

International Telecommunications Union (ITU). (2014). *Smart sustainable cities: An analysis of definitions*. http://www.itu.int/en/ITU-T/focusgroups/ssc/Documents/Approved_Deliverables/TR-Definitions.docx (01/09/2020).

Kitchin, R. (2014). The real-time city? Big data and smart urbanism. *GeoJournal, 79*(1), 1–14.

Kourtit, K., & Nijkamp, P. (2012). Smart cities in the innovation age. *Innovation: The European Journal of Social Science Research, 25*(2), 93–95.

Kummitha, R. K. R., & Crutzen, N. (2017). How do we understand smart cities? An evolutionary perspective. *Cities, 67*, 43–52.

Lee, J. H., Hancock, M. G., & Hu, M. C. (2014). Toward an effective framework for building smart cities: Lessons from Seoul and San Francisco. *Technological Forecasting and Social Change, 89*, 80–99.

Leydesdorff, L., & Deakin, M. (2011). The triple-helix model of smart cities: A neo-evolutionary perspective. *Journal of Urban Technology, 18*(2), 53–63.

Lombardi, P., Giordano, S., Farouh, H., & Yousef, W. (2012). Modeling the smart city performance. *Innovation: The European Journal of Social Science Research, 25*(2), 137–149.

Longworth, N., & Osborne, M. (2010). Six ages toward a learning region—A retrospective. *European Journal of Education, 45*(3), 368–401.

Marsal-Llacuna, M. L., Colomer-Llinàs, J., & Meléndez-Frigola, J. (2015). Lessons in urban monitoring taken from sustainable and livable cities to better address the Smart Cities initiative. *Technological Forecasting and Social Change, 90*, 611–622.

Meijer, A., & Bolívar, M. P. R. (2016). Governing the smart city: A review of the literature on smart urban governance. *International Review of Administrative Sciences, 82*(2), 392–408.

Mora, L., Bolici, R., & Deakin, M. (2017). The first two decades of smart-city research: A bibliometric analysis. *Journal of Urban Technology, 24*(1), 3–27.

Nam, T., & Pardo, T. A. (2011). Conceptualizing smart city with dimensions of technology, people, and institutions. In *Proceedings of the 12th annual international digital government research conference: Digital government innovation in challenging times* (pp. 282–291). ACM.

Neirotti, P., De Marco, A., Cagliano, A. C., Mangano, G., & Scorrano, F. (2014). Current trends in Smart City initiatives: Some stylised facts. *Cities, 38*, 25–36.

OECD. (2018). *Housing dynamics in Korea: Building inclusive and smart cities*. OECD Publishing. https://doi.org/10.1787/9789264298880-en

Paskaleva, K. A. (2011). The smart city: A nexus for open innovation? *Intelligent Buildings International, 3*(3), 153–171.

Schaffers, H., Komninos, N., Pallot, M., Trousse, B., Nilsson, M., & Oliveira, A. (2011). Smart cities and the future internet: Toward cooperation frameworks for open innovation. In *The future internet assembly* (pp. 431–446). Springer.

Shelton, T., Zook, M., & Wiig, A. (2015). The "actually existing smart city". *Cambridge Journal of Regions, Economy and Society, 8*(1), 13–25.

Söderström, O. (2014). *Cities in relations: Trajectories of urban development in Hanoi and Ouagadougou*. John Wiley & Sons.

Söderström, O., Paasche, T., & Klauser, F. (2020). Smart cities as corporate storytelling. In *The Routledge companion to smart cities* (pp. 283–300). Routledge.

Teece, D. J., Pisano, G., & Shuen, A. (1997). Dynamic capabilities and strategic management. *Strategic Management Journal, 18*(7), 509–533.

Townsend, A. M. (2013). *Smart cities: Big data, civic hackers, and the quest for a new utopia*. WW Norton & Company.

UNDP. (2021). *Handbook on smart urban innovations*. https://www.undp.org/publications/handbook-smart-urban-innovations

Vanolo, A. (2014). Smartmentality: The smart city as disciplinary strategy. *Urban Studies, 51*(5), 883–898.

Yang, C., Huang, Q., Li, Z., Liu, K., & Hu, F. (2017). Big Data and cloud computing: innovation opportunities and challenges. *International Journal of Digital Earth, 10*(1), 13–53.

Yigitcanlar, T., & Lee, S. H. (2014). Korean ubiquitous-eco-city: A smart-sustainable urban form or a branding hoax? *Technological Forecasting and Social Change, 89*, 100–114.

Yigitcanlar, T., Kamruzzaman, M., Buys, L., Ioppolo, G., Sabatini-Marques, J., da Costa, E. M., & Yun, J. J. (2018). Understanding "smart cities": Intertwining development drivers with desired outcomes in a multidimensional framework. *Cities, 81*, 145–160.

Yigitcanlar, T., Kamruzzaman, M., Foth, M., Sabatini-Marques, J., da Costa, E., & Ioppolo, G. (2019). Can cities become smart without being sustainable? A systematic review of the literature. *Sustainable Cities and Society, 45*, 348–365.

Zupic, I., & Čater, T. (2015). Bibliometric methods in management and organization. *Organizational Research Methods, 18*(3), 429–472.

Zygiaris, S. (2013). Smart city reference model: Assisting planners to conceptualize the building of smart city innovation ecosystems. *Journal of the Knowledge Economy, 4*(2), 217–231.

Open Access   This chapter is licensed under the terms of the Creative Commons Attribution 4.0 International License (http://creativecommons.org/licenses/by/4.0/), which permits use, sharing, adaptation, distribution and reproduction in any medium or format, as long as you give appropriate credit to the original author(s) and the source, provide a link to the Creative Commons license and indicate if changes were made.

The images or other third party material in this chapter are included in the chapter's Creative Commons license, unless indicated otherwise in a credit line to the material. If material is not included in the chapter's Creative Commons license and your intended use is not permitted by statutory regulation or exceeds the permitted use, you will need to obtain permission directly from the copyright holder.

4 IR Technologies to Facilitate Planning in Smart Cities of the Future

Densil A. Williams and Curtis B. Charles

Abstract Urbanization is becoming a grave concern for city planners. Globally, as the urban population continues to grow and places greater demands on resources, citizens require quick, efficient, and cost-effective services, among other things. Critically, city planners must provide relevant infrastructural support to make their city run effectively. These include but are not limited to potable water, electricity, roads and bridges, housing, transportation, and information and communications technologies, among other things.

To address this increasing demand in their cities, planners have turned to the use of technology to better assist them in the delivery of services to improve the lives of their citizens and stimulate the growth of their local and national economies. This chapter provides an overview of the types of technological solutions that are available to planners to assist them in facilitating the fast and efficient delivery of services to their citizens. These technological solutions include but are not limited to big data, cloud computing, the Internet of Things (IoT), artificial intelligence (AI), and machine learning (ML), as exemplars. City policymakers will be able to use the knowledge generated from this chapter to derive policies around the employment of 4IR technologies to assist their cities in delivering better and more efficient services to citizens.

Keywords Smart Cities · 4IR Technologies · Service Delivery · Innovation · Planning

D. A. Williams (✉)
The University of West Indies, Five Islands Campus,
Five Islands Village, Antigua and Barbuda
e-mail: densil.williams@uwimona.edu.jm

C. B. Charles
Academic Affairs, The UWI Five Islands Campus, Five Islands Village, Antigua and Barbuda
e-mail: curtis.charles@uwi.edu

© The Author(s) 2024   17
F. Belaïd, A. Arora (eds.), *Smart Cities*, Studies in Energy, Resource and
Environmental Economics, https://doi.org/10.1007/978-3-031-35664-3_2

1 Introduction

While there is a plethora of practitioner literature on the concept of a smart city, there is a dearth of academic work on the subject. Indeed, as it relates specifically to the area of Fourth Industrial Revolution (4IR) technologies that city planners generally tap into to assist them with designing their smart cities, very little theoretical or empirical academic work exists on the subject. Furthermore, despite the large body of work on these technologies and how they can assist city planners in designing their smart cities, there is no single article that pulls all these technologies into a document so that policymakers can peruse, compare, and contrast ideas as they seek the most appropriate technology for their smart city. This chapter aims to fill that gap in the academic literature by providing a single document that outlines all the possible 4IR technologies that city planners can access to build their smart city. The chapter will become a major reference point for policymakers interested in designing smart cities in their country.

Indeed, the technologies highlighted below will go a far way in assisting city planners in using the vast amount of data to make informed decisions on the future of their cities. For instance, policymakers will be able to tap into technologies such as the Internet of Things (IoT) to stimulate the use of advanced sensors and wireless communication in all kinds of physical objects. Critically, the use of sensor technologies will be able to create substantial volumes of data, thus providing a fine-grained digital view of the physical world. With these data now in hand, smart city planners can now use smart systems that optimize the use of infrastructure and resources in a more informed way (Ward et al., 2015).

To achieve the research objective as stated above, the remainder of the chapter is organized as follows: the next section will provide a brief background to the concept of *smart cities* and show some of the major challenges of designing such a city. The section will also provide a definition of what is meant by a smart city so that readers can be aware of how this new development differs from just having a city as an urban construct. Subsequently, the chapter will provide a list of the 4IR technologies that are heavily used to drive and facilitate the concept of a smart city. It will, among other things, show some of the issues ranging from the connection of citizens to commerce, access, and use of information communications technologies (ICT), information sharing, upgrading of transportation systems and other infrastructure, etc., that these technologies have used to make a smart city possible. The chapter will end with some concluding remarks and point to a direction for how city planners can incorporate these 4IR technologies into their planning process to facilitate the Smart Cities of the Future.

2 Smart Cities and the New Technologies

Geoffery West, a renowned physicist, is a household name in the work on smart cities. He predicts that the planet will be dominated by cities over the next couple of years. This is because urbanization has been expanding at an alarming rate over the last 200 years, according to West. Urbanization also comes with significant problems. These include but are not limited to environmental damage due to high levels of pollution, health problems, diseases, financial problems to keep budgets within limits, energy crises to keep moving, transportation systems, factories, households, etc. Finding a smart way to overcome these problems is exactly at the heart of the concept behind smart cities as more of the world is moving toward urbanization.

The data on urbanization put into context the great need for cities to become smarter for the future. Indeed, West (2011) quoted some very sobering statistics on the subject. He noted that every week for the foreseeable future, by 2050, more than one million people will be added to cities. Furthermore, one of the largest populations on earth, China, is expected to build 300 new cities in the next two decades. The United States, which a few centuries ago had very few urban areas, is now 82% urbanized.

With this mass urbanization and with the future growth of the world population, which is predicted to reach the 10 billion mark by the end of this century, and more persons wanting to live in cities, the problem of planning and effectively managing cities will become much greater and more sophisticated. The use of technologies will enable planners to better plan and design effective strategies to address mass urbanization, which comes with significant problems. The 4IR technologies will provide a strong enabler to help city planners better address the management of the various issues in their cities. Technology will be at the heart of smart cities.

2.1 What Are Smart Cities?

The most cogent definition of the concept of a smart city is espoused by Microsoft, one of the world's leaders in technological advancement. It noted that a "smart city is an urban area that uses an array of digital technologies to enrich residents' lives, improve infrastructure, modernize government services, enhance accessibility, drive sustainability, and accelerate economic development." This definition encapsulates the need for digital twenty-first century technologies as the backbone of the cities of the future. These technologies will allow city planners and managers to gain a fulsome view of their city's operations, infrastructure, and necessary service delivery demanded by their citizens.

Indeed, as Microsoft noted, digital technologies can help city planners design solutions to the following:

- Protect and connect with residents and businesses.
- Improve accessibility for all people in the community.
- Support businesses and fuel economic growth.
- Share information with the public.
- Streamline government operations.
- Deliver user-friendly community services.
- Provide reliable, intelligent infrastructure.

- Drive environmental sustainability.
- Promote cross-agency collaboration.
- Upgrade public transportation.
- Manage city resources to avoid waste.
- Collect and analyze data to obtain valuable insights.

There is no doubt that as the urban population continues to grow, citizens will demand more significant levels of service and want those to be delivered on time, with greater levels of efficiency and at a cost-effective fee as well. City managers and planners will therefore require sophisticated technologies that will assist them in planning and overseeing the operations of their city so that they can deliver on the benefits of city life for residents and businesses alike.

2.2   Benefits of Smart Cities

The practitioner literature is replete with examples of how citizens and businesses can benefit from smart cities and generally from urban centers. West (2011) noted that cities are generally vacuum cleaners that suck up creative people, generate cutting-edge ideas, and drive wealth creation and economic growth. These benefits normally drive people toward cities compared to those who remain in rural areas. With the increase and improvements in digital technologies, urban centers have become much more smart and, as such, can provide innovative services and solutions to overcome the general problems associated with urbanization.

With the use of digital technologies, smart cities normally deliver:

- Improved quality of life for citizens and businesses alike through shorter commutes to work, safer streets, green spaces, and increased economic opportunities.
- Better services such as modern utilities, intelligent infrastructure such as transportation, banking services, healthcare, etc. The technology is helping cities to better streamline their transportation system to have a more modern multimodal logistics operation.
- Stronger economic growth because businesses are typically drawn to these cities for the creative talent of citizens, better communication systems, reliable infrastructure, and more sophisticated consumers. Indeed, Porter (1990), in his diamond model of competitiveness, spoke to the need for sophisticated buyers and sellers as strong enablers for the competitiveness of an economy. Cities are known to be vital enablers of economic competitiveness precisely because they have a larger pool of more sophisticated customers and suppliers who demand more from their businesses and force them to innovate and grow (Williams & Morgan, 2013).

2.3 Innovate or Die

Deloitte (2022) predicts that over 55% of the world's population now lives in urban areas, and by 2050, this percentage will increase to over 68%. With this massive growth in urbanization in such a short space of time, if cities do not innovate, they will definitely face the strong possibility of collapse as the demands of citizens and businesses become increasingly stronger and more sophisticated.

To start the innovation journey, cities must first have a clear vision for their future and a clear road map to implement the same. Without a starting vision, innovation will not be possible, as there will not be an environment to facilitate creativity and resources directed to support the innovation agenda. With a clear vision and the right leadership, cities can start the innovation journey to deliver higher quality of life, economic value creation, and long-term sustainability. This leadership entails a collection of personnel with the appropriate skill set, the emotional intelligence to motivate the people to reach a higher level of creativity, and the necessary supporting technologies. These features, according to Deloitte, are critical characteristics of a smart city leader (Deloitte, 2022).

The smart city will deliver a quality of life that enhances every aspect of its citizens' existence, including but not limited to safer streets, green spaces, and efficient means of commute, among other things. Indeed, what makes the city smart compared to other urban living is that the smart city provides the best of urban living while minimizing the hassle of city living. A part of smartness is the need for businesses to be creative and generate new economic value, create high-value-added jobs, and drive an innovative economy. In addition, being conscious of the environment and sustainably using natural resources is a crucial characteristic of the smart city concept.

To drive the growth and innovation needs in smart cities, data will be critical. Smart cities will need to have data to properly plan and design strategies to overcome the general problems that come with urbanization. Given the quantity and quality of data in urban centers, to capture the information and use it to generate creative solutions and deliver a high quality of life, economic innovation, and sustainable consumption of resources, the support of technologies will be very important. This chapter focuses on critical 4IR technologies that can be used to facilitate the development of the Smart Cities of the Future. These technologies are important because the cities of the future will be led by smart automation powered by ICTs as opposed to over 200 years ago when the steam engine was the driver of the Industrial Revolution process. The next section of this chapter brings together a suite of 4IR technologies that can be used to facilitate and support the drive toward smart cities.

3   4IR Technologies for Smart Cities

The history of the modern industrialization agenda can be traced back to 1782, with the steam engine as the major technology to drive industrialization of cities and countries. Engineering science was critical to designing the technology to drive power generation and mechanical automation. Fast forward to 131 years after the steam engine technology, which powered the Industrial Revolution, the Second Industrial Revolution was powered by the conveyor belt, the newest technology to drive the automation process. This is a period in human history where mobility was a key driver of the Industrial Revolution; thus, the conveyor belt became an important technology to move items from one place to another in a more efficient way. Within four decades of this new development in the industrialization process, human thinking evolved, and more technologically efficient solutions were being designed to drive the Industrial Revolution. By 1954, electronic automation led by computers became the driving force in the Third Industrial Revolution. Just a mere 172 years after the steam engine, the technological efficiency of the Industrial Revolution process has improved tremendously, with electronics being the driving force behind the new phase of the Industrial Revolution.

Nevertheless, as humans' innovative and creative skills have advanced, newer technologies and smarter ways of working have become the norm. A mere six decades after the electronics revolution, the Fourth Industrial Revolution arrived with cyber and physical systems driving smart automation, which is the current phase of the Industrial Revolution. Indeed, with smart automation coming after 200 years of steam engine technology, this current wave of the Industrial Revolution will need advanced and sophisticated technological solutions to drive innovation at the country and city levels. This wave will deliver a higher quality of life, economic value added, and a sustainable future.

The rapid advancement in science and technology over the last 200 years has led to the development of a number of sophisticated technological breakthroughs, which can assist city planners in further advancing the quality of life of their citizens and businesses and taking the hassle out of urban living. This chapter provides a comprehensive but not exhaustive list of these new technologies, which are part of the 4IR drivers that can facilitate the movement of urban centers to smarter cities.

3.1   The Technologies

As the Fourth Industrial Revolution unfolds, smart cities are leveraging the pace and scope of ground-breaking scientific and technological advances to better assist in their planning and designing strategies to meet the demands of citizens and businesses. These technologies include but are not limited to artificial intelligence and machine learning, big data analytics, cloud computing, the Internet of Things (IoT), robotics, cobots, and intelligent automation. These emerging 4IR disruptive technologies would enable smart cities of the future to rapidly adapt to change and provide a flexible structure and operational approach that facilitates interaction with a wide range of stakeholders, including alignment with industry, government, and global citizens.

3.1.1 Big Data

Big Data are defined as having volume (e-commerce, mobility, and social media that generate large amounts of data), velocity (generating new data at a rapid pace), and variety (data in many different formats: emails, documents, images, videos, etc.) (Kitchin, 2014). This 4IR emerging technology is the fuel that would operationalize the efficiency and effectiveness of smart cities of the future. According to du Sautoy (2019), 1 exabyte (10^{18} bytes) of data is created on the internet every day, roughly the amount of data that can be stored on 250 million DVDs. To add, humankind now produces in 2 days the same amount of data that was generated from the dawn of civilization until 2003. The plummeting prices of computer processing power will continue to fuel the prominence of big data intelligence on a scale unimaginable just a few years ago for the future design and operation of cities.

For architects and urban planners, collecting and analyzing data to inform design decision-making have always been a natural stage in the early phase of designing cities. However, until now, this early stage of the design process called programming has been limited to a manual process. Grounded in research and decision-making, programming is a solution-seeking and problem-solving process in which city planners seek to understand the design problem they are trying to solve. To solve the complex design problems of smart cities, urban planners employ this human-centered programming process to interview hundreds of potential users to gain a deeper personal understanding of their experiences to inform the design solution. This process also includes research on hundreds of city design precedents, building and zoning codes, building materials, pedestrian and vehicular patterns, energy consumption, etc.

The rise of big data intelligence can now reduce the collection and analysis from months to weeks. Urban planners can leverage their firm's Big Data Intelligence cloud infrastructure to ingest the above structured and unstructured disparate data sets. They can employ machine learning (ML) and artificial intelligence (AI) to gain deeper insights by extracting themes, patterns, and trends from the data. This might *not be humanly possible*. Furthermore, algorithms can learn from past precedent performances and make predictions about what 4IR emerging technologies would achieve operational excellence in the design and human-centered experiences of smart cities of the future.

As observed earlier, big data is the 4IR technology that fuels the efficiency and effectiveness of smart cities of the future. If we continue to witness 1.3 million people moving into cities every week (West, 2011), Smart Cities of the Future will need big data analytics to keep pace with the massive amounts of data being collected by sensors created by the IoT and shared from disparate mobile devices. For example, transportation professionals would be able to quickly access more up-to-date and comprehensive data for every road in cities every day of the year. Entire cities would become a transportation data set full of unexploited potential. City planners and engineers would be able to use computer algorithms and ML to gain deep insights into data to generate answers to questions they did not even know they could ask (Streetlight, 2022).

Big Data Deep-Insight Analytics dashboards powered by AI that analyze data from sensors and IoT devices for pattern recognition would enable city dwellers and operators to gain real-time urban informatics for answering questions such as:

- What will create congestion in Cities of the Future? (Al Nuaimi et al., 2015)
- Where would congestion occur a month from today?
- How long would the congestion last?
- Which alternative routes would I be able to take to reach my destination without losing time?
- What would be the impact of traffic flow on connecting residents to key city centers during the labor day weekend?
- For this year's Tour de St. Johns, which city routes would separate cyclists from cars and pedestrians?

3.1.2 Cloud Computing

If big data is the fuel that would operationalize the efficiency and effectiveness of smart cities of the future, cloud computing is the container that holds the fuel. Cloud computing's cost-effectiveness, scalability, elasticity, reliability, security, and global availability make this 4IR technology essential for storing, processing, and mining the huge amounts of big data that cities of the future would generate (Cryptopas, 2020). To engage with their citizens more effectively and actively, governments of future smart cities would be able to rent computing resources from Amazon, Microsoft, or Google to leverage intelligent data generated from disparate sources to improve the performance of health, transportation, energy, education, and water services, leading to higher levels of comfort for their citizens (Mohamed et al., 2015). In recent years, businesses seeking to gain a competitive edge have been quick to embrace this model where the cloud provider is responsible for the physical computing hardware and for keeping it up to date. These cloud computing services include infrastructure-as-a-service (IaaS), platform-as-a-service (PaaS), and software-as-a-service (SaaS) (Microsoft, 2022).

In the words of Marc Andreessen (2011), "Software Is Eating the World." This Software-as-a-Service 4IR cloud computing service is scaffolding the critical infrastructure for Smart Cities of the Future shaping digital marketplaces in connecting citizens to commerce and services—the Digital Platform Economy (Microsoft, 2022). The most impactful companies of the past decade have been software platforms: Airbnb; Alibaba; Alphabet (Google) Amazon; Apple; and many more that do not start with the letter "A". According to McFadyen (2021), these software platform businesses have transformed entire industries: Amazon vs. retail; Airbnb vs. hotels; Apple vs. record companies; Uber vs. taxis; Instacart vs. grocers; DoorDash vs. restaurants; Craigslist vs. newspaper classifieds, etc. According to McKinsey, more than 30% of global economic activity—some $60 trillion— could be mediated by the digital platform economy in 6 years (McKinsey Global Institute, 2017).

To add, a 2018 Accenture study of the readiness index for Government as a Platform (GaaP) indicated that three in four citizens globally say government needs to tackle complex issues by collaborating with citizens, companies, and nongovernmental organizations (NGOs). Furthermore, 60% of citizens globally would themselves take an active role in personalizing government services, and seven in ten start-ups/entrepreneurs globally find that collaboration with public agencies is key to their companies' innovation activities (Le Masson, 2018). The term "Government as a Platform" is used to refer to the whole ecosystem of shared application programming interfaces (APIs) and components, open standards, and canonical data sets, as well as the services built on top of them and governance processes that keep the wider smart city system safe and accountable (Pope, 2019).

In the Smart Cities of The Future digital platform economy, the government as a platform acts as an intermediary to facilitate collaboration, connect citizens to commerce providers, and coordinate to deliver next-generation public services. This can be done through the orchestration of the IoT cloud services that receive, transmit, and monitor information signals from interconnected nodes (sensors) throughout cities. For example, in Spain, ICT company Telefonica has installed sensors that are attached to refuse containers to report, in real-time, how full they are—which means refuse collectors do not have to waste time traveling to bins that are only half full. It also means that key performance indicators (KPIs) can be more closely tied to bottom-line impacts, such as how many bins are close to overflowing and will not be emptied within the next few hours (McKinsey Global Institute, 2018). However, sensors alone are not sufficient. Smart Cities of the Future would need a mature software IoT platform to manage the sensors, receive and process data, and make these data available to smart solutions through application program interfaces (Dubbeldeman & Ward, 2020). According to Kirwan and Zhiyong (2020), the Software-as-a-Service (SaaS) 4IR cloud computing service is scaffolding the critical infrastructure for Smart Cities of the Future shaping digital marketplaces in connecting citizens to commerce and services—the Digital Platform Economy.

3.1.3   Internet of Things (IoT)

The Internet of Things refers to the massive use of advanced sensors and wireless communication in all kinds of physical objects. The wide-scale use of sensor technology creates massive volumes of data providing a fine-grained digital view of the physical world. These data can be used by smart systems that optimize the use of infrastructure and resources (Ward et al., 2015). To add, pervasive computing in Smart Cities of the Future would entail a vision of the world in which computing is not limited to tablets, smartphones, and laptops. The realization of this vision, called the "Internet of Things," is the ever-expanding collection of connected devices that capture and share data (Ornes, 2016). As city governments begin to unlock the full potential of urban data platforms, AI, smart devices, and interconnectivity, the need for IoT will grow exponentially, leading to efficient problem solving, smart mobility, sustainability, and more (Appleton, 2021). For example, public transport routes can be adjusted in real time according to demand, and intelligent traffic light systems can be used to improve congestion (Marr, 2020). In smart cities, each building can be outfitted with applied intelligence smart-energy meters with human-like capabilities that mimic cognitive functions—learning from volumes of data to increase the efficiency and effectiveness of the electric grid.

These applied pervasive intelligence agents include swaths of tiny sensors that will track everything in the city, from steps and calories to humidity and light (Scientific American, 2014). In addition, with a projection of 27.1 billion connected IoT devices in 2025, former Google and Alphabet executive chairman Eric Schmidt said it best when he made this bold IoT prediction: "[T]he Internet will disappear. There will be so many IP addresses, devices, sensors, things that you are wearing, and things that you are interacting with that you will not even sense. It will be part of our presence all the time" (Thomas, 2022). In sum, the IoT enables more efficient use of resources, improvements to services and safety, and creates a greater sense of connectedness within emerging smart cities (Butler & Lachow, 2016).

As we reimagine Smart Cities of the Future, it must be borne in mind that, currently, the most important generator of city data is a familiar tool: the ubiquitous mobile phone. These devices are, in effect, personal sensing devices that are becoming more powerful and more sophisticated with each product iteration (Pentland, 2015). Each mobile phone personal sensing device leaves breadcrumbs of who you called; who called you; where you are; how you move; how much you spent and where you spent; who you are with; who else is around; your health data; browsing, email, and apps history. This kind of data capture is critical for planning in smart cities, and highly complex and sophisticated technologies will need to be deployed to help city planners make sense of these data.

3.1.4   Artificial Intelligence (AI) and Machine Learning (ML)

We live in a technology marvel where computers continue to speed up while the price of processing power continues to drop precipitously, doubling and redoubling the capacity of machines. This is driving the advance of ML—the ability of computers to learn from data—and the push for AI (Aoun, 2017). In addition, AI technologies can perceive, learn, and reason to extend the capabilities of people and organizations, making them a pivotal 4IR technology enabler for smart cities of the future. As such, AI and ML 4IR technologies would be employed to extract patterns from live and historical data to diagnose and predict what measurable actions city officials should take to improve data-informed decision-making. Let us take energy systems as an example. It will be observed that in a system where energy spikes tend to occur, AI and ML can learn where they usually occur and under which circumstances, and this information can be used for better management of the smart power grid (Choudhary, 2019).

Furthermore, smart cities of the future could employ AI to optimize city services such as garbage collection, vehicle and pedestrian traffic, electricity use, and parking space based on myriad algorithms. However, "being smart" is more than just technology; it is about creating liveable, equitable, and sustainable cities where 4IR technology is first and foremost an enabler to support solutions for addressing urban challenges such as depletion of resources (financial and environmental), inequality, and climate change (Comer, 2016).

Similarly, in securing smart cities of the future, an intelligent network of sensors can capture hackers' behavior and patterns of attacks, while ML could use the large metadata generated from the sensors to create training data to be used to predict when the right indicator will be deployed to protect the cyber-physical infrastructure before a cyberattack happens (Charles, 2016). Sending the right indicators to protect the right machines before a cyberattack would be equivalent to healthcare professionals predicting what part of your body would be attacked by a disease and dispatching a protective medication to that specific area of the body before the disease attack occurs. The volume of patient data and scientific knowledge has increased to a level that cannot be understood or handled by humans anymore without the help of technology. As such, AI and cognitive computing will be applied to assist health professionals in interpreting medical data to establish the right diagnosis and define the most effective treatment. In addition, the increasing data volume combined with new 4IR technologies such as big data analytics, AI, and ML would create opportunities for better risk assessment to assist financial service providers who are highly dependent on their ability to estimate risks (Dubbeldeman & Ward, 2020). The ultimate goal is for AI to assist in the self-regulation of cities of the future as living systems. If we think of city governance as the process of triage, then self-regulating smart cities would use AI and ML to monitor systems in real time and anticipate problems.

3.1.5 Robots, Co-bots, and Intelligent Automation

Robotics, intelligent automation, and drones are some of the fastest-growing 4IR technologies. When coupled with IoT, these would be the vanguards in reimagining how we design, manage, live, work, and engage with each other—harmoniously with machines—in sustainable smart cities of the future. Furthermore, the McKinsey Global Institute (MGI) confirmed that AI can be combined with complementary technologies such as robotics to provide integrated solutions, including autonomous driving, robotic surgery, and household robots that respond to stimuli (Woetzel et al., 2018). A major 4IR innovation of relevance for Smart Cities of the Future is robots acquiring cognitive and social abilities. They use AI and speech recognition to interact with people through natural language. They also recognize and respond to human emotions and express their own emotions (Dubbeldeman & Ward, 2020). For example, robots now analyze stocks, write in def and informative pros, and interact with customers (Gray, 2016). In China, cobots—machines that can work in factories safely alongside human beings—are upending that country's vaunted manufacturing sector, allowing fewer laborers to be vastly more productive (Aoun, 2017). Indeed, according to the market research firm, Mordor Intelligence, algorithmic trading accounted for approximately 60–73% of the overall United States equity trading in 2020. This is a critical area for the future of smart cities (Mordor Intelligence, 2022).

4IR Cognitive Robotic Process Automation (RPA) tools and solutions leverage AI technologies such as Optical Character Recognition (OCR), Text Analytics, and ML to improve the experience of the workforce and customers. For example, predictive analytics can enable a robot to make judgment calls based on the situation that presents itself, while ML can enable the system to learn, expand capabilities, and continually improve on its own (NICE, 2022). Intelligent Automation is the next logical step Robotic Process Automation (RPA) is growing toward. Ideal for automating complex systems for smart cities of the future, intelligent automation is an advanced form of RPA that combines technologies such as structured data interaction (SDI), RPA, ML, natural language processing (NLP), natural language generation (NLG), AI decision systems, chatbots, and more (10×DS, 2020).

4   Securing the City's Critical Infrastructure

Securing the city's critical infrastructure from manmade and natural catastrophic threats would require deliberate planning to mitigate disruption in services. In addition, planners must consider that every intelligent node depends on complex, interconnected cyber systems—if not fortified—that can be the weak link through which the adversary can attack the city's critical infrastructure systems. A city's critical infrastructure sectors can include the chemical sector; commercial facilities sector; communication sector; critical manufacturing sector; dams sector; defense industrial base sector; emergency services sector; energy sector; financial services sector; food and agriculture sector; government facilities sector; healthcare and public health sector; information technology sector; nuclear reactors, materials, and waste sector; transportation systems sector; and water and wastewater systems sector (Ustun, 2021). It is clear that cyberattacks on any, some, or all of these critical infrastructure sectors can have direct impacts on things such as life-sustaining medical devices, industrial control systems running a power grid, a smart sensor indicating the malfunction of a plane's engine, or tools that examine water contamination. When planning future cities, it is more important than ever to bridge cybersecurity and operational risks to effectively protect critical infrastructure and business operations (LogRhythm, 2022).

Cyber-physical systems (CPS) are at the center of the unification of what have always been distinct physical and virtual worlds. While the convergence of our physical and virtual worlds is not conceptually new, it is the capability that CPS possess that creates one of the greatest intellectual and technical challenges of our time (Trevino, 2019). For example, a cyber incident or attack affecting the systems that control nuclear facilities or dams could be devastating if it resulted in the flooding of a city downstream from a dam or an explosion that spreads radiation over a wide radius (McCarthy et al., 2009).

Overall, while the 4IR technologies are critical to shaping and facilitating the city of the future, a fundamental issue that must be borne in mind is the security and safety of the technologies. Any breach or attack on these technologies will pose a huge risk to the efficient operations of these cities. Therefore, before employing these technologies, policymakers must perform the necessary risk analyses to assess which technology will give the best outcome.

5 Conclusion and Ways Forward

This chapter provides a comprehensive but not exhaustive list of Fourth Industrial Revolution technologies that are crucial to driving the operationalization of smart cities in the future. As urbanization increases and more of the population starts living in urban centers, the demand on city planners becomes more onerous as citizens and businesses require fast and efficient services, higher quality of life, and greater economic value-added and a more sustainable future. Put simply, citizens will want the benefits of urban living with less hassle of normal city life. The smart city aims to generate hassle-free living while giving citizens and businesses the benefits of city life. To effectively achieve this goal, planners will need sophisticated and complex planning tools to assist them in making the best decisions on all aspects of city life from efficiency in transportation to the consumption of natural resources in the most sustainable way.

Planning will, therefore, require a vast amount of data to generate critical information to aid in the decision-making process. To adequately capture these data, process it, and make sense of the information, the Fourth Industrial Revolution technologies itemized and explained in this chapter provide a strong portfolio of tools that planners can access to assist them in making better and more robust decisions in the execution of their smart city. Aggregating these technologies into one place where policymakers can compare and contrast their benefits and drawbacks is a strong value-added, which this chapter brings to the extant literature that focuses on technologies that can drive smart cities of the future. The technologies highlighted above will go a long way in assisting city planners in using the significant amount of data gathered through Internet of Things technologies to make informed decisions on the future of their cities. Data that will drive the development of smart cities, however, given the volume that is to be generated, human capacity will not be able to process the same. It will require algorithms powered by these 4IR technologies to process and make sense of the large volume of data to assist city policymakers in decision making. The insights from this chapter will be useful to help in the selection of the most appropriate technologies for this purpose.

References

Al Nuaimi, E., Al Neyadi, H., Mohamed, N., et al. (2015). Applications of big data to smart cities. *Journal of Internet Services and Applications, 6*, 25. https://doi.org/10.1186/s13174-015-0041-5

Andreessen, M. (2011). Why is software is eating the world. *Wall Street Journal*. https://www.wsj.com/articles/SB10001424053111903480904576512250915629460

Aoun, J. (2017). *Robot-Proof*. The MIT Press.

Appleton, J. (2021, May 11). *What is IoT and why is it important for smart cities*. Emerald Publishing Limited.

Butler, R. J., & Lachow, I. (2016). *Smart cities and the internet of things: Benefits, risks, and options*. New America. http://www.jstor.com/stable/resrep10510.5

Charles, C. (2016). *The national cyber sensing initiative using graph theory & cognitive computing*. Small Business Innovation Research (SBIR) Program proposal.

Choudhary, M. (2019). *Six technologies crucial for smart cities*. Shorturl.at/knKQV.

Comer, A. (2016). *Planning smarter cities*. BuroHappold Engineering. burohappold.com

Cryptopas. (2020, July 31). *What are smart cities and the role of blockchain in smart cities*? https://cryptotapas.com/role-of-blockchain-in-smart-cities/

Deloitte. (2022). *Smart cities of the future: From vision to reality*. https://www2.deloitte.com/global/en/pages/public-sector/solutions/gx-smart-cities-of-the-future.html. Accessed 7 May 2022.

du Sautoy, M. (2019). *The creativity code: How AI is learning to write, paint and think*. The Belknap Press of Harvard University Press.

Dubbeldeman, J., & Ward, J. (2020). The role of software platforms in smart cities. International *Journal of Information Management, 53*, 102276. https://doi.org/10.1016/j.ijinfomgt.2020.102276

Exponential Digital Solutions (10xDS). (2020). *Robotic process automation vs. intelligent automation*. https://10xds.com/blog/rpa-vs-intelligent-automation/

Gray, A. (2016, May 24). Mastercard to start trailing pepper the robot in pizza hut. *Financial Times*. http://www.ft.com/content/2b78d806-20f2-11e6-aa98-db1e01fabc0c

Kirwan, C., & Zhiyong, F. (2020). *Smart cities and artificial intelligence: Convergent systems for planning, design, and operations*. Elsevier.

Kitchin, R. (2014). Big Data, new epistemologies and paradigm shifts. *Big Data & Society, 1*, 2053951714528481. [CrossRef].

Le Masson, B. (2018). *Government as a Platform: 2018 GaaP Readiness Index*. Accenture.

LogRhythm. (2022). *Overcoming OT and IT security challenges*. LogRhythm.

Marr, B. (2020). *The smart cities of the future: 5 ways technology is transforming our cities*. Forbes. https://urlcc.cc/6vt9y

McCarthy, J. A., Burrow, C., Dion, M., & Pacheco, O. (2009). *Cyberpower and critical infrastructure protection: A critical assessment of federal efforts*. University of Nebraska Press; Potomac Books.

McFadyen, T. (2021). *Marketplace best practices: Transforming commerce in the platform economy*. McFemale Digital Publications.

McKinsey Global Institute. (2017). *Artificial intelligence: Implications for China*. McKinsey & Company.

McKinsey Global Institute. (2018). *Smart cities in Southeast Asia*. Produced for World Cities Summit 2018 in Collaboration With the Center for Livable Cities.

Microsoft. (2022). *Smart cities: The cities of the future*. https://www.microsoft.com/en-us/industry/government/resources/smart-citie. Accessed 7 May 2022.

Mohamed, A., Alenezi, A., & Khan, M. A. (2015). The role of cloud computing in smart cities. *International Journal of Computer Science and Network Security, 7*(9), 18–25. https://doi.org/10.5815/ijcsns.2015.09.04

Mordor Intelligence. (2022). *Algorithmic trading market – Growth, trends, COVID-19 impact, and forecasts (2022–2027)*. https://www.mordorintelligence.com/industry-reports/algorithmic-trading-market

NICE. (2022). *What is cognitive robotic process automation?* https://www.nice.com/guide/rpa/what-is-cognitive-rpa

Ornes, S. (2016). The internet of things and the explosion of interconnectivity. *Proceedings of the National Academy of Sciences of the United States of America, 113*(40), 11059–11060.

Pentland, A. (2015). *Social physics: How social networks can make us smarter*. Penguin Books.

Pope, R. (2019). *Playbook: Government as a platform*. Ash Center for Democratic Governance and Innovation, Harvard Kennedy School.

Porter, M. E. (1990). *Competitive advantage of nations*. The Free Press.

Scientific American. (2014). The city of the future is here. *Scientific American, 310*(5), 44–51. https://doi.org/10.1038/scientificamerican0514-44

Streetlight. (2022). The potential of machine learning for smart cities. *The Journal of Smart Cities, 5*(2), 679–692. https://doi.org/10.1109/JSC.2022.3157034

Thomas, M. (2022). *27 Top internet of things examples you should know*. https://builtin.com/internet-things/iot-examples

Trevino, M. (2019). *Cyber physical systems: The coming singularity* (Vol. 8, No. 3, pp. 2–13). Institute for National Strategic Security, National Defense University. PRISM. https://www.jstor.org/stable/10.2307/26864273

Ustun, D. (2021). *National interest exception for travel: What are 16 critical infrastructure sectors in US?* https://ustunlawgroup.com/national-interest-exception-for-travel-what-are-16-critical-infrastructure-sectors-in-us/

Ward, J., Kim, J., & Dey, A. K. (2015). Smart city sensors: A survey. *ACM Computing Surveys (CSUR), 48*(1), 1–34.

West, G. (2011). *The surprising math of cities and corporations*. www.ted.com

Williams, D. A., & Morgan, B. (2013). *Competitiveness of small nations: What matters?* (pp. 1–187). Arawak Publications.

Woetzel, J., et al. (2018). *Smart cities: Digital solutions for a more livable future*. McKinsey Global Institute.

Open Access This chapter is licensed under the terms of the Creative Commons Attribution 4.0 International License (http://creativecommons.org/licenses/by/4.0/), which permits use, sharing, adaptation, distribution and reproduction in any medium or format, as long as you give appropriate credit to the original author(s) and the source, provide a link to the Creative Commons license and indicate if changes were made.

The images or other third party material in this chapter are included in the chapter's Creative Commons license, unless indicated otherwise in a credit line to the material. If material is not included in the chapter's Creative Commons license and your intended use is not permitted by statutory regulation or exceeds the permitted use, you will need to obtain permission directly from the copyright holder.

Financing of Smart City Projects

Isam Shahrour

Abstract This chapter deals with financing smart city projects, which generally constitutes a significant barrier to smart city transformation. The chapter is organized into four sections. The first section presents the emergence and evolution of the smart city concept, and then it discusses the barriers facing this concept with an emphasis on financing smart city projects. The second section highlights the city challenges and the need for a smart city transformation to respond to citizens' demand for modernized services and the legally binding international treaties on climate change. The third presents the smart city concept with an emphasis on the capacity of the city to address city challenges and the architecture of the smart city system. The last section describes the need and financing for the main sectors of the smart city: communication infrastructures, data infrastructures, public infrastructures and services, and private infrastructures and services. The chapter shows that considering city challenges and the city's role in global challenges, smart city transformation becomes a must. Smart city financing could be covered by regional and green funds, savings related to reducing energy consumption, and increased competitiveness gained by digital transformation.

Keywords Smart City · Financing · Barriers · Citizens · Green deal · Climate agreement · Infrastructure · Data · Buildings · Mobility · Funds · Competitiveness

1 Introduction

The smart city concept has emerged as a powerful instrument to address urban challenges in the past two decades. In the first period, smart city projects focused on the safety and efficiency of energy infrastructures. The smart grid concept emerged as an efficient solution to address the risk of electrical blackouts. The use of smart sensors in the smart grid proved to be efficient in detecting electrical failure early and

I. Shahrour (✉)
Laboratoire de Génie Civil et géo-Environnement, Université de Lille, Lille, France
e-mail: Isam.Shahrour@univ-lille.fr

© The Author(s) 2024
F. Belaïd, A. Arora (eds.), *Smart Cities*, Studies in Energy, Resource and Environmental Economics, https://doi.org/10.1007/978-3-031-35664-3_3

taking preventive actions to avoid failure and limit its rapid extension in the grid. The success of the smart grid created significant promise for the use of smart technology to improve the efficiency and safety of complex urban systems such as urban mobility, water supply, sanitation, building management, and municipal waste collection. Considering the urgent need for the green transition and the development of the digital economy, regional and national authorities launched extensive programs to boost smart city implementation. For example, the European Union launched initiatives to "bring together cities, industry, and citizens to improve urban life through more sustainable integrated solutions" (EIP-SCC, 2020). China launched an ambitious program concerning the construction of 500 smart city pilot projects (Deloitte, 2018; Hu, 2019). India has initiated a smart city program to develop 100 smart city projects (India Smart City, 2015).

Smart city implementation has been facing many barriers. Some of them are related to a lack of experience and capacity building. Others are associated with the absence of regulations and the disruptive transformation in urban management. The challenge for the city managers is how they could transform the silo-based management organization into a transversal and shared management model.

The issue of financing also appeared as a fundamental barrier with many questions (Blanck & Ribeiro, 2021; He et al., 2020; Mishra et al., 2019): How can the digital infrastructure of the smart city be financed? How can transversal expenses be shared with the urban "silo"? How can new investments be found in a challenging period? What is the economic return of the investment? In addition, we must consider other benefits of the smart city, such as building a sustainable and resilient city with a positive impact on social, economic, and environmental development. Who should pay for these benefits?

This chapter attempts to address financing barriers through an approach that starts with identifying the expectations of smart city transformation. This phase will be followed by exploring the boosters for embracing smart city implementation and their consequences for finance mobilization. The last part proposes a framework for financing smart city projects.

The contribution of this chapter concerns the identification of financial barriers facing smart city projects and the ways to address these barriers with emphasis on national and regional and green funds, savings in resource consumption, increased competitiveness gained by digital transformation, and the development of new services in the smart city environment.

2 City Challenges

2.1 Overview

The smart city emerged first as an efficient technology to improve the safety and efficiency of energy, transport, and water infrastructures and facilitates. Then, it was extended to services related to these infrastructures through the involvement of citizens as end users of these services.

Although the concepts of smart infrastructure and smart services were promising, their implementation faced significant barriers such as the lack of experience and capacity building, inadequate regulation, incompatibility with the silo management model, lack of funds, and the highly economical risk for the private sector. The objective of the smart transformation of urban infrastructures and services appeared very complex to achieve without ambitious regional and national supporting programs.

In recent years, two phenomena have boosted the need for smart city transformation. The first is related to the expectation of the population for digital and ecological adaptations of services and governance modes. The second is associated with the European and national respect for the legally binding international treaty on climate change, called the Paris Agreement (UNFCCC, 2016).

2.2 Citizens Demand Ecological and Digital Services

The rapid development of digital technology, particularly the internet, social media, and smartphones, created a world of connected people. The emergence of the Internet of Things (IoT) extended the human-connected world to that of things. Today, citizens live in an intensively connected world and consider this world to be the new normal. Consequently, they expect the public and private sectors to embrace digital transformation through its intensive use in services, management, and governance processes. This expectation results in massive pressure on the public and private sectors. In a competitive economy, companies could lose business or eventually disappear if they do not respond to the population's expectations. The challenge for the public sector is also significant because the absence of digital transformation could deteriorate the quality and efficiency of public services. It could also result in a transfer of some of these services to the private sector, increasing social inequity. The public sector is also subjected to the population's demand for increased participatory governance.

The rise in the population's awareness of the green transition also results in increasing pressure on the public and private sectors to embrace the green transition. The increasing social and environmental concerns of citizens and public authorities have resulted in high pressure on both private and public organizations to embrace and implement policies related to social and environmental responsibility.

Since the digital and green transformations of services and governance modes face tremendous barriers, the public and private sectors need to develop and implement technical and social innovations to achieve these transformations.

2.3   *Legally Binding International Treaty on Climate Change, Paris Agreement*

Due to global warming and climate change challenges, cities' transformation into sustainable and smart cities has become the top priority of regional, national, and local authorities. Indeed, since the city is responsible for 80% of greenhouse gas emissions, it has a significant role in achieving the goal of the legally binding international treaty on climate change (Paris Agreement on climate change) to keep global warming below 2 °C in 2100. The Paris agreement also aims to strengthen countries' capacity to address climate change's impacts through appropriate financial flows, innovation, and enhanced capacity building.

The achievement of the goal of the Paris agreement requires a massive cut in greenhouse gas emissions. Recently, the E.U. announced an ambitious commitment for reducing greenhouse gas emissions by at least 55% by 2030 compared to 1990 and becoming climate-neutral by 2050. The European climate law will translate this commitment into a legal obligation. The climate goal of the E.U. requires a profound transformation in sectors concerned by high gas emissions, such as buildings, transportation, energy production, and industry. The goal is to reduce the energy consumption in these sectors, particularly fossil energy, and to increase the use of renewable energy. This mandatory goal will help mobilize public and private resources, accelerate the update of laws and regulations, and boost innovations.

3   Smart City

3.1   *How the Smart City Could Help Respond to the City's Challenges*

The smart city is based on extensive monitoring of urban systems and the environment using smart sensors and crowdsourcing. The collected data could be completed by data from other valuable sources, as illustrated in Table 1 (Shahrour & Xie, 2021). Comprehensive urban data have high economic value because they can help companies understand market expectations and then adapt their products and services to these expectations or develop innovative services (OECD, 2019). As "data" becomes seen as the "new oil," cities could create new financial resources from these data.

Analyzing collected data using engineering tools and artificial intelligence algorithms allows real-time tracking of urban infrastructures, services, and the environment. It also allows rapid detection and localization of abnormal events and the ability to take adequate response actions. Consequently, intelligent cities help improve city safety and security.

Data analysis also permits exploring the performances of several management scenarios and selecting the best one to optimize the management of both

Table 1 Data for the smart city (Shahrour & Xie, 2021)

Urban system	Source	Data
Urban infrastructures	The city administration, urban services providers, facility managers	Digital model including geo-referenced data for architectures and components (GIS, BIM, etc.), functioning data (traffic, congestion, consumptions, flow, pressure, quality, tension, frequency, temperature, humidity, accessibility)
Urban environment	The city administration, environmental and weather agencies, NGOs, urban services providers, citizens, public authorities	Indicators concerning air pollution, quality of water and soils, biodiversity including green areas, biological species, public health indicators, as well as safety and security
Urban services	The city administration, urban services providers (transport, water, energy, municipal wastes.), citizens, companies,	Indicators concerning the quality, availability, affordability, and risk, of urban services (mobility, energy, and water supply, telecommunication, municipal wastes, sanitation, health, education, cultural, sportive, and artistic activities, etc.)
City stakeholders	citizens, policymakers, urban services providers, and socioeconomic actors	Data for citizens concerning urban indicators (urban services, strategies, significant projects, impact analysis, finance, etc.) Data from citizens, including feedback and evaluation about urban services, city functioning, quality of life, and improvement suggestions
Socioeconomic activities	The city administration, public authorities, social activity managers and providers, economic actors	Indicators concerning type and distribution of socioeconomic activities, building capacity, innovative industrial power, city attractiveness, availability and use of cultural and sportive facilities, commercial and industrial land, and labor availability

infrastructures and services. The share of urban data permits understanding the interaction and interdependency among these systems and then developing comprehensive management considering both the security and efficiency of urban systems. The smart city system also provides extensive data to citizens. Consequently, it enhances governance transparency and participation. Innovative companies can use available data to develop new services and businesses and reinforce their competitiveness.

Globally, smart cities help transition to digital society and respond to citizens' expectations to access modernized services. It also helps to optimize the functioning of urban infrastructures and services, emphasizing the reduction of energy consumption, greenhouse gas emissions, and air pollution. Consequently, it helps authorities achieve the climate goal.

Since the smart city concerns many interdependent infrastructures, services, the environment, and stakeholders, it is very complex to estimate the investment and running costs and their share. Therefore, we need a good understanding of the smart city's functioning and related costs. This issue is discussed in the following section.

3.2   Smart City Architecture

The smart city system is operated through a central platform called a smart platform, which coordinates the connectivity and functioning of the smart city components and data flow and operation (Shahrour & Xie, 2021). The platform architecture includes different layers (Silva et al., 2018; Alvear et al., 2018). For example, Fig. 1 shows a smart city architecture based on six layers.

The first layer corresponds to urban physical systems, including infrastructures, environment, services, and stakeholders.

The second layer is the sensing layer, which operates data collection and transmission to the smart platform. This layer includes the following components: (i) Internet of Things (IoT), including sensors, cameras, and RFID (Bandyopadhyay & Sen, 2011; Mulligan & Olsson, 2013); (ii) data from public authorities such as weather, air quality, and traffic information; (iii) open data from public authorities; and (iv) crowdsourcing, which corresponds to data transmitted by citizens.

The third layer concerns data transmission to the smart city platform. It combines wired and wireless communications protocols. The former uses phone lines, ethernet cables, fiber optic, and PLC lines (Power Line Communication). The latter refers to various technologies such as Wi-Fi, radio, LP-WAN, including LoRa and Sigfox, Bluetooth, and 2G, 3G, 4G, and 5G connections (Ejaz et al., 2016). Generally, wireless technology is used at the local level to transmit data to a local gateway, which then transmits data to the smart city platform via wired or wireless communication protocols (Fig. 2).

The fifth layer is the control layer. It includes IoT devices such as actuators, switches, breakers, valves, electronic motors, pumps, and locks. It also provides communication tools with citizens and other stakeholders for data collection and diffusion.

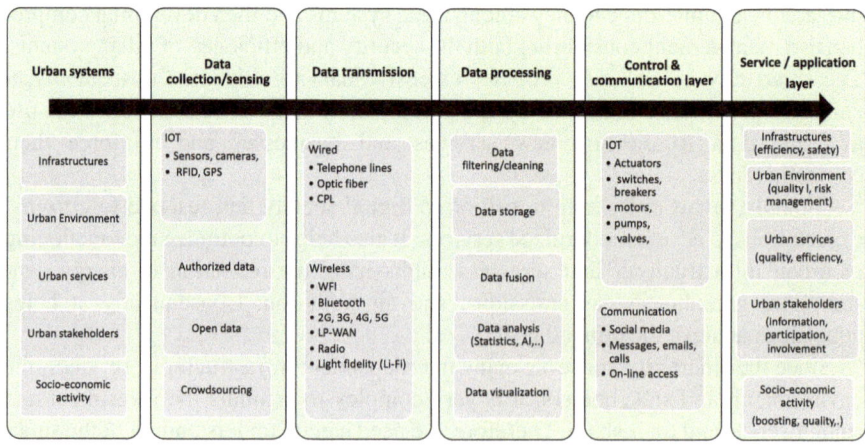

Fig. 1 Smart city architecture based on six layers (Shahrour & Xie, 2021)

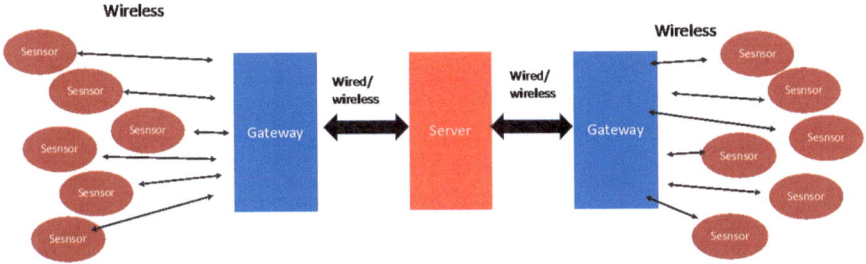

Fig. 2 Data communication layer (Shahrour & Xie, 2021)

The last layer corresponds to smart services such as (i) the optimal and safe management of urban infrastructures, (ii) the survey of the urban environment, and (iii) the enhancement of the quality of urban services, for example, optimization of public transport and traffic, real-time mapping of parking availability, rapid detection of water leakage or contamination (Farah & Shahrour, 2017; Saab et al., 2019), optimization of the stormwater system (Abou Rjeily et al., 2017), early detection of electrical outages, and optimization of the public lighting system. It could also include smart tools for managing utilities hosted in shared spaces (Shahrour et al., 2020).

This layer includes interactive tools to inform citizens about city news, projects, events, realizations, and challenges to reinforce their involvement in participatory governance.

4   Finance of the Smart City

4.1   Overview

The financing of the smart city covers both investment and operating costs. This section focuses on the financing of the investment because it constitutes the principal barrier to smart city financing. Figure 3 illustrates the smart city infrastructures according to the operating sector and owner. It includes four sectors: communication infrastructures, data infrastructures, public infrastructures and services, and private infrastructures and services.

The following subsections present the financial needs and resources for each sector for its smart transformation, emphasizing the French context.

4.2   Communication Infrastructures

The first sector concerns the communion infrastructures that ensure data transmission among smart city systems. Data transmission uses wired and wireless protocols. The former includes phone cables and internet cables, particularly fiber optic

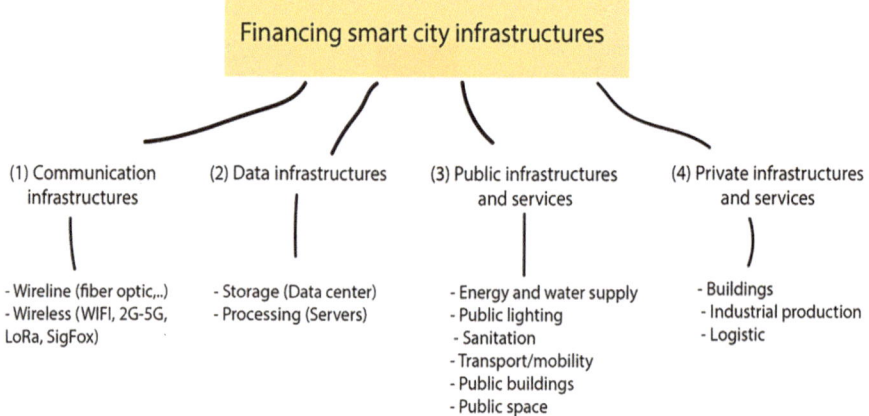

Fig. 3 Finance of smart city infrastructures (Shahrour & Xie, 2021)

cables. The owner of the wired infrastructure depends on the country. In France, the private sector owns this infrastructure, but it benefits from public aid to cover low-density populated areas. The private sector mainly owns wireless infrastructure. The income of related services covers the finance of the communication infrastructure.

4.3 Data Infrastructures

The second sector concerns data storage and processing. The private sector developed high-capacity data centers that offer customized cloud services to store data and data processing. The income of provided services covers the financing of these data centers. Some public institutions have their own data centers, but they were designed to operational needs. Creating a smart city data center requires gathering or connecting existing public data centers and servers into a data center that can host and process the intensive amount of smart city data. It requires public investments, which could be covered by services related to the smart city and businesses connected to data. The data infrastructure could also benefit from a public–private partnership with an appropriate economic model.

4.4 Public Urban Infrastructures and Services

The third part concerns public urban infrastructures and services such as those related to water and energy distribution, sanitation, waste collection, transport, public lighting, public space, and public building. The responsibility of these infrastructures depends on the national codes. In France, cities are responsible for these infrastructures. Their management could be assumed by cities or delegated to external operators. The smart transformation of the energy distribution infrastructure is a must for operating security and management issues. Societies in charge of energy distribution have already launched national programs for installing smart meters. The smart transformation of the water distribution could be included in the water delegation contract. End users will then ensure the finances.

The smart transformation of stormwater infrastructures is a must in cities subjected to high flood risk. It could benefit from national and local funds related to flood fighting. Insurance companies should also contribute to this program.

The smart transformation of public lighting results in a reduction of approximately 75% in energy expenses. Consequently, this transformation could be easily financed by energy savings.

The smart transformation of public buildings constitutes the biggest challenge for cities because of the poor energy performance of buildings and the enormous financial need for their green transformation. The recent French tertiary decree (Legifrance, 2019) has fixed the following obligations for energy savings in public and private tertiary buildings: 40% by 2030, 50% by 2040, and 60% by 2050. Buildings'green transformation is also a concern of the European Green Deal (E.U., 2019) and the European Climate Law (E.U., 2021). Considering the enormous financial need to achieve the goal, European and national authorities should support this program through climate funds, which are precious for the first stages of the buildings' green program. Energy savings obtained in the first stages could be used to achieve the subsequent stages of the program.

4.5   Private Infrastructures and Services

The private sector's responsibility concerns three main domains: buildings, industrial production, and logistics. The smart transformation of the industrial and logistic sectors is necessary for regulatory and competitiveness issues. It could also anticipate a social demand for green industry. The private sector could benefit from European and national green funds to achieve the smart transformation of these sectors. Companies can also use the funds' return on energy savings and competitiveness in the first stages to finance the latter stages.

The green building transformation program constitutes a major challenge for the private sector because of the enormous financial need to achieve this program. However, since this program contributes to national and European climate strategies (Legifrance, 2019; E.U., 2019, 2021), it can benefit from European and national climate funds for the first stages of the program.

5 Conclusion

This chapter discussed the financing of smart city projects, which constitutes a significant barrier to smart city transformation. Analysis shows that the smart city concept constitutes an efficient and adequate response to citizens' demand for modernized and green services and for national and regional authorities to respect legally international treaties on climate change. The implementation of smart city projects requires finances for communication infrastructures, data infrastructures, public infrastructures and services, and private infrastructures and services. However, some of these finances are common with other programs, such as energy security supply, health protection, green industry, green buildings, and green mobility. Smart city projects could also benefit from the economic return of energy savings and increased performance and competitiveness. Cities could use data as a financing source. Finances could also be generated by developing new services in the smart city environment. Cooperation between the private and public sectors constitutes an efficient way to finance smart city projects with a mutual benefit for these sectors, society, and the environment.

Since finance constitutes a significant challenge for smart city projects, policymakers should explore direct and indirect financing sources and integrate smart city projects into existing and future infrastructure and urban social projects. This integration reduces the cost of urban projects and contributes to encouraging private finances and social involvement.

References

Abou Rjeily, Y., Abbas, O., Sadek, M., Shahrour, I., & Chehade, F. H. (2017). Flood forecasting within urban drainage systems using NARX neural network. *Water Science and Technology, 76*, 2401–2412.

Alvear, O., Calafate, C. T., Cano, J.-C., & Manzoni, P. (2018). Crowdsensing in smart cities: Overview, platforms, and environment sensing issues. *Sensors, 2018*(18), 460. https://doi.org/10.3390/s18020460

Bandyopadhyay, D., & Sen, J. (2011). Internet of things: Applications and challenges in technology and standardization. *Wireless Personal Communications, 58*, 49–69. https://doi.org/10.1007/s11277-011-0288-5

Blanck, M., & Ribeiro, J. L. D. (2021). Smart cities financing system: An empirical modeling from the European context. *Cities, 116*(2021), 103268. https://doi.org/10.1016/j.cities.2021.103268

Deloitte. (2018). *The challenge of paying for smart cities projects.* Available in-line https://www2.deloitte.com/au/en/pages/about-deloitte/articles/challenge-paying-smart-cities-projects.html (Accessed 16/05/2022).

EIP-SCC. (2020). *European innovation partnership for smart cities and communities.* https://ec.europa.eu/info/eu-regional-and-urban-development/topics/cities-and-urban-development/city-initiatives/smart-cities_en#european-innovation-partnership-on-smart-cities-and-communities (Accessed February 20, 2020).

Ejaz, W., Anpalagan, A., Imran, M. A., Jo, M., Naeem, M., Qaisar, S. B., & Wang, W. (2016). Internet of Things (IoT) in 5G wireless communications. *IEEE Access, 2016*(4), 10310–10314. https://doi.org/10.1109/ACCESS.2016.2646120

E.U. (2019). *European Green Deal,* European Commission. https://europeanclimate.org/the-european-green-deal/ (Accessed 15/5/2022).

E.U. (2021). *European Climate Law: Framework for achieving climate neutrality and amending Regulations.* Regulation (EU) 2021/1119 of the European Parliament and of the Council. https://eur-lex.europa.eu/legal-content/EN/TXT/?uri=CELEX:32021R1119 (Accessed 15/5/2022).

Farah, E., & Shahrour, I. (2017). Leakage detection using smart water system: Combination of water balance and automated minimum night flow. *Water Resources Management, 2017*(31), 4821–4833. https://doi.org/10.1007/s11269-017-1780-9

He, Z., Liu, Z., Wu, H., Gu, X., Zhao, Y., & Yue, X. (2020). Research on the impact of green finance and fintech in smart city. *Complexity, 2020*, 6673386. https://doi.org/10.1155/2020/6673386

Hu, R. (2019). The State of Smart Cities in China: The Case of Shenzhen. *Energies, 12*(22), 4375. https://doi.org/10.3390/en12224375

India Smart City. (2015, June). *Smart cities mission statement & guidelines*, Government of India Ministry of Urban Development. http://smartcities.gov.in/upload/uploadfiles/files/SmartCityGuidelines(1).pdf (Accessed 20 February 2020).

Legifrance. (2019). *Obligations d'actions de réduction de la consommation d'énergie finale dans des bâtiments à usage tertiaire.* Décret n° 2019–771 du 23 juillet 2019, https://www.legifrance.gouv.fr/jorf/id/JORFTEXT000038812251 (Accessed 15/5/2022).

Mishra, A. K., George, H., & Mohring-Harwitz Theorems. (2019). Lessons for financing smart cities in developing countries. *Environment and Urbanization ASIA, 10*(1), 13–30. https://doi.org/10.1177/0975425318821797

Mulligan, C. E., & Olsson, M. (2013). Architectural implications of smart city business models: An evolutionary perspective. *IEEE Communications Magazine, 51*, 80–85. https://doi.org/10.1109/MCOM.2013.6525599

OECD. (2019). *Digital innovation: Seizing policy opportunities.* Retrieved from http://www.oecd.org/innovation/digital-innovation-a298dc87-en.htm

Saab, C., Shahrour, I., & Hage Chehade, F. (2019). Risk assessment of water accidental contamination using smart water quality monitoring. *Exposure and Health, 2019*, 281. https://doi.org/10.1007/s12403-019-00311-1

Shahrour, I., & Xie, X. (2021). Role of Internet of Things (IoT) and crowdsourcing in smart city projects. *Smart Cities, 4*, 1276–1292. https://doi.org/10.3390/smartcities4040068

Shahrour, I., Bian, H., Xie, X., & Zhang, Z. (2020). Use of smart technology to improve management of utility tunnels. *Applied Sciences, 2020*(10), 711. https://doi.org/10.3390/app10020711

Silva, B. N., Khan, M., & Han, K. (2018). Toward sustainable smart cities: A review of trends, architectures, components, and open challenges in smart cities. *Sustainable Cities and Society, 38*, 697–713. https://doi.org/10.1016/j.scs.2018.01.053

UNFCCC. (2016). *United Nations Framework Convention on Climate Change (UNFCCC). The Paris Agreement.* Available on-line: https://unfccc.int/sites/default/files/resource/parisagreement_publication.pdf (Accessed 16/05/2022).

Open Access This chapter is licensed under the terms of the Creative Commons Attribution 4.0 International License (http://creativecommons.org/licenses/by/4.0/), which permits use, sharing, adaptation, distribution and reproduction in any medium or format, as long as you give appropriate credit to the original author(s) and the source, provide a link to the Creative Commons license and indicate if changes were made.

The images or other third party material in this chapter are included in the chapter's Creative Commons license, unless indicated otherwise in a credit line to the material. If material is not included in the chapter's Creative Commons license and your intended use is not permitted by statutory regulation or exceeds the permitted use, you will need to obtain permission directly from the copyright holder.

Smart Cities: Development and Benefits

Razan Amine

Abstract With rapidly increasing urbanization, new ways of living and policies are racing to catch the train of modernization while meeting the changing demands of growing populations. Following the introduction, the second part of this chapter describes the pillars of smart cities as a response to growing worldwide urbanization. The third part details benefits across areas, including economic, demographic, environmental, social, and health. The fourth part sheds light on the nature of governance required for this transition into intelligent and interconnected cities. However, on the path of transition, governments face significant challenges— outlined in part fifth. Finally, given the great need for data-driven policy, the final section suggests ways of smart data collection that meet the standards of research and governance in smart cities. The last part concludes.

Keywords Cities transformation · Benefits of Smart Cities · Cities governance · Urban policy

1 Introduction

What are the features of a city that lie at the intersection of citizen welfare, environmental sustainability, and technological advancement? This chapter describes both the need for and the pillars of smart cities in a general context, in reference to some examples from various regions of the world. First, the chapter discusses in detail the pillars of a smart city. Relatedly, it offers the resulting benefits of a smart city in various dimensions. Moving into practice, the chapter outlines the features of governance required to establish the development of smart cities. Expectedly, the chapter then delves into the challenges of smart cities, as the transition depends on the initial level of development that the city has been characterized with. As part of smoothing challenges, data play an immense role in

R. Amine (✉)
Cambridge University, The Old Schools, Trinity Ln, Cambridge, UK

© The Author(s) 2024
F. Belaïd, A. Arora (eds.), *Smart Cities*, Studies in Energy, Resource and Environmental Economics, https://doi.org/10.1007/978-3-031-35664-3_4

informing policies that in turn foster the features of smart cities. This chapter contributes to a growing literature on smart cities and further motivates the need for smart cities as a solution to the problems created by urbanization.

2   Growing Worldwide Urbanization and Its Challenges

With rapidly increasing urbanization, new ways of living and policies are racing to catch this train of modernization and meet the changing demands of growing populations. Urbanization leads to a set of challenges that span all city areas. First, the orientation toward maximization of production has led to the exhaustion of natural resources. Firms and businesses are motivated by competition and profits, which in turn leads to focusing on revenues as opposed to environmentally friendly production. The challenge here is that the government's role is needed to further foster environmental taxation on firms and on polluters. Second, mass production through industrialization has exacerbated pollution levels of air and water, further harming the environment and in turn increasing the number of diseases. This is also driven by the massive competition and focus on profit maximization. Third, from a social perspective, urbanization has contributed to higher crime levels due to indirectly incentivizing disorder, chaos, stemming from a greedy mentality of production and possession. Fourth, urbanization has contributed to very high levels of pollution. This in turn motivates the usage of environmental taxes, which are a powerful way to curb down these dismal practices because the problem is that firms are not interested in internalizing environmental costs unless they face a tax on their profits. Of course, ideally, by introducing a value system, firms would be more interested in taking care of environmental sustainability, but that would be a long-term goal. By imposing the "polluter pays" principle on polluting firms or businesses or even individuals, governments can force the private sector to internalize the cost of harming the environment. Morocco and Tunisia have already introduced these taxes, but the extent of their true impact is unclear. Moreover, all the countries of North Africa continue to implement fossil fuel subsidies, a policy once aimed at protecting low- and middle-class citizens, which, given that high-income households are more likely to own cars, effectively amount to an upper-class license to pollute. As a result of these harmful impacts of urbanization, the health of individuals is put at stake, where water pollution, air pollution, and natural resource exhaustion in turn contribute to more diseases and lower life expectancy in the longer term and deteriorate the quality of life in both the short and the longer term. One other dimension could be increasing inequality, and this is reasoned by the fact that rich people focus on production while imposing a negative externality through harming the environment for everyone else and exhausting the resources of future generations.

For example, in the MENA region, an above-average population growth rate of 1.56% per year (vs. 1.1% globally) and a high speed of urbanization leads to the high importance of smart city solutions. While in 1960, less than 40% of the MENA population was living in cities, in 2020, this rose to above 60% and is expected to significantly increase further (World Bank, 2022).

Addressing the challenges of high population growth and urbanization, urban agglomerations need to be designed intelligently, enabling sustainable living in "smart cities."

3   Benefits of Smart Cities

This section sheds light on the multifold benefits, on numerous levels, of smart cities. First, let us define a smart city, although there is no one universal definition. A broad definition of a smart city is the use of information and communications technology (ICT) in governance and daily life in a city. In fact, the first use of the phrase "smart city" dates back to 1990 and has been associated with globalization, technology, and creativity. It can also be defined as an instrumented, interconnected, and intelligent city (Harrison et al., 2010). A third definition of smart cities involves outlining their characteristics, including but not limited to economy, governance, environment, people, and mobility. Each definition includes elements of technical structure, application domain, system integration, and data processing. It is important to distinguish between the following three overlapping but different terms to characterize a modern city: a digital city that is based on technology, an intelligent city that depends on artificial intelligence, and a smart city that focuses on the user-friendly adaptation of technological solutions. All definitions of smart cities point to the intersection between knowledge and technology to enhance sustainable development, thus making smart cities resilient and inclusive to all "users" or citizenst, as well as adaptable to shocks.

The concept of a smart city was first implemented in the United States at the time of President Barack Obama. Numerous initiatives followed, including the Digital Agenda initiative in Europe, i-Japan strategy in Japan in 2015, and Intelligent Nation 2015 plan in Singapore (Yin et al., 2015).

There are multiple benefits for smart cities, offering solutions for urbanization-caused challenges and new approaches for optimal policy making. These benefits become visible when shining light on the varieties of application domains of smart cities. First, smart cities make government work more efficient. It enhances e-governance, including services such as e-taxation and online documentation of public documents. It also boosts the role of the government in emergency response, transparency, and public safety (Yin et al., 2015). Relatedly, smart cities focus on users, i.e., citizens, to ease their lifestyle, including public transportation that reduces traffic, ample high-quality education and health services, and enhances social cohesion by boosting well-being (Yin et al., 2015). Third, smart cities are designed to make businesses more prosperous by diversifying efficient and need-based production in various sectors starting with agriculture and injecting innovation into entrepreneurship, marketing, and management. Fourth, smart cities enhance environmental protection. Smart cities are ideally based on renewable energy sources and sustainable water supply, reducing all types of pollution (Yin et al., 2015). Relatedly, to achieve environmental sustainability, smart cities implement resilient infrastructure, including the industrial structure and sewage system, necessary for the long-term prosperity of the city.

For example, in the MENA region, a new city called "The Line" is envisioned to grasp all mentioned benefits through its smartly planned infrastructure and even going beyond. As a new city, it is possible to design it on a line that is planned to optimize transportation inefficiencies, minimizing the time needed for personal mobility and business logistics. Another example is Abu Dhabi's Masdar City, which is often seen as a frontrunner for smart city development, reaching interconnectedness and minimal environmental pollution through large greenfield investments (Ringel, 2021). This top-down government-steered transition to smart cities is typical for the smart city approach in the MENA region (with dedicated initiatives of high prominence in Casablanca, Algiers, Cairo, Kuwait, Doha, Dubai, Abu Dhabi, and many more). Top-down steering can be highly effective for efficient, homogeneous interconnected infrastructure based on a smart electricity grid and allows for precise long-term planning—although this is dependent on the management ability to coordinate the implementation, as described in Sects. 4.3 and 4.4.

While the multitude of benefits of smart cities are attracting governors worldwide, transitioning into a smart city requires dedicated policy initiatives. The following section outlines necessary pillars and governmental curtail investments to realize the benefits of a smart city.

4   Transition Pillars and Governance Required

This section describes the features and transition pillars of smart cities as a response to the growing worldwide urbanization and the nature of governance required for this transition into intelligent and interconnected cities. There is no set consensus on what makes up a smart city, besides the general idea of their sustainability and resilience. Azevedo Guedes et al. (2018) developed a four-step approach to understand what the drivers boosting the intelligence of cities are. First, bibliographic research allows them to identify key literature on smart cities. Second, they scan the literature for core drivers that increase the intelligence of cities. Third, they survey experts to understand the importance of drivers and list them by priority. Finally, they quantitatively summarize the collected data. The results show 20 factors that increase the intelligence of cities, 15 of which are related to city governance and 5 to technology. To begin with the governance drivers, urban planning that enhances the cleanliness and sustainability of the environment and thus contributes to well-being bolsters the intelligence of a city. It is also essential for the government to invest in proper management of the city's infrastructure, including sewage system and sanitation. Curing problems is essential for developing smart cities, but another impact-driven approach is the prevention of risks by building urbanization strategies that minimize the impacts of disasters. Furthermore, to establish the feature of sustainability, it is fundamental for governments to incorporate in their public policies the effective management of resources to bolster the well-being of citizens. As part of environmental sustainability and the enhancement of the well-being of the people, smart urban mobility through clean individual and group modes of transport is needed. This extends to efficiently transporting and managing goods in smart supply chains, storage, and sustainable packaging. The well-being of people also includes the essential availability of a high-quality healthcare system. As part of enhancing the lifestyles of citizens, public safety measures should be put in place not only to reduce but also to prevent violence and crime. This is part of a general list of regulations that enforce order and guidance in the city. That in turn is also part of the complex process of public policy construction. A complementary approach to regulation is self-regulation, which in a way promotes the city to be self-sufficient in enforcing discipline through a rooted value system that naturally motivates citizens to adopt ethical standards to guide their lifestyles. A smart city by default implies innovation and creativity for the development of business and culture. On the business front, deepening networks of stakeholders and partnerships enable, in turn, a surge in innovation through greater interconnectivity of ideas. As new solutions arise, they require funding, so cooperation between public and private partnerships bridges the gaps between ideas and their feasibility. In addition to the relationship between the public and private sectors, it is useful to also understand all other forms of relationships that affect the city (Azevedo Guedes et al., 2018).

In terms of technology, which is a major skeleton of smart cities, indeed involves the usage of information and communications technologies (ICT). One other technological component includes smart energy management grids and smart buildings that reduce energy spending through natural components such as natural lighting and temperature as opposed to nonrenewable resources. Moreover, smart logistical applications can be developed that allow the usage of tools such as radio frequency identification and electronic routing of goods. Relatedly, the use of technology alone is not sufficient; rather, it is also important to integrate it into the production process and labor markets (Azevedo Guedes et al., 2018).

The complementarity of all these factors combined to offer a long but smooth process of the transition of cities to smart cities. As an example, national programs such as Qatar's TASMU Program provide a policy framework to enable investments into a smooth transition process, including 114 digital use cases across the five priority sectors of transport, logistics, environment, health care, and sports.

5   Challenges of Smart Cities

This section flips the coin to look at the challenges to building smart cities, which this section aims to outline. Like all other aspects of our modern technology-dependent lives, smart cities are prone to downsides by design and implementation challenges. First, technology dependence leads to an increase in the need for materials to manufacture access devices as well as infrastructure and electricity to run the interconnected infrastructure. For example, interruptions in the supply chains of rare metals and semiconductors have obstructed the rapid expansion of smart infrastructure, while electricity cuts remain a risk to the real-time usage of the capabilities of smart cities. Second, the magnitude of data collected increases privacy risks for individuals as a consequence of the sensors necessary to gather data. In a similar fashion, interconnectedness leads to information technology (IT) security risks for companies, which need to increase their cybersecurity spending to protect sensitive company data and technology-enabled operation flows. For a deeper picture of privacy and security challenges, see Braun et al. (2018).

Beyond these problems inherent to technology and data-driven life in smart cities, implementation challenges slow down the transition to an interconnected, sustainable urban life. Silva et al. (2018) remark that inclusiveness through the easy usability of smart city infrastructure is particularly difficult to reach in cities with fast-growing and diverse populations of citizens and firms.

In the context of the MENA region, congestion, scarcity of resources, and waste management demand intelligently designed cities. By first laying a foundation with aspirations (e.g., "The line" in Saudi Arabia or Tunis and Cairo's mostly conceptual stage in the smart city transition), citizens, businesses, and investors are pooled toward jointly designing smart city life. The second stage curtails the convergence of a city toward the outlined vision through a dedicated plan and investment efforts. GCC countries have pushed their "signature cities" into the convergence phase with successful implementations and large-scale investments. Finally, the transformation phase encompasses network-enabled utilities, security services integrations, and smart transport available to all inhabitants (Yahia & Shokeir, 2020). Ideally, a transformation aligned to the needs of all stakeholders is thus reached: business, citizen, and nature. There are risks and challenges in all stages: aligning on a vision might be difficult with diverse and changing populations in cities across the MENA region. Successful convergence is dependent on significant governmental coordination and stable ground for investments, which cannot (yet) fully be found across the MENA region. In addition, Ringel (2021) finds that poor management is seen as the biggest barrier to successful convergence to smart cities. Lastly, the transformation is dependent on the continuous availability of sustainable energy, i.e., the ability to harness MENA's abundant solar energy capacity.

6   Smart Data Collection

Finally, given the great need for data-driven policy, this section suggests the type of smart data collection that meets the standards of research and governance in smart cities. There are existing data technologies, including cloud computing, "big data," data visualization, and "internet of things," among other technologies that have facilitated the transmission of information across parts of the world. The motivation for this big data revolution lies in the ability to measure indicators for better predictions and informed decision-making. Decision-making can span a variety of areas, including but not limited to business, policy, and research. Big data also has its own managerial challenges; the right people should be hired for each process of assembling mass data, analyzing and finding patterns in the data, and making the most informative decisions about the data (McAfee & Brynjolfsson, 2012). The "internet of things," through remote sensing, eases access to managing infrastructures of a city in real time, such as water supply, roads, and transport networks, from a distance (Kopetz, 2011). Data visualization, elaborated on by Yuan et al. (2012), acts as a bridge between information and analysis through proper illustration of the data and data patterns. Moreover, mobile computing facilitates the access of data by users at any time from any place, boosting the interconnectivity and delivery of real-time data.

In fact, Rong et al. (2014) approach the development of smart cities as being data-oriented. They summarize it into layers of data acquisition, representing data sources, data transmission, data visualization and storage that incorporates data cleaning and maintenance, and data processing. The last layer consists of a support service layer, where the stored and visualized data are used as common platforms for user access.

According to Ringel's (2021) survey of key stakeholders of smart cities in MENA, the acceptance of smart city infrastructure is generally higher than in comparable settings in Europe. Integration with global corporations and citizens' willingness to share data for the sake of modernization reduce the barriers to a fast transition to smart cities. This implies a high potential of smart data collection and data-driven decision-making in MENA cities.

Despite the tremendous burst of the data world, we can still go "smarter" about it. The area of emotional intelligence data through biological indicators is still underdeveloped. To maximize unbiased data collection and what "is" rather than what "might be", and in order to objectively estimate the most policy-relevant variables, new ways are yet to be developed to grasp emotional data on attitudes and preferences, for example.

7 Conclusion

Smart cities are growing worldwide, offering a wide range of solutions to problems created by urbanization while accelerating the worldwide boom of fast information and data-driven governance. While there are common benefits of smart cities everywhere, the implementation of smart cities differs widely depending on the country and existing urban structure. The development of smart cities should be extremely tailored to the needs of every city. To harness a proper socioeconomic model for the city, the combined efforts of researchers, policy makers, and the government are greatly needed. The first step is to conduct a needs assessment for the city. This step could be achieved with the help of expert researchers and urban planners. Second, the needs should then be communicated to policy makers that can then design policies that match the needs. Third, these recommendations must be transferred to the government to ensure the implementation of policies that align with the findings of researchers and policy makers. In addition, the government should design environmentally friendly policies. For example, environmental fiscal reform, including environmental taxation, has immense positive impacts on developing a smart city. Relatedly, it also has implications on long-term poverty alleviation, through direct measures such as palliating environmental burdens such as water and air pollution that affect the lives of the poor, as well as indirectly by freeing up public finance space to invest in anti-poverty programs (OECD, 2005).

Further research is still needed on how to make the concept of smart cities very specific in terms of implementation, depending on the city's level of development, needs assessment and evaluation, which leads to impact-driven smart strategies. Many challenges remain, starting from the lack of simplicity to identify what each city needs in the first place. In addition, political bureaucracy remains a burden. Nonetheless, despite all the challenges and the long process, in the longer term, smart cities can have significant positive implications on the standards of living of people and their mobility.

References

Azevedo Guedes, A., Carvalho Alvarenga, J., dos Santos Sgarbi Goulart, M., Rodriguez y Rodriguez, M., & Pereira Soares, C. (2018). Smart cities: The main drivers for increasing the intelligence of cities. *Sustainability, 10*(9), 3121. https://doi.org/10.3390/su10093121

Braun, T., Fung, B. C. M., Iqbal, F., & Shah, B. (2018). *Security and Privacy Challenges in Smart Cities, and Society.* https://doi.org/10.1016/j.scs.2018.02.039

Harrison, C., Eckman, B., Hamilton, R., Hartswick, P., Kalagnanam, J., Paraszczak, J., & Williams, P. (2010). Foundations for smarter cities. *IBM Journal of Research and Development, 54*(4), 1–16. https://doi.org/10.1147/jrd.2010.2048257

Kopetz, H. (2011). *Real-time systems* (pp. 307–323). Springer.

McAfee, A., & Brynjolfsson, E. (2012). Big data: The management revolution. *Harvard Business Review, 90*, 60–68.

OECD. (2005). *Environmental fiscal reform for poverty reduction.* https://doi.org/10.178 7/9789264008700-en

Ringel, M. (2021). Smart city design differences: Insights from decision-makers in Germany and the Middle East/North-Africa Region. *Sustainability, 13*(4), 2143. https://doi.org/10.3390/ su13042143

Rong, W., Xiong, Z., Cooper, D., et al. (2014). Smart city architecture: a technology guide for implementation and design challenges. *Network Technology Applications, 11*, 56–69.

Silva, B. N., Khan, M., & Han, K. (2018). Toward sustainable smart cities: A review of trends, architectures, components, and open challenges in smart cities. *Sustainable Cities and Society, 38*, 697–713. https://doi.org/10.1016/j.scs.2018.01.053

World Bank Group. (2022). World Bank Open Data | Data. 2022. https://data.worldbank.org/

Yahia, I. B., & Shokeir, R. (2020). Moving toward Smart Cities: Insights from the MENA Region. *IJWBC, 16*(1), 1. https://doi.org/10.1504/ijwbc.2020.10026217

Yin, C. T., Xiong, Z., Chen, H., Wang, J. Y., Cooper, D., & David, B. (2015). A literature survey on smart cities. *Science China Information Sciences, 58*(10), 1–18. https://doi.org/10.1007/ s11432-015-5397-4

Yuan, Y. M., Qin, X., Wu, C. L., et al. (2012). Architecture and data vitalization of smart city. *Advanced Materials Research, 403*, 2564–2568.

Open Access   This chapter is licensed under the terms of the Creative Commons Attribution 4.0 International License (http://creativecommons.org/licenses/by/4.0/), which permits use, sharing, adaptation, distribution and reproduction in any medium or format, as long as you give appropriate credit to the original author(s) and the source, provide a link to the Creative Commons license and indicate if changes were made.

The images or other third party material in this chapter are included in the chapter's Creative Commons license, unless indicated otherwise in a credit line to the material. If material is not included in the chapter's Creative Commons license and your intended use is not permitted by statutory regulation or exceeds the permitted use, you will need to obtain permission directly from the copyright holder.

Cities: The New Form of International Environmental Governance

Jamila El Mir

Abstract Traditionally, international environmental agreements have taken the form of national bilateral or multilateral commitments made by the national governments of the respective parties. With the increased urbanization that happened in the twentieth century, the world has seen a growing number of collaborations happening at the city level between local governments. This movement began in the early 1900s and grew considerably in number starting in the 1990s, reflecting the growing prominence of cities as global actors in environmental action and sustainability transitions. This chapter examines the different dynamics cities apply with their host national governments and international organizations. It covers the key tools within their hands that empower them to become global actors. The chapter then explores the key blind spots cities display as global actors, highlighting the limitations that they face in acting as a parallel platform for international environmental governance.

Keywords International environmental governance · Cities · City alliances · Transnational municipal networks

1 Introduction

Traditionally, international environmental agreements have taken the form of national bilateral or multilateral commitments made by the national governments of the respective parties. This is now changing, and cities are taking a front seat in international environmental discussions. With the increased urbanization that happened in the twentieth century, the world has seen a growing number of

---

The original version of this chapter was revised. The correction to this chapter is available at https://doi.org/10.1007/978-3-031-35664-3_20

J. El Mir (✉)
MENA Environmental and Sustainability Policy and Strategy Expert, Government of Dubai, Dubai, United Arab Emirates

© The Author(s) 2024, Corrected Publication 2024
F. Belaïd, A. Arora (eds.), *Smart Cities*, Studies in Energy, Resource and Environmental Economics, https://doi.org/10.1007/978-3-031-35664-3_5

collaborations happening at the city level beginning from the early 1900s and growing considerably in numbers starting from the 1990s, reflecting their growing prominence as global actors in environmental action and sustainability transitions. This chapter aims to explore the extent to which and the limitations of their role as global environmental actors.

Their share in the overall environmental footprint justifies this: according to the UN Habitat, cities are responsible for 78% of the world's energy consumption and generate over 60% of global greenhouse gas emissions (United Nations). Additionally, sustainable transition action plans are local by nature and most often fall under the responsibility of the local authorities. This gives them an opportunity to determine local sustainability and environmental targets. Cities can, however, be hindered by the availability of funding as well as a lack of supporting regulatory and fiscal powers, amongst other factors. The ability to shape their local agenda as well as the common challenges cities face have acted as a motivation to form city alliances and coalitions focused on sustainability and climate action, be it by cities themselves or by national governments and international organizations, as will be explored below.

To answer the question above, the chapter explores the enablers cities have that empower them as actors on the international scene, and the various pathways cities' alliances have to adopt and influence global environmental action. It starts by looking at the (re)emergence of cities as international players and by identifying the key enablers to their growing prominence, including the expansive reliance and use of data. It then explores the different types of city alliances and the dynamics of the influence at play between the local, national, and international levels. We contend that while national alliances still form the majority of city networks, international alliances naturally provide greater visibility and leverage for cities as key players in international environmental governance. This is followed by an overview of the key tools cities' alliances use to have global influence. Finally, the limitations cities still face today as actors of international environmental governance are discussed. These consist mostly of blind spots in representing the global community and the difficulty of enforcing accountability mechanisms within their platforms.

2 Why Cities: An Overview of Cities' Role in International Governance

2.1 Background: The Re-emergence of Cities on the Global Scene

Cities were originally the center of governance, standing at the helms of the Roman, Greek, and Babylonian empires, for example. To this point, Andrew Bodiford cunningly notes that the word 'citizenship' has its roots in belonging to a city, rather than a state, reflecting the city's placement at the core of government and managing the populace (Bodiford, 2020). However, our modern understanding of international governance has national states as its key players. It is based on the model adopted following the Peace of Westphalia, ending a 30-year war between Germany, Spain, and the Dutch state in 1648 (Bodiford, 2020; Patton, 2019). To be acceptable by all parties, the two ensuing Peace treaties recognized the equal sovereignty of all treaty members and agreed on the principles of state sovereignty and freedom of action and alliances by the states previously subordinated to the Roman Empire (Hassan, 2006). In a way, this turning point in international relations was a hint of what would happen over four centuries later: local governments want more autonomy and empowerment to act and deliver on their local sustainability agendas. This has taken place following increased urbanization (Sharpe, 1988).

Indeed, the continued growth of cities (Fig. 1) has resulted in an increased demand for local social and infrastructure services by local communities, strongly fuelled by industrialization and trade. This in turn led to greater devolution of powers from central governments to local governments. Urban communities are also gaining diversity in ethnicities and cultural backgrounds. A study of the metropolitan and micropolitan areas in the United States found that they had experienced an increase in diversity in 98% and 97% of the areas, respectively, between 1980 and 2010 (Barret et al., 2014). This requires being in close contact with the community to understand the changes in demography and the implications on demand for community facilities, infrastructure services and social services, as well as identifying growing areas of risks such as safety and security and social welfare. An assessment undertaken by the Organization for Economic Co-operation and Development (OECD) confirms that subnational government spending largely focuses on the services mentioned above, namely, education (21.8%), general public services (20.3%), economic affairs and transport (13.8%), social protection (12.5%), health (9.4%), and housing and community amenities (8.8%) (Allain-Dupré, 2016).

To this effect, cities have been noticeably active in promoting decentralization and voicing their needs to national and international institutions, often using partnerships with local and international organizations. The UN-Habitat has also

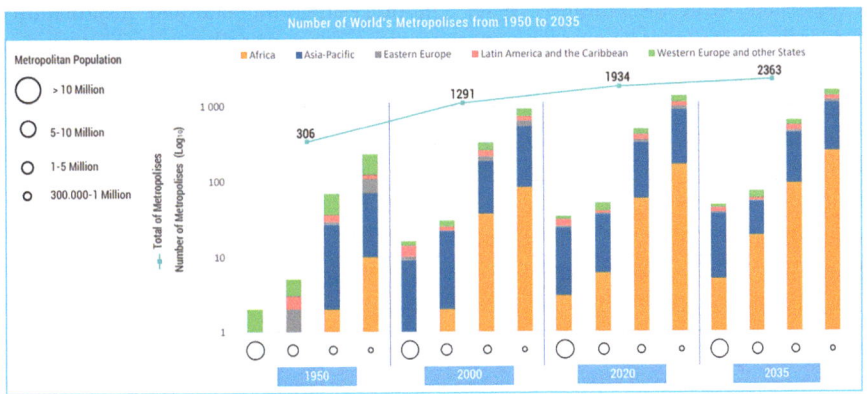

Fig. 1 Number of world's metropolises from 1950 to 2035. (Source: UN-Habitat (2020))

been active in promoting decentralization as a means of enhancing sustainability outcomes at the local level (Nijman, 2016). Professor Janne Nijman assesses the growing role of cities as global actors and notes that "decentralization is changing the relationship between the city and the state, and between the city and the global level. It empowers the city, locally as well as globally" (Nijman, 2016).

2.2 A Necessary Partner for the Global Agenda

2.2.1 Engines for the Global Economy

Cities are also the main engine for the global economy, with more than 80% of the global gross domestic product (GDP) being generated in cities (World Bank, 2020). This positions them as a focal point of foreign investors, as well as key drivers in national and global economies. The role of cities in the global economy is an area that has been a subject of study as early as 1915 with Patrick Geddes and that has been much popularized in the past two decades by the work of Saskia Sassen, who defines global cities as the nodes of the global economy, linked by finance, trade, and communication routes that transcend the borders of their host countries (Acuto, 2011; Sassen, 1991).

Beyond the role of cities in the global economy, it is also important to acknowledge that cities "are entrepreneurial cities because they purposefully undertake political and economic activities"; as such, they "are not passive elements in a world of flows, but are also hubs, motors and magnets of these, and thus key agents of the world system" (Acuto, 2011). Nijman concurs with cities' important role as a testbed for global innovation and change (Nijman, 2016). The size of the city's economy is not the only factor to consider but rather its ability to move away from business as usual and succeed in doing so. This capacity to innovate is particularly important in the context of international environmental governance. The willingness to act as a testbed for environmental transition allows other cities to learn from their experience, while the boldness to set ambitious environmental and sustainability targets beyond national commitments encourages others to follow suit. The trend of increased adoption of environmental action by cities is reflected in the CDP's annual 'A List', which displays the cities that stand out in environmental action and transparency, particularly looking at climate change mitigation and adaptation actions. The number of cities listed grew from 88 cities in 2020 to 95 cities in 2021.[1] Indeed, there are many cities, such as Copenhagen and Stockholm in Europe and Auckland in New Zealand, that have established themselves as environmental leaders, although they are not megacities per se.

---

[1] It could be argued that the increase in number of cities in the 'A List' is a result of a bigger number of cities reporting on the CDP portal and not necessarily a rise in ambition amongst cities. However, as discussed below, monitoring and reporting is in itself an enabler for cities to act as agents for global environmental change.

2.2.2 Agents for Sustainable Transitions

Cities have become a necessary partner for the sustainability agenda. Agenda 21, developed during the United Nations Environment and Development Convention in Rio de Janeiro in 1992, highlighted the importance of cities as key actors in ensuring sustainable development at a global scale. The recognition of their role in a document that was endorsed by over 178 representatives of heads of states is a testament of the breadth of acknowledgment of their importance. This was followed by a surge in closer collaboration between international organizations, such as the United Nations, the World Health Organization and the World Bank, and cities' networks to inform international policy or implement international agreements (Nijman, 2016). More recently, the decision of the Conference of Parties adopting the Paris Agreement to the United Nations Framework Convention on Climate Change (UNFCCC) referred to "the efforts of all non-Party stakeholders to address and respond to climate change, including those of civil society, the private sector, financial institutions, cities and other subnational authorities", asking them to "scale up their efforts and support actions" in the fight against climate change (Sands & Peel, 2018, p. 934). Cities have therefore permeated through the ranks of international environmental governance to gain a seat in the first row of key actors mentioned in what can be considered the most critical multilateral environmental agreement of our time.

The ascent of cities to the global scene has allowed them to become the loci of a new platform to discuss topics of international importance and common interest, amongst which the environment is finding its way in the lead (Acuto & Rayner, 2016). This platform is operating parallel to the traditional states' platforms, which have historically been the main interlocutors for international organizations, such as the United Nations and the World Bank, on topics related to environmental governance.

3 Enablers for Cities to Shape Their Growth

3.1 Planning, Operations, and Regulations

As alluded to above, cities have the responsibility of catering to the needs of their communities, and they typically oversee local development (OECD, 2017). They are also best placed to implement pilot projects to test new solutions (Acuto, 2011; Acuto & Rayner, 2016; Nijman, 2016; OECD, 2017). They have a unique opportunity to introduce change at scale based on validation and lessons learned from local pilot projects – an opportunity that does not exist at the state level.

The OECD performed an extensive study in 2017 looking at the governance of land use in its member countries. The study concludes that, despite a growing complexity of the governance and stakeholder array in land-use planning, its implementation remains mainly the responsibility of the local government. OECD's study succinctly depicts the task cities have at hand: "planning in the city combines statutory functions, balances public and private interests, offers a vehicle for economic development and seeks to strengthen local democracy through participatory processes" (OECD, 2017, p. 54).

While it is not the intention of this chapter to study in detail the powers cities possess regarding planning, service operations, and regulations, it is important to note that the city always has a role to play in shaping these aspects locally. For example, they have a say in determining local detailed land uses, in deciding on levers impacting mobility choices such as parking fees and local building development and energy efficiency guidelines, all of which have a direct impact on sustainability performance and the carbon footprint within their boundaries (OECD, 2017; World Bank, 2021). As explained below, these powers provide them with additional levers to enhance their environmental performance, such as managing municipal budgeting and city procurement, as well as obtaining data that are critical to understanding their baseline and defining environmental action.

3.2   Budgeting and Investment Decisions

A growing focus of the sustainability transition of cities and specifically relating to the climate transition is access to finance. There are several levers that cities can use to access finance for sustainability programmes and projects. In its report entitled *The State of Cities Climate Finance*, the World Bank highlights the role cities can play in securing climate finance and splits it across two main levers, the first looking at the city as a 'provider' and the second looking at the city as a 'steward', further detailed below (World Bank, 2021):

- *The city as a provider:* the city, as a consumer of goods, can use its purchasing power to influence the market transition and reduce its operational environmental footprint. As a service provider, it can also leverage its role in designing and operating these services (including municipal buildings) to reduce the environmental footprint of the services it provides to the broader community (such as waste management and landscaping).
- *The city as a steward:* As mentioned above, the city can impact sustainability transitions by leveraging its regulatory and policy-making powers.

Intersecting these two roles, the city also acts as a 'fundraiser' to finance the services it controls (World Bank, 2021). Beyond raising its own funds through local municipal fees and charges, it can receive funds from the subnational and national government as well as raise debt.

The World Bank identifies the key contributing sectors to greenhouse gas emissions (GHGs) in cities and the typical aspects related to each that the local government controls and pays for. These act as its main point of leverage to direct local budget allocation towards reducing its environmental footprint. These are listed in Table 1:

Table 1 Direct city functions and expenditure responsibilities in key sectors responsible for GHGs in urban areas

Sources of GHGs in urban areas	Direct city functions and expenditure responsibilities
Transport	Street, bike, bridge network extension, and maintenance Parking management Public transport route planning Fast electric vehicle (EV) charging network Green vehicle fleet procurement Public transit fleet procurement and operation
Buildings	Planning and design standards, construction permitting Social/public housing Trunk infrastructure connections Parks and green spaces City-owned assets (public buildings, land, property)
Energy	Management of electricity distribution network (grid) Street lighting (including energy efficient or solar powered street-lamps) Energy consumption of city-owned assets (rooftop solar on city-owned buildings)
Waste	Water, wastewater, and SWM strategic plans Solid waste collection and street cleaning Water supply systems Storm and waste-water management
Industry	Planning, location, and design standards for industrial zones/estates Business licensing Trunk infrastructure, roads, and grid connections

Source: Reproduced from World Bank (2021)

3.3 Partnerships

On the other hand, with the growing size of cities and the diversity of their communities, there is a growth in the complexity of governance (OECD, 2017). This is compounded by the increasing sophistication of urban services' delivery with the advent of digitization and technological advances. To have any effective sustainability interventions, there is a need to ensure the ability to implement the intervention itself, which is greatly dependent upon stakeholder buy-in and capabilities as well as clear governance for the initiatives at hand. Local governments therefore have a critical role to play in communicating with local stakeholders to ensure this buy-in and establish clear governance, and they are uniquely placed to do so. The public sector is often best placed to do that, as it is a not-for-profit organization that holds decision-making power, thereby giving confidence to stakeholders that the topic is of importance, there is hope for fruitful outputs for the common good, and there is the capability to implement. The Paris Fonds Verts initiative is a point in case whereby the City of Paris established a Green Fund that is co-financed and co-managed with the private sector, with the objective of financing local projects that align with the city's climate action plan (Ville de Paris, 2019). The Fund had managed to raise 200 million euros in its first round of funding in 2018. A smaller-scale example includes the pilot project of extended producer responsibility in Az-Zarqa town in Jordan hosted under the Ministry of Environment in collaboration with the local municipality and representatives from the stakeholders involved in the different stages of the waste lifecycle, namely, waste management companies, food and beverage companies, schools, businesses and consultants. The project's lessons learned and outcomes were then fed into the design of the national EPR scheme (El Mir et al., 2021). In both cases, local authorities were the key instigator of the initiative, and it was linked to a long-term vision: the carbon transition in the case of Paris and establishing a national policy in the case of Az-Zarqa. This gave the stakeholders from the private sector confidence that efforts put will lead to fruitful outcomes while also giving them the opportunity to sit on the table and shape these outcomes.

3.4 Digitization and Data

Least but not least, a powerful tool that cities have is data: data about their communities, their services and assets, and the flow of resources within their geography. This gave rise to the concept of a "smart city" and several related game-changing concepts, such as the Internet of Things (IoT), blockchain, and artificial intelligence. These were spoken of as a vehicle to walking us into the next era of the global economy, the fourth industrial revolution (OECD, 2020). OECD defines a smart city as one that applies "initiatives or approaches that effectively leverage digitalization to boost citizen well-being and deliver more efficient, sustainable and inclusive urban services and environments as part of a collaborative, multistakeholder process" (OECD, 2020). In addition, true to this definition, the concept of smart cities brought the promise of improved social, environmental, and economic outcomes as a result of increased data availability and use. While there are many reports grounding the expectations of what smart cities they can deliver (Baykurt & Raetzsch, 2020; Weekes, 2019), the fact that there has been a colossal jump in the focus on and availability of data is undeniable (Fig. 2).

There are many contributing factors that led to this step change in cities' investment in digital infrastructure and data collection. While those will not be covered in detail in this chapter, it is worth highlighting several key factors that, in our view, are related to the sustainability transition. First, the growing focus on the climate transition has led to a global upskilling in emissions reporting, which in turn requires in-depth understanding and monitoring of emissions and performance across the key sectors at hand. Cities are best placed to oversee the emissions from these respective sectors given their role in delivering and monitoring these same sectors within their boundary. This was further reinforced by the detailed formulation of the United Nations Sustainable Development Goals in 2015, which cover 17 goals and 231 unique key performance indicators, also instigating the need to monitor, measure, and report on a broad set of topics.

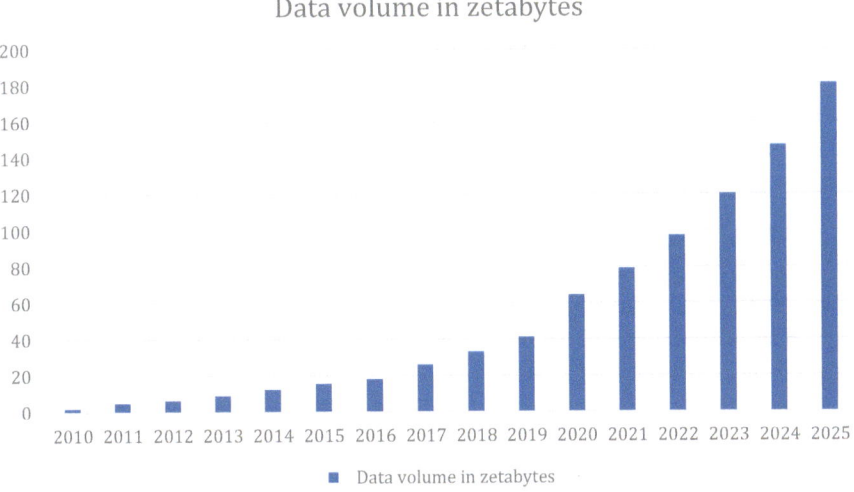

Fig. 2 Volume of data/information created, captured, copied, and consumed worldwide from 2010 to 2025. (Source: Satista, based on International Data Corporation data (www.idc.com), and Statista forecast for 2021–2025 (https://www.statista.com/statistics/871513/worldwide-data-created/)

Parallel to the focus on climate or sustainability transition, there has been an increased focus on community wellbeing and happiness over the past decade (Carnegie Trust UK, 2016). This prompted the need to define KPIs to measure wellbeing to develop action plans that can be monitored and the impact of which can be measured. The OECD, for example, developed a suite of KPIs to measure wellbeing that it applied to 395 subnational regions, finding it necessary to measure at the local level rather than stopping at the national level to better understand the state of wellbeing in its member countries (OECD, 2014a). As such, data not only became a tool to improve the performance of local services but also became a central tool to assess performance against global agendas.

Several platforms were developed, building on this recent recognition of the role of data. These include most notably the World Council on City Data (WCCD) and the Carbon Disclosure Project (CDP). WCCD was created to respond to this need for harmonized, city-level data to be able to compare and assess performance against set KPIs, notably KPIs set by the International Organization for Standardization (ISO) in relation to their standards for sustainable, safe, and resilient cities. WCCD now has over 100 member cities. CDP shares a similar purpose focused on standardizing reporting on climate action and had 1128 cities reporting on the platform in 2021.

4 Cities Alliances: A Voice in Front of National and International Audiences

The past three decades have seen a significant jump in cities' networks and alliances (refer to Fig. 3). To understand the role they are playing in international environmental governance, it is important to understand what underpins international governance. As stated by Adonova et al., governance is intentional and structured and aims to steer the members towards a set objective that consists of the provision or delivery of a certain public good, following agreed principles and procedures (Andonova et al., 2009). Acuto and Rayner tried to map the 'causes' on which cities' alliances were focused, looking at 170 cities' networks. They found that 29% of the networks focused on environmental causes, followed by culture, poverty, energy, peace-building, gender, and other causes (Acuto & Rayner, 2016). All of these topics relate to the public good, thus confirming that cities' alliances are contenders to play an important role in international governance, not least in the environmental sphere. Acuto and Rayner also mapped the different types of city organizations and identified that out of the 170 networks assessed, approximately 49% were national, 21% were regional, and 29% were international. To take a deeper look into the role of cities in international environmental governance, we will look separately into national and international cities' alliances, focusing mainly on the direction of the influence of these alliances between the local, national, and international spheres.[2]

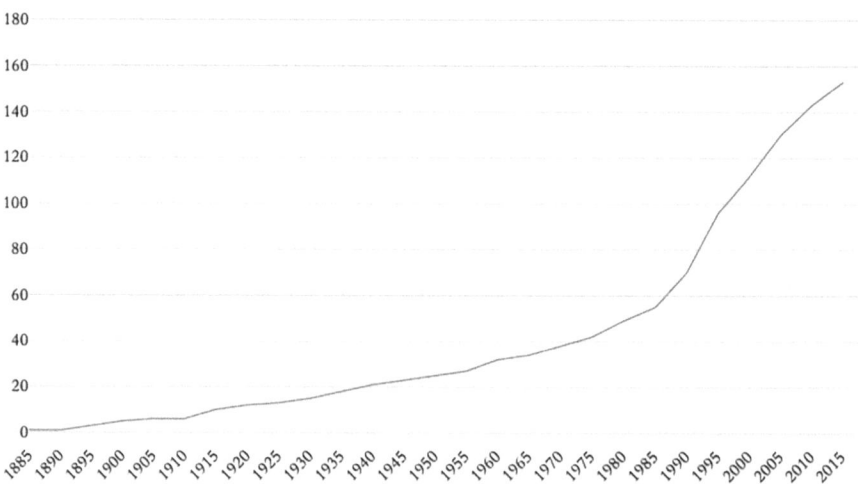

Fig. 3 Numbers of networks per year, 1885–2015. (Source: Acuto and Rayner (2016))

---

[2] There is much to analyze to comprehensively assess the role cities' alliances are playing in international governance including membership requirements, form of alliance, objectives, and diversity of members and the growing role of nonstate actors including the private sector, to name a few

4.1 National Alliances

Partnership between cities from the same country presents clear benefits to member cities, as it strengthens their position in front of the national government. This is particularly the case with regard to common aspects impacting them, such as fiscal autonomy, regulatory powers, and influence over national policies and development plans. National coalitions help member cities better formulate requests for national policy and governance changes that empower local governments and address their specific needs (Somanathan, 2001). Indeed, they represent the larger portion (49%) of all cities' alliances: according to Acuto and Rayner, this reflects the relative and enduring importance of states and national governments to cities.

A powerful example of cities' alliances at the national level includes The League of the Cities of Philippines (LCP), which was established in 1991 by law. It has a clear remit to enhance collaboration between cities of the country, to strengthen decentralization in city governance, to act as a forum for discussion and feedback on matters impacting city governments and to foster collaboration and dialogue between national government and local governments with the aim of improving the welfare of the local communities and implementing supporting development programmes (The League of the Cities of Philippines). Putting to practice the ability of cities to act as testbeds for innovation and sustainability transitions, the LCP began working on developing city development strategies in three cities and has since upscaled its work in this area to more than 19 cities. Another important project the LCP is working on consists of establishing a national database of harmonized datasets provided by cities, linking back to the importance of data at the city level mentioned above. This will allow improved knowledge sharing, informed decision-making, and enhanced collaboration between the public, private, and academic sectors (The League of the Cities of Philippines).

An interesting example of a city's alliance consists of the US Mayors Climate Protection Agreement. The agreement was formed by mayors in the United States to internalize the commitments of the Kyoto Protocol after President Bush had refused to endorse it at a national scale (Nijman, 2011). The agreement was unanimously endorsed by the US Mayors Conference, an organization established as early as 1933 with the aim of magnifying the voice of cities as well as enabling knowledge-sharing amongst its member cities totalling over a thousand cities. There are national alliances of local governments that are also set up by the national government with the aim of implementing an international agenda. This is the case for the Associations of Local Authorities and Regions in Denmark, Finland, Iceland, Norway, and Sweden, which act as key partners in the implementation and monitoring of the United Nations Agenda 2030 (Agenda 2030 at the Local Level).

---

parameters. Many of these aspects have been assessed to a certain degree by leading authors on the topic including Michele Acuto, Steve Rayner, Harriet Bulkeley and Christie Swiney and Janne Nijman. For this article, however, we are mainly interested in understanding the nature of influence that cities are having in international environmental governance and will therefore focus on the direction of this influence between the local, national, and international levels.

The 100 smart cities formed by the Indian government in 2015 present yet another form of national city alliance. Its main objective is to drive the digitization of the systems of member cities as part of a national program to enhance the performance of local governments and the quality of life of their communities (Smart Cities Mission). The national government set a roadmap for the member cities to implement by developing their local action plans in consultation with the public and by applying for funding of recommended initiatives on a competitive basis.

The examples above illustrate how national cities' alliances may serve various objectives and how their relationship with the national and international levels can be subject to different dynamics. The LCP demonstrates a primarily bottom-up approach whereby the local government uses the alliance as a channel to influence the national government. The 100 Smart Cities initiative by the Government of India, on the other hand, serves as a conduit for the national government to implement a national policy at a local level. The US Mayor Conference is, in turn, an example of how a city coalition enables local governments to bypass national government in adopting and implementing international policies. Finally, the associations of local governments in the Nordic countries present an example where the city alliance is formed by the national government to implement an international agenda (Fig. 4).

It is worth noting that most often, the flow of influence is not singular and that the alliances established allow for a fluid discussion between the stakeholders and the management body of the organization. This is all the truer in light of the proliferation in the number and form of national alliances focused on sustainability issues and climate action, which include stakeholders from civil society, local governments, and private sector players.[3]

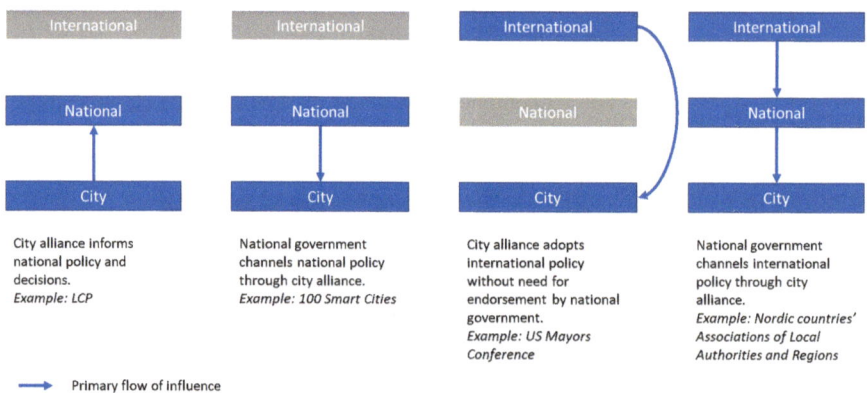

Fig. 4 Primary flow of influence of national cities' alliances. (Source: Author's assessment)

---

[3]The private sector is playing an increasing role itself in international environmental governance by either joining TNMs or by joining international environmental agreements and initiatives such as the Race to Zero. Their role in international environmental governance also warrants in-depth analysis, it is however not within the objective of this chapter.

4.2   International Alliances

Following the increasing powers given to local governments in the 1970s and 1980s, new cities' alliances of an international nature began forming in the late 1990s and early 2000s at scale. Three of the most prominent examples active in the environmental space include ICLEI, which was created in 1990 by the United Nations, UCLG, which was formed in 2004, and the C40 Climate Leadership Network of Cities, which was formed in 2005 and focused on climate action. It is worth noting, however, that the first international transnational municipal network (TNM) was formed in 1913 in the city of Ghent in Belgium and was called l'Union Internationale des Villes. It formed a nucleus for a burgeoning of other municipal networks that expanded geographically from Europe to include North America. This bourgeoning slowly reached other geographies. In 2004, many of these TNMs finally coalesced into the UCLG, forming the largest network of subnational governments in the world, with over 240,000 members in over 140 states, representing over half of the global population (United Cities and Local Governments, 2021). Today, UCLG acts as the main representative of cities and subnational governments in front of the United Nations and has been able to directly input into several multilateral agendas (Swiney, 2020). C40, which was established in 2005 by the Mayor of London Ken Livingstone, has since grown into an organization of 96 cities, mostly megacities above three million residents, representing approximately 25% of the global GDP (C40 Cities). C40 distinguishes itself by the detailed performance-based membership requirements it applies, which are centered around developing climate action plans that are compliant with the Paris Agreement (C40 Cities). These TNMs are joining hands to magnify their impact on the global environmental agenda. The most notable example is the Race to Zero initiative backed by the United Nations. The C40 joined hands with ICLEI, GCoM, WRI, UCLG, CDP, and the World Wildlife Fund for Nature (WWF) to mobilize over 1000 cities to commit to halving their emissions by 2030 and reaching carbon neutrality by 2050 in time for COP 26 in Glasgow last year.

Transnational municipal networks (TNMs) give manifestation to new forms of partnerships that state alliances and agreements cannot, such as public–private partnerships (Acuto & Rayner, 2016). Acuto and Rayner's study highlighted that private sector players are instigators of approximately 19% of existing networks, while intergovernmental organizations are behind 23% of TNMs created. In this way, cities become the platform for the political influence of non-state actors and give them (especially private sector players) an entry point to become active players in international governance (Nielsen & Papin, 2021). One such cities' network is the 100 Resilient Cities (100RC) founded by The Rockefeller Foundation in 2013 (until 2019), with the aim of helping cities enhance their resilience to increasing urban stresses by sharing knowledge and supporting them in developing their local resilience strategies as well as creating a role for and appointing a Chief Resilience Officer (The Rockefeller Foundation, 2022).

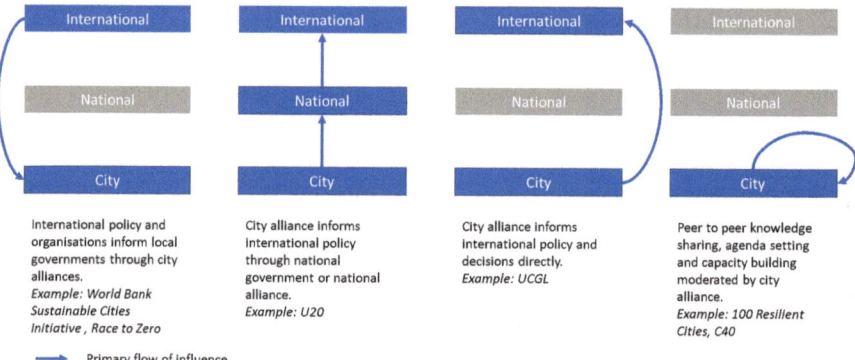

Fig. 5 Primary flow of influence of international cities' alliances. (Source: Author's Assessment)

Acknowledging the growing importance of cities in the international arena, international organizations are also establishing their own city-focused networks and initiatives. Examples include the UN-Habitat Cities and Climate Change Initiative and the World Bank with its Sustainable Cities Initiative. Although not necessarily acting as formal TNMs, such initiatives require the mobilization of cities on a specific agenda through specific action, linking back to Andonova's definition. From this angle, cities appear as a necessary partner (and therefore target) for international organizations to achieve agenda-specific goals. Another dynamic of TNM consists of the U20, which is a TNM gathering cities from the member countries of the G20. It aims to influence the position of national governments within the G20 on global matters to better represent the priorities of cities and help magnify their voices in front of the international community.

International city alliances therefore display different flows of influence between the local, national, and international compared to national city alliances (Fig. 5). The main difference is their greater reach to the international level, whether as direct or indirect influencers or influenced directly by international organizations, which is to be expected given their nature.

5   Soft Tools for a Global Influence

In addition to the enablers cities have within their boundaries, cities' alliances make use of the fruits of collaboration to magnify the voice of local governments in front of their national and international stakeholders.

5.1   Knowledge Sharing and Capacity Building

The primary objective of cities' alliances is knowledge sharing, which in turn helps achieve other objectives, such as capacity building, adopting evidence-based decision-making, and developing proposals for technical and financial support (Barbi &

De Macedo, 2019). Indeed, a study by Bulkeley et al. found that although 93% and 88% of the 60 transnational climate groups they studied worked on knowledge sharing and capacity building, respectively, only 13% of the groups were solely focused on these two objectives (Bulkeley et al., 2012), all the rest having additional objectives. One point to keep in mind is the importance of data in enabling knowledge sharing, which was covered in Sect. 3.4 and is further detailed in the Sect. 5.2 below.

Knowledge sharing and capacity building, equipped with data from self-monitoring, allows cities to become global advocates on the set agenda. Acting as testbeds for environmental transition, they become teachers to other cities by sharing lessons learned, success stories, and best practices. To this effect, C40 membership includes two categories of cities: megacities, which must have at least over three million inhabitants, and innovator cities, which can benefit the rest of the cities by their lessons learned and innovative approaches to solving climate change challenges.

Another aspect highlighted by Acuto and Rayner's study consisted of how different configurations of the networks gave more or less room for non-state actors to be active. They argue that multitiered networks (where there were two layers or more) allow for more sector-specific interventions that then open the doors for the private sector to present solutions tailored to the needs of the cities concerned while also allowing cities to benefit from knowledge exchange and 'pooled procurement'. This has been seen, for example, in the fields of low-emission vehicles and renewable energy within the organizations of C40 and ICLEI (Acuto & Rayner, 2016).

5.2   Monitoring and Reporting

Many of the TNMs require or encourage some form of monitoring, with varying frequencies for reporting. C40 and the Global Covenant of Mayors together with ICLEI, for example, require periodic reporting on climate action and greenhouse gases although the CDP. The 100RC required profiling the cities' performance across over 52 KPIs through 156 questions. Indeed, Acuto and Rayner's study, which considers 170 cities' networks, identifies that 45% of the outputs of the networks consist of regular reports followed by joint pilots and policies representing 38% of their collective outputs (Acuto & Rayner, 2016). This reflects a regular flow of information from member cities to the managing team of the relevant network.

Monitoring action at a city level has also transcended the sphere of network requirements as it becomes a tool for cities to place themselves on the map of international environmental players. Platforms such as the WCCCD and the Race to Zero welcome the willing provided there is commitment to report on data and action in a specified format and at a regular basis, regardless of their membership with other TNMs. The example of CDP's 'A List' mentioned above is a case in point.

It is worth noting that CDP's reporting system was updated in 2019 in collaboration with ICLEI to respond to ICLEI's member subnational governments' request to have a unified reporting system (Staden, 2019). CDP has grown to become the default reporting system for several international alliances, including the Global Covenant of Mayors, ICLEI, C40, to name a few. As such, TNMs help address the challenge of standardized monitoring and reporting on global matters (such as climate change) by setting uniform requirements. Inspired by this success of standardizing reporting efforts, the Japanese Ministry of Environment also established a partnership with CDP, which resulted in 200 new Japanese cities reporting on the platform in 2021.

5.3   Rule and Target Setting

Until recently, cities had no perceived role in setting international law. Swiney argued that it was only in 2006 that the role of cities in international law began to be fully explored through the work of Yishai Blank. Blank's work specifically highlights two means through which cities are gaining prominence as key players in international environmental law: first, cities are adopting international environmental agreements at a local level regardless of commitments made by the national government (as per the example of the US Conference of Mayors Agreement). Second, they gain status as global actors by doing so. Swiney does not fully agree with Blank's claim that cities are now objects of international regulation, contending that there is no rule setting from the international to the local level directly. While her stance is justified, it is nonetheless true that there is a growing direct relationship between the international and local levels, as seen in Sect. 4.

This relationship is not one where the international environmental agenda is mandated (in a top-down manner). It is rather one where international agreements are either informed (through the U20 and the UCLG, for example) or where international environmental policies are applied at the local level without direct intervention by the national government (US Mayors Conference and the Nordic Associations for Local Authorities, for example). On a broader level, cities' alliances tend to set internal targets. The study by Bulkeley et al. highlights that 60% of the assessed transnational climate groups had some form of target setting and 23% established mandatory rules (Bulkeley et al., 2012). In this sense, they adopt internal mechanisms and targets to act upon their selected agenda. This helps create a parallel platform for environmental action at the international level that is defined and implemented by cities.

6   Main Limitations of Cities

Despite their power over the global sustainability transition, cities do have their own limitations. The European Commission highlights that despite their growing responsibilities, "this political recognition has not always been accompanied by an adequate level of autonomy, capacity development and financial resources, leaving their empowerment incomplete" and thereby not providing them with the tools needed to fulfil their responsibilities.

6.1 Legislative Powers

6.1.1 Local Policy

One of the recurring challenges cities face in pushing their ambition in the sustainability transition is their limited regulatory powers. Cities often fall under the jurisdiction of states where it is the state government that has regulatory powers, as is the case in the United States of America, Canada, Australia, India, Brazil, and many other countries around the world. This limits their ability to decide upon environmentally driven tariffs targeting waste or high-emission transport, for example (World Bank, 2021), as well as environmental standards and mandated programmes (such as vehicle emissions standards or extended producer responsibility schemes).

On an even broader level, planning at the local level is often subject to overarching strategies set by the regional or national level that can limit the role cities play in pursuing aggressive sustainability policies. The absence of coordination at the regional and national levels can also result in reduced efficiencies and incompatible infrastructure development between neighbouring cities (OECD, 2017). In the absence of direct control over these matters, local governments can work on establishing local pilots that inform state governments and lobby for the required change, such as in the case of Az-Zarqa municipality and the work done by the LCP mentioned above.

6.1.2 International Policy

Related to regulatory powers is the question of the ability to enforce commitments through city coalitions. The enforcement of international agreements lacks strong levers except for a couple of agreements that have developed sophisticated non-compliance mechanisms starting with the 1987 Montreal Protocol to the 1985 Vienna Convention on the Protection of the Ozone Layer as well as the Kyoto Protocol. In the sphere of international environmental law, it is preferred to avoid disputes and then reach the stage of dispute settlement mostly through arbitration or adjudication. There are clear dispute avoidance mechanisms that are broadly referred to in the United Nations Charter article 33, and these include monitoring, enquiry, conciliation, mediation, reporting, and compliance technical and financial support (United Nations Environment Programme, 2001). The International Law Commission (ILC) further developed draft articles on Responsibility of States for Internationally Wrongful Acts codifying the states' responsibilities in case of non-compliance, namely, consisting of stopping the non-compliance and restituting, compensating and/or satisfying the party or parties wronged by the wrongful act (International Law Commission, 2001).

Cities and cities' alliances have not reached the same level of maturity in dealing with non-compliance. While many alliances and networks have put in place some form of monitoring and compliance assistance mechanisms (technical and sometimes financial assistance), such as in the case of C40 and the Global Covenant of Mayors, more in-depth accountability in the case of non-compliance is lacklustre if not absent. Perhaps one of the most stringent memberships of international cities' alliances lies in the leadership standards of C40, which renders members inactive for a year after not meeting essential commitments of the network, followed by revoking their membership should the non-compliance not be rectified. This is, however, an exception rather than the norm. This points to a weakness when comparing state responsibility to city or local government responsibility towards international commitments and objectives: the mechanisms of enforcement and accountability of the latter remain limited and toothless and mostly rely on monitoring and reporting.

6.2   Budgets and Finance

While there is an increasing trend in the decentralization of government responsibilities to state or local governments, the transition has not taken place in all aspects. Indeed, in her analysis of decentralization across countries for the OECD, Dorothée Allain-Dupré highlights that fiscal decentralization has not taken place to the same extent as overall decentralization (Allain-Dupré, 2016). This can limit the delivery of important enabling infrastructure for sustainable transition, such as infrastructure to enhance low-emissions transport (walking, cycling and public transport) or increase the efficiency and reduce the carbon footprint of the energy grid.

According to the Cities Climate Finance Leadership Alliance, the specific barriers cities face in accessing finance for climate action and sustainable transitions include "limited or restricted regulations in cities regarding private sector participation, low or no credit rating, limited capacity to structure bankable climate-smart projects, or lack of consistency in policy resulting from changing mayoral election cycles" (The Cities Climate Finance Leadership Alliance, 2021). The Alliance further highlights that there is an overall shortage in climate finance, specifically finance for cities in developing countries, on the one hand, while there remains a sizeable gap in financing adaptation, on the other hand, with it only receiving 9% of overall obtained finance. Additional challenges highlighted by the Stockholm Environment Institute in regard to climate finance in Swedish cities include the lack of a common framework for sustainable finance between the national level and local governments as well as the low appetite for risk-sharing in financing sustainability initiatives and access to finance covering operational costs that are considered high risk (Waltré et al., 2022).

To address these challenges, some of the TNMs do help member cities access finance such ICLEI, C40, and UCLG. They do so by building the capacity to structure financial proposals within the city government or providing them with supporting tools that facilitate decision-making related to procurement options. They also do so by providing linkages with donor organizations and raising the voices of cities in front of these donors in platforms such as the annual UNFCCC Conference of Parties. However, Bulkeley et al. found that less than 12% of overall city climate networks work on facilitating access to finance per se, highlighting considerable limitations of TNMs in addressing the financial shortage of cities. In this regard, states still present greater securities to financing organizations and donours, as they can present greater guarantees and have better credit rating compared with subnational governments (OECD, 2014b).

6.3   Infrastructure Networks

The efficiency of key infrastructure networks and services is impacted by whether it is planned at a local or regional level. This includes waste management, energy, water, wastewater, transportation infrastructure, and services and communication networks. A study conducted by the Inter-American Development Bank on Latin America and the Caribbean highlighted that efficiency gains were negatively correlated with the size of the city, pointing to two main reasons. The first reason consisted of the assumption that smaller cities would experience greater growth and therefore greater demand for infrastructure services, and the second reason considered greater limitations in local skills for efficient infrastructure planning compared to the regional level (Serebrisky et al., 2017).

On this point, the Executive Director of the Colorado Smart Cities Alliance, Tyler Svitak, resonates clearly: *"Smart regions are the new smart cities," "Without a regional, collaborative approach across jurisdictions, new technology solutions will never reach the scale or standardization required to improve complex civic issues"* (Collier, 2021). The need to coordinate at a regional scale to obtain increased efficiencies in infrastructure investment and operation applies to most utilities' networks, such as transport, water, energy, waste and wastewater, and communications. The history of the electricity grid in the United States of America is an example of 'regionalization' resulting from the need for more efficient and coordinated infrastructure planning. Indeed, according to the Energy Information Administration, the country had over 4000 individual utilities at the beginning of the 1900s, which slowly merged into three main interconnected networks over the past century, driven by a need for improved efficiency and resilience of the networks, reducing the cost of operation and oversupply of electricity.

Svitak mentions the word standardization, which is crucial to developing assessments and action plans at scale, whether at a regional, national, or international level. In this regard and as mentioned above, cities' alliances requiring monitoring can act as catalysts for standardizing data collection and reporting, thereby enhancing the ability to compile and analyse data at scale. On the other hand, alliances that do not capture regional governments are less influential in ensuring improved coordination in infrastructure planning and operation at a regional scale.

6.4 Blind Spots

6.4.1 The Non-urban

The increasing urbanization and growing visibility of cities at the international level begs questions about the fate of rural communities. Where and how are they represented? Benoît Bréville, French journalist who focuses on the role of cities in the global scene, has hinted at this blind spot through his writings (Bréville, 2020). He specifically warns of entering a vicious cycle whereby cities lobby the World Bank and the United Nations and their peer international organizations for policies and funding that are best suited for them, thereby fuelling further urbanization and exacerbating the problems inherent to cities without allowing a balanced approach to growth and allocation of international support to non-urban areas. It is worth noting that in 2018, 79% of the poor globally lived in rural areas (United Nations Statistics Division, 2019).

When thinking deeper about the responsibility of cities towards non-urban areas, one must think of cities as the main consumers globally. For example, cities consumed 70% of all food produced in 2020 (FAO, 2020), and they hosted 56.2% of the global population, while smallholder farmers produced approximately 50–70% of global food commodities and nutrients (Giller et al., 2021). Artisanal and small-scale mining (ASM) activities, dominated by rural communities, generate approximately 20% of the global gold supply, 80% of the global sapphire supply, 26% of global tantalum production, and 25% of tin, which are both important for the production of electronics. Overall, ASM is estimated to account for 15–20% of "precious metals, gems, building materials and (mostly) nonfuel minerals" (IGF, 2018), while approximately 70–80% of ASM activities are informal and not regulated. Finally, ASM is the largest source of mercury leak into the environment globally (IGF, 2018), leading to considerable environmental damage to exposed communities.

What this shows is that cities cannot play a representative and holistic role in international environmental governance if they do not look at the impacts of their local activities and consumption on communities beyond their boundaries. Late British social scientist and geographer Doreen Massey advocated for this throughout her work. While quite an evolved concept in international environmental law and embodied in the "polluter pays" and common but differentiated responsibilities' principles, it is still a burgeoning concept in its application to cities (Massey, 2004). It has manifested mainly through attempts to look at consumption-based greenhouse gas emissions, which C40 has looked at and which the Carbon Disclosure Project is now recording, although there is no mandatory requirement to do so. In April 2022,

Sweden announced its intent to report on consumption-based emissions, although there are several key challenges that need to be overcome, not least the lack of an internationally agreed carbon emissions calculation methodology (Hebron, 2022).

6.4.2 The Small Urban

Related to the point above, another blind spot of cities' alliances consists of small and medium cities, which typically have larger budgetary, skills, and resource constraints than larger cities (Häußler & Haupt, 2021).

Häußler and Haupt explored the possibility and usefulness of establishing a domestic city network for small and medium cities (below 100,000 inhabitants) in Germany to focus on adaptation. Their findings highlighted the need for this to be championed at a regional level to shoulder the resources required to manage such networks and to enable useful and impactful knowledge-sharing and capacity-building.

We contend that by forming regional or state networks, small and medium cities could raise their voice towards national governments on par with the voice of larger cities. The challenge of accessing the international arena, however, would remain given their limited global environmental and economic footprint. As illustrated in Fig. 5, larger cities have found a pathway to influence international stakeholders directly both bottom-up and top-down.

A possible mitigation measure to address this blind spot could be to scale-up representation from a city level to a regional level, as indirectly hinted at by Svitak. UCLG and ICLEI, amongst other TNMs, have done so; however, there has not been a detailed assessment of how well this has given equal representation to the smaller cities and their specificities compared to the larger cities.

7 Conclusion

The combination of the size of cities and the supporting services required to ensure a minimum quality of life for urban communities, added to the concentration of economic activities and their relative agility to implement actions, is what makes cities a key player in facilitating sustainable transitions. This has led to the empowerment of cities as influencers of the global environmental agenda. In Swiney's words, we witness "the rising soft power of cities in global governance". With the increased emancipation of cities from the national level, cities are gaining powers to act locally and shape their own development through the establishment of local action plans and policies, using their purchasing powers to influence the local market and reduce their environmental footprint. They also have the power to mobilize their local stakeholders for increased impact when needed. Cities' alliances in turn provide member cities with increased visibility at the national and international scale, thereby reinforcing their role as global actors in the environmental transition. They do so by magnifying the voices and efforts of local governments by compiling and converting local data into global assessments while providing cities with additional support and leverage when needed.

We have looked at the influence that both national and international cities' alliances can have at the local and international level and the different dynamics that can exist between the alliances and the national and international stakeholders. Cities that are members of international alliances, or TNMs, have greater visibility and influence on international players and international agreements. This has crystalized through the specific mentioning of cities in Agenda 21 and the Paris Agreement to the UNFCCC. The development of city-focused programmes by international agencies such as the United Nations and the World Bank is further testament to their importance as players in the international environmental agenda.

There are, however, current limitations that cities have that prevent them from becoming a parallel platform for international environmental governance on par with states. These mainly consist of the limitations in forming as strong of an international accountability and liability framework as states have been able to in the realm of international law as well as the blind-spot cities have toward the non-urban. They are not capable of putting forward a holistic approach to a just and fair environmental transition unless they address their consumption-induced impacts on non-urban and global communities. They fully have the potential to do so, and one could argue that it is perhaps partially their responsibility to move in this direction given their growing voice and influence on the international scene, using 'local economic strategy' aiming towards 'global responsibility' to use Massey's terminology. Remaining challenges, and areas for future development, include a more detailed mapping of these blind spots in the international debate on environmental and sustainable transition and the development of appropriate means of representation – for example, through cities platforms, existing international organizations, new dedicated networks – so that no one is left behind.

Bibliography

Acuto, M. (2011). Finding the global city: An analytical journey through the 'invisible college'. *Urban Studies, 48*(14), 2953–2973.

Acuto, M., & Rayner, S. (2016). City networks: Breaking gridlocks or forging (new) lock-ins? *International Affairs, 92*(5), 1147–1166. https://doi.org/10.1111/1468-2346.12700

Allain-Dupré, D. (2016). *Decentralization trends in OECD countries: A comparative perspective for Ukraine*. OECD. https://www.oecd.org/regional/regional-policy/Decentralisation-trends-in-OECD-countries.pdf. Accessed 10 Mar 2022.

Andonova, L., Betsill, M., & Bulkeley, H. (2009). Transnational climate governance. *Global Environmental Politics, 9*(2), 52 73. https://doi.org/10.1162/glep.2009.9.2.52

Barbi, F., & De Macedo, L. (2019). Transnational municipal networks and cities in climate governance: Experiments in Brazil. In J. Van der Heijden, H. Bulkeley, & C. Certomà (Eds.), *Urban climate politics: Agency and empowerment* (pp. 59–79). Cambridge University Press. https://doi.org/10.1017/9781108632157.004

Barret, A. L., Iceland, J., & Farrel, C. (2014). Is ethnoracial integration of the rise? Evidence from metropolitan and micropolitan America since 1980. In *Diversity and disparities: America enters a new century* (pp. 415–456). Russell Sage Foundation. Chapter 13.

Baykurt, B., & Raetzsch, C. (2020). What smartness does in the smart city: From visions to policy. *Convergence, 26*(4), 775–789. https://doi.org/10.1177/1354856520913405

Bodiford, A. (2020). *Cities in international law: Reclaiming rights as global custom.* 23 CUNY L. Rev. 1. The CUNY Law, Office of Library Services at the City University of New York. https://academicworks.cuny.edu/clr/vol23/iss1/1

Bulkeley, H., Andonova, L., Backstrand, K., Betsill, M., Compagnon, D., Duffy, R., et al. (2012). Governing climate change transnationally: Assessing the evidence from a database of sixty initiatives. *Environment and Planning C: Government and Policy, 30*, 591.

Carnegie UK Trust. (2016). *Sharpening our focus: Guidance on wellbeing frameworks for cities and regions.* Carnegie UK Trust. ISBN: 978-1-909447-47-9. https://d1ssu070pg2v9i.cloudfront.net/pex/pex_carnegie2021/2016/09/09130209/Sharpening-our-Focus.pdf

FAO. (2020). Food and Agriculture Organisation Website. Accessed June 2022. Urban Food Agenda | Food and Agriculture Organization of the United Nations (fao.org).

El Mir, J., Elgendy, K., & Khamlichi, H. (2021). *Circular economy in cities of the MENA region: Prospects and challenges for material circularity.* Friedrich-Ebert-Stiftung. https://library.fes.de/pdf-files/bueros/amman/18984.pdf

Giller, K. E., Delaune, T., Silva, J. V., Descheemaeker, K., van de Ven, G., Schut, A. G. T., van Wijk, M., Hammond, J., Hochman, Z., Taulya, G., Chikowo, R., Narayanan, S., Kishore, A., Bresciani, F., Teixeira, H. M., Andersson, J. A., & van Ittersum, M. K. (2021). The future of farming: Who will produce our food? *Food Security, 13*, 1073–1099. https://doi.org/10.1007/s12571-021-01184-6

Hassan, D. (2006). Rise of the territorial state and the treaty of Westphalia. *Yearbook of New Zealand Jurisprudence, 9*, 62–70. ISSN 1174-4243.

Häußler, S., & Haupt, W. (2021). Climate change adaptation networks for small and medium-sized cities. *SN Social Sciences, 1*, 262. https://doi.org/10.1007/s43545-021-00267-7

Intergovernmental Forum on Mining, Minerals, Metals and Sustainable Development (IGF). (2018). *Global trends in artisanal and small-scale mining (ASM): A review of key numbers and issues.* IISD.

International Law Commission. (2001). *Draft articles on responsibility of states for internationally wrongful acts,* Supplement No. 10 (A/56/10), chp.IV.E.1

Massey, D. (2004). The responsibilities of place. *Local Economy, 19*(2), 97–101. https://doi.org/10.1080/0269094042000205070

Nielsen, A. B., & Papin, M. (2021). The hybrid governance of environmental transnational municipal networks: Lessons from 100 Resilient Cities. *Environment and Planning C: Politics and Space, 39*(4), 667–685. https://doi.org/10.1177/2399654420945332

Nijman, J. E. (2011). The future of the city and the international law of the future. In S. Muller et al. (Eds.), *The law of the future and the future of the law* (pp. 213–229). Torkel Opsahl Academic EPublisher.

Nijman, J. E. (2016). Renaissance of the City as Global Actor: The role of foreign policy and international law practices in the construction of cities as global actors. In G. Hellmann, A. Fahrmeir, & M. Vec (Eds.), *The transformation of foreign policy: Drawing and managing boundaries from antiquity to the present* (pp. 209–239). Oxford University Press. https://doi.org/10.1093/acprof:oso/9780198783862.003.0010

OECD. (2014a). *How's life in your region? Measuring regional and local well-being for policy making.* OECD Publishing. Accessed at http://www.oecd.org/gov/how-s-life-in-your-region9789264217416-en.htm

OECD. (2014b). *Cities and climate change, national governments enabling local action.* OECD Publishing. https://www.oecd.org/env/cc/Cities-and-climate-change-2014-Policy-Perspectives-Final-web.pdf

OECD. (2017). *The governance of land use in OECD countries: Policy analysis and recommendations.* OECD Publishing. https://www.oecd-ilibrary.org/urban-rural-and-regional-development/the-governance-of-land-use-in-oecd-countries_9789264268609-en;jsessionid=BD63RhKN0gPUu7Egj0VQzyRG.ip-10-240-5-25. Accessed 25 Feb 2022.

OECD. (2020). *Smart cities and inclusive growth: Building on the outcomes of the 1st OECD roundtable on smart cities and inclusive growth.* OECD. https://www.oecd.org/cfe/cities/OECD_Policy_Paper_Smart_Cities_and_Inclusive_Growth.pdf. Accessed 25 Jan 2022.

Patton, S. (2019). The peace of Westphalia and it affects on international relations, diplomacy and foreign policy. *The Histories, 10*(1), Article 5. https://digitalcommons.lasalle.edu/the_histories/vol10/iss1/5

Sands, P., & Peel, J. (2018). *Principles of international environmental law* (4th ed.). Cambridge University Press. https://doi.org/10.1017/9781108355728

Sassen, S. (1991). *The global city*. Princeton University Press.

Serebrisky, T., Suárez-Alemán, A., Pastor, C., & Wohlhueter, A. (2017). *Increasing the efficiency of public infrastructure delivery: Evidence-based potential efficiency gains in public infrastructure spending in Latin America and the Caribbean.* Inter-American Development Bank. Infrastructure and Energy VI Series.

Sharpe, L. J. (1988). The growth and decentralization of the modern democratic state. *European Journal of Political Research, 16*, 365–380.

Somanathan, E. (2001). Empowering local government: Lessons from Europe. *Economic and Political Weekly, 36*(41), 3935–3940. http://www.jstor.org/stable/4411235

Swiney, C. (2020). The urbanization of international law and international relations: The rising soft power of cities in global governance. *Michigan Journal of International Law, 41*(2), 227.

The Cities Climate Finance Leadership Alliance. (2021). *The state of cities climate finance part 1: The landscape of urban climate finance.* Climate Policy Initiative. https://citiesclimatefinance.org/wp-content/uploads/2021/06/Part-1-1-The-Landscape-of-Urban-Climate-Finance-FINAL.pdf

UN-Habitat. (2020). *Global state of metropolis 2020 – Population data booklet.* United Nations Human Settlements Programme. HS Number: HS/013/20E. https://unhabitat.org/sites/default/files/2020/09/gsm-population-data-booklet-2020_3.pdf

United Nations Environment Programme. (2001). *Dispute avoidance and dispute settlement in international environmental law: Compilations of documents.* https://wedocs.unep.org/20.500.11822/29653

Waltré, N., Sjöström, E., Agerström, M., Vanhuyse, F., & Requena Carrion, A. (2022). *The role of private market capital in financing sustainable cities: Investor and municipal views in a Swedish context.* Stockholm Environment Institute, Stockholm School of Economics, Cleantech Scandinavia.

Weekes S. (2019). *Smart cities: Beyond the hype. How far away are truly intelligent cities that improve quality of life for citizens?* Smart Cities World. https://www.smartcitiesworld.net/whitepapers/whitepapers/smart-cities---beyond-the-hype#:~:text=Less%20than%20a%20third%20of%20cities%20%2831%20per,by%20SmartCitiesWorld%2C%20in%20association%20with%20Interact%20by%20Signify

World Bank. (2021). *State of cities climate finance 2021 part 2: The enabling conditions for mobilizing urban climate finance.* © World Bank. License CC BY 3.0 IGO.

Websites

100 Resilient Cities. The Rockfeller Foundation. https://www.rockefellerfoundation.org/100-resilient-cities/

Agenda 2030 at the Local Level. https://nordregioprojects.org/agenda2030local/. Accessed 15 Mar 2022.

Alliance Team. (2021, October 4). *The national smart coalitions partnership unites 100+ governments across seven regional smart cities consortiums.* Colorado Smart Cities Alliance. https://coloradosmart.city/the-national-smart-coalitions-partnership-unites-100-governments-across-seven-regional-smart-cities-consortiums/

Bréville B. (2020, May). *Quand les grandes villes font sécession.* Le Monde Diplomatique. https://www.monde-diplomatique.fr/2020/03/BREVILLE/61548

C40 Cities. https://www.c40.org/cities/. Accessed 1 Mar 2022.

C40 Cities. *Financing the green transition.* https://www.c40.org/what-we-do/influencing-the-global-agenda/financing-the-green-transition/

Collier, C. (2021, October 8). *Smart regions unite to form the national smart coalitions partnership.* Smart Cities Connect. https://smartcitiesconnect.org/smart-regions-unite-to-form-the-national-smart-coalitions-partnership/

Hebron E. (2022, April 4). *Sweden heeds Greta's call to target 'consumption-based emissions' in world-first.* Euronews. https://www.euronews.com/green/2022/04/13/sweden-heeds-gretas-call-to-target-consumption-based-emissions-in-world-first

The League of the Cities of Philippines. https://lcp.org.ph/10/about-us

The League of Cities of the Philippines Mandate, League of Cities of the Philippines Mandate. https://web.archive.org/web/20090204010713/http://lcp.org.ph/au_mandate.htm

McBride J., & Siripurapu A. (2021, May 14). *How does the U.S. power grid work?* Council on Foreign Relations. https://www.cfr.org/backgrounder/how-does-us-power-grid-work

Smart Cities Mission. *Ministry of Housing and Urban Affairs, Government of India.* https://smartcities.gov.in/. Accessed 13 Feb 2022.

Staden M. (2019, February 11). *CDP and ICLEI: Introducing streamlined climate reporting.* CDP. https://www.cdp.net/en/articles/cities/cdp-and-iclei-introducing-streamlined-climate-reporting

The Rockefeller Foundation. (2022). *100 Resilient Cities.* https://www.rockefellerfoundation.org/100-resilient-cities/#:~:text=Quick%20Take&text=In%202013%2C%20The%20Rockefeller%20Foundation,part%20of%20the%2021st%20century. Accessed 15 Jan 2022.

United Cities and Local Governments. (2021). *Centenary of the International Municipal Movement 1913–2013.* https://www.uclg.org/en/centenary. Accessed 2 Mar 2022.

United Nations. *Climate action.* https://www.un.org/en/climatechange/climate-solutions/cities-pollution#:~:text=According%20to%20UN%20Habitat%2C%20cities,cent%20of%20the%20Earth's%20surface. Accessed 10 Apr 2022.

United Nations Statistics Division. (2019). *End poverty in all its forms everywhere.* https://unstats.un.org/sdgs/report/2019/goal-01/. Accessed 10 Apr 2022.

Urban Development, Overview. World Bank. (2020, April 20). https://www.worldbank.org/en/topic/urbandevelopment/overview#1

US Department of Energy. *Clean Cities Coalition Network.* https://cleancities.energy.gov/about/. Accessed 20 Feb 2022.

US Energy Information Administration. *Homepage – U.S. Energy Information Administration (EIA).* Accessed 28 Mar 2022.

Ville de Paris. (2019, February 1). *Paris, une collectivité engagée dans la finance verte et durable.* https://www.paris.fr/pages/une-finance-verte-et-responsable-5686\

Open Access This chapter is licensed under the terms of the Creative Commons Attribution 4.0 International License (http://creativecommons.org/licenses/by/4.0/), which permits use, sharing, adaptation, distribution and reproduction in any medium or format, as long as you give appropriate credit to the original author(s) and the source, provide a link to the Creative Commons license and indicate if changes were made.

The images or other third party material in this chapter are included in the chapter's Creative Commons license, unless indicated otherwise in a credit line to the material. If material is not included in the chapter's Creative Commons license and your intended use is not permitted by statutory regulation or exceeds the permitted use, you will need to obtain permission directly from the copyright holder.

Smart Energy Cities: The Role of Behavioral Interventions in Reducing Electricity Demand in Buildings in Principality of Monaco

Fateh Belaïd, Mira Toumi, and Nathalie Lazaric

Abstract With energy powering the most attractive aspects of urban environments in modern society, from health, transportation, and comfort to information, business, and leisure, energy cities are perfectly positioned to design the smart city of the future by leveraging the energy foundations of the city. This chapter focuses on the emerging concept of energy cities through the lens of sustainable behaviors and their role in alleviating climate change. We use the results of a randomized control trial experiment implemented in Monaco to illustrate our arguments on the role of behavioral intervention in empowering citizens on the importance of saving energy. The results will offer a vision of what steps cities are taking to increase environmental awareness and the role of individual behaviors in tackling climate change.

Keywords Energy Cities · Smart and Sustainable Cities · Behavioral change · Residential energy use

1 Introduction

Modern cities and human activities cause significant climate change issues as well as energy and mobility challenges and need to take initiatives to find sustainable solutions. Currently, over half of the world's population lives in an urban

F. Belaïd (✉)
King Abdullah Petroleum Studies and Research Center, Riyadh, Kingdom of Saudi Arabia
e-mail: fateh.belaid@kapsarc.org

M. Toumi
GREDEG, University of Côte d'Azur, Valbonne, France

N. Lazaric
GREDEG, University of Côte d'Azur, Valbonne, France

University of Gothenburg, Göteborg, Sweden

© The Author(s) 2024
F. Belaïd, A. Arora (eds.), *Smart Cities*, Studies in Energy, Resource and Environmental Economics, https://doi.org/10.1007/978-3-031-35664-3_6

environment, and by 2050, this figure is projected to exceed two-thirds (United Nations, 2019). This accelerating urbanization process has resulted in many challenges, including intensive energy consumption, high-carbon GHG emissions, environmental pollution, social inequality, and traffic congestion. This is a long list of challenges included in the Agenda 2030 Sustainable Development Goals, putting the development of energy-efficient, more sustainable, and smart cities at the top of the agenda in the coming years (United Nations, 2022). Here, we examine the role of behavioral intervention in shaping residential electricity consumption within the contexts of "smart cities" and "smart energy cities" (SEC).

The smart cities concept underlines an urban development area using digital and information and communication technologies (ICT) solutions to improve traditional networks and services. Smart cities are designed to address challenges and improve the workability, quality of life, and sustainability of cities. Although the design of the smart city has traditionally focused on technology, smart devices, and urban infrastructure, it now goes beyond the technical-centric nature of using digital solutions for the efficient use of resources. Over recent years, the concept has been expanded to incorporate socioeconomic aspects (Pira, 2021). Accordingly, this paradigm shift allows the smart city approach to expand its potential impacts on the economic, social, and environmental dimensions.

The recent phase of smart cities, the so-called smart city 3.0, rather than adopting a technology-driven or city-driven model, focuses on developing co-creation models involving citizens in developing efficient and practical solutions. This new concept focuses on the inhabitant's role and involvement in addressing community issues and assisting municipality managers in identifying effective and reliable solutions for various city challenges, including social, economic, and environmental problems. The new paradigm strongly suggests that a sustainable future will rely on a combination of innovative technology—improving new technologies and system performance—and promoting more environmentally friendly behavior (Sovacool et al., 2022).

The concept of "smart energy cities," which builds on the concept of smart cities, has been developed recently to recognize the prominent role of energy in the built environment (Thornbush & Golubchikov, 2021). SEC ideas are anchored both in the expansion of "smart cities" concepts and in a sustainability framing. The SEC concept has grown to depict digitally enhanced, zero-carbon cities. Accordingly, Übelmesser et al. (2020) proposed the following definition: "SEC is a concept at the core of the smart city, that uses technology, including information and communication technologies (ICT), to address the challenges of increasing urban energy demand and climate change, while ensuring the quality of life of its citizens … the SEC uses ICT to integrate different domains, resulting in a holistic view of the energy system" (p. 1). Thus, the SEC paradigm is strongly aligned not only with the smart cities concept but also with the climate-neutral city concept and its variants, such as low-carbon, net-zero cities, and postcarbon cities (Thornbush & Golubchikov, 2021).

With many of the world's major cities pursuing important initiatives to improve citizens' urban life and achieve sustainability and climate goals and with energy powering the most attractive aspects of urban environments in modern society, from health, transportation, and comfort to information, business, and leisure, energy cities are perfectly positioned to design the smart city of the future by leveraging the energy foundations of the city. Thus, SEC will serve as an example for other cities going forward. They will harness cutting-edge technologies, such as green energy, superfast telecommunications, autonomous transportation, and artificial intelligence. Using these technologies, they will become desirable and comfortable places to live, work, and relax.

Based on the underlying premise of "smart energy cities," this chapter discusses the role of behavioral insights in shaping the future of urban living. Then, it uses results from a randomized controlled trial (RCT) to illustrate the role of behavioral intervention in increasing the impact of household actions to save energy. This will offer valuable insights and inform policymakers on the importance of behavioral intervention in shaping cities' sustainability. Behavioral interventions can be broadly defined to encompass interventions for which no command-and-control regulations or financial incentives are involved, e.g., providing information, goal setting, invoking values and norms, engaging, and restructuring choice options, or so-called nudges (Lazaric & Toumi, 2022).

The setting of the analysis in this chapter is the Principality of Monaco, a sovereign city-state located on the French Riviera in Western Europe. The uniqueness of our context and experimental framework makes the statistical analysis and the evaluation criteria we have implemented compelling for a variety of reasons. First, despite the growing literature in the field of smart cities, there have been no studies linking behavioral intervention within the concept of "smart energy cities." Second, the analysis takes advantage of a recent field experiment to illustrate how behavioral interventions may reduce the contribution of individuals' activities to carbon emissions and climate change.

This chapter contributes to the literature on SEC in the following ways. It emphasizes the emerging energy city concept and the critical role of behavioral insights in shaping the future of urban living. These findings can help city policymakers leverage behavioral intervention to reduce energy demand in cities to achieve goals toward environmental carbon neutrality. It may also add to the behavioral literature and stimulate studies to rethink how to implement behavioral interventions efficiently and incorporate behavioral insights to improve sustainability in cities.

The remainder of this chapter is structured as follows. Section 2 reviews the literature on behavioral economics in reducing energy demand in buildings. Section 3 focuses on Monaco as a smart energy city and presents Monaco's smart city initiatives. Section 4 presents and discusses the experimental design and Smartlook project's main results, while Sect. 5 provides some concluding remarks and policy implications.

2   Behavioral Economics and Policymaking

To meet challenging climate and sustainability goals, the ambitious curtailment of GHGs is needed worldwide. A low-carbon world will rely on a combination of green technological innovation and sustainable behaviors such as energy sobriety and waste reduction. There is also an unequivocal consensus about the critical role that behavioral change may play in decarbonizing the building sector (Maréchal, 2010; Sovacool et al., 2022). Today, behavioral change interventions are widely implemented in a range of public policy settings with the goal of moving individuals in the desired direction, e.g., toward more sustainable lifestyles, more eco-friendly practices, and more responsible financial decisions. For example, Moran et al. (2020) show that a consumer-oriented policy can reduce GHG emissions by 25% in the European context, and Asmare et al. (2021) show that providing information via a web portal lowers electricity consumption by 8.6% in Lithuania. Thus, a considerable part of CO2 emissions can be reduced or eliminated with lifestyle changes, thus leading to a positive impact on fighting climate change, reinforcing energy security, and ensuring affordable energy access (Belaïd, 2022a, b, 2024).

Governments have used traditional economic tools and other behavioral solutions to foster environmentally friendly practices. Depending on the target behavior and context, traditional tools such as regulations and taxes can be an efficient response to reducing emissions. However, in some specific situations, they will not be enough to drive effective behavioral change or, even more, generate counterproductive behaviors. For instance, in terms of residential energy consumption, it is challenging to implement a law to force people to reduce the electric heating temperature by 1° to save 7% of their energy bill. Indeed, residential energy consumption is a multifaceted sociotechnical process shaped by a variety of interdependent factors (Belaïd, 2016, 2017). In addition, the complexities of consumers' lifestyles and the role of individuals' behavior in the energy demand process have contributed to ambiguities and partial comprehension of residential energy use patterns (Belaïd et al., 2020, 2021; Belaïd & Rault, 2021). Many studies have documented that there is a large gap between theoretical and observed energy consumption. This is due mainly to difficulties in capturing the behavioral aspects of domestic energy use (Bakaloglou & Charlier, 2019; Bakaloglou & Belaïd, 2022). Therefore, policymakers and governments are increasingly becoming concerned about effectively changing consumer behaviors. They rely on behavioral sciences, now widely recognized as a source of alternative or complementary tools to empower citizens toward the greenest sustainable behaviors.

A critical element of the pro-environmental behavior question is intrinsic human nature. In fact, since it has been proven that individuals are not entirely rational, policymakers and stakeholders should address this bounded rationality and cognitive biases (Maréchal, 2010). In fact, contrary to traditional rational choice analysis, behavioral economics has shown that individuals often rely on heuristics and are easily influenced by their cognitive biases. An interesting illustration is the claim "It only happens to the others" when seeing the consequences of climate change. This sentence, often heard in Western countries, can be explained by the abstract frame of the environmental problem due to psychological distance. This implies that humans are more interested in the present than in the future, in what has an impact on them (or her) rather than what might impact others, and in what happens close to them rather than far away. This starting point for anchoring environmental issues is tricky since it pushes citizens to procrastinate on their potential actions. Behavioral tools then raise opportunities for the awareness that by now, as members of the worldwide community, "we are all the others."

The most famous tool, coming from behavioral economics insights, is the "Nudge." In their popular book *Nudge, Improving Decisions about Health, Wealth and Happiness* (2008), Richard Thaler and Cass Sunstein define a nudge as "any aspect of the choice architecture that alters people's behavior in a predictable way without forbidding any options or significantly changing their economic incentives. To count as a mere nudge, the intervention must be easy and cheap to avoid" (Thaler & Sunstein, 2008, p. 6). Relying on bounded rationality theory, the authors suggest that a considerable part of the decision-making pattern is the result of cognitive boundaries, biases, or habits and that this pattern may be "nudged" toward better options by integrating insights about the former boundaries and biases in ways that promote a more preferred behavior rather than obstruct it. After considerable hype around the concept for several years, the efficiency and ethical outcomes of nudging started to be discussed and questioned to conclude that nudging is not the magic wand able to solve all problems in all settings. From this questioning emerged other tools such as "boosts." In fact, while nudges shape human decisions by changing the choice setting and the encountered information, they boost aim to foster human competencies and motivation by working on his (or her) skills and knowledge. In addition, presenting accurate information by changing the options humans are exposed to "boosts" work to empower an individual to make better decisions with respect to his (or her) personal goals and preferences (Hertwig & Grüne-Yanoff, 2017).

The success of Thaler and Sunstein's book, combined with the considerable body of materials, academic articles, and university programs, made the young behavioral science field gain such legitimacy that several so-called "nudge units" appeared all over the world, while some behavioral economists were involved in policymaking. At the European level, the United Kingdom has pioneered the use of behavioral economics for policymaking with the *Behavioral Insight Team (BIT)*. The BIT nudge unit is a governmental team dedicated to implementing soft methods since its creation in 2010 under the advice of Richard Thaler. The BIT advises the government on several subjects, such as pandemic management or actions to fight global warming. BIT's actions are numerous around the world, and their success has led to the creation of nudge units in other countries. In France, the *French behavioral team*, the DITP (Direction Interministerielle de la Transformation Publique), arose under Emmanuel Macron's presidency and Bercy's supervision. Since 2020, the team has grown with the creation of the Transformation of the Public Service Ministry. In the Principality of Monaco, no specific governmental behavioral team is dedicated. However, led by the government impulse and the success of projects held in their neighborhood, local utilities such as SMEG (Société Monégasque d'Electricite et de Gaz) are inspired and motivated to adopt new behavioral insights to create new ideas for policy development. This is the reason why SMEG in the Principality of Monaco has initiated a partnership with the University Côte d'Azur on new behavioral tools for increasing awareness of opportunities to reduce electricity consumption. Monaco is among the innovating smart cities in southern Europe as an environmentally committed region at the cutting edge of technology. The municipality uses various tools and policy interventions to act in this field. Behavioral tools are one of these options.

3 Experimental Research on Smart Cities

3.1 Smart and Energy Cities Research

A growing interest in smart cities and energy consumption is observed in policy-making and academic areas. Some empirical evidence of the growing interest in smart and energy city studies can be found in Scopus data with the yearly number of publications in social sciences. We observe two similar rising trends.

Figure 1 shows that the number of articles published each year in social sciences on smart cities and energy remained constant from 2006 to 2009. After 2011, there was an increasing number of publications until 2019, when it reached a maximum of 138 articles. This growing interest of governments and several founding agencies explains this rising trend. Figure 2 displays the number of projects referring to a sponsor within the considered period and shows that the main sponsor is the

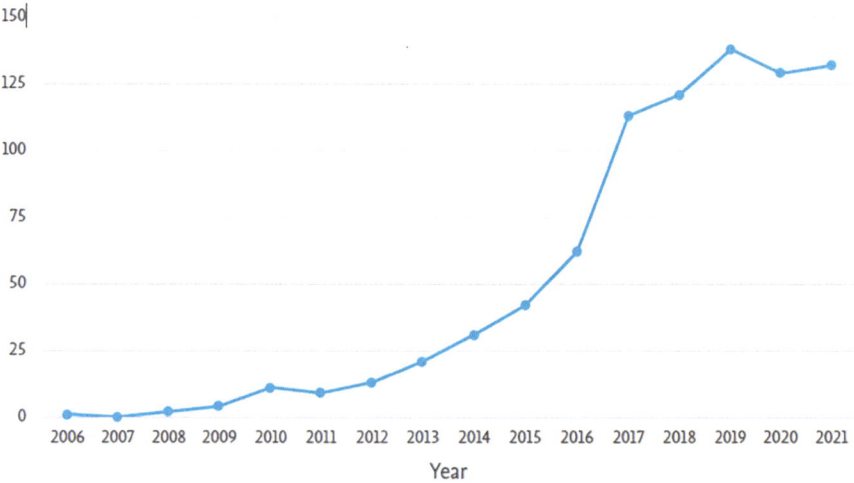

Fig. 1 Smart and energy city studies over time and disciplines. (*Source*: Scopus – Authors' calculation)

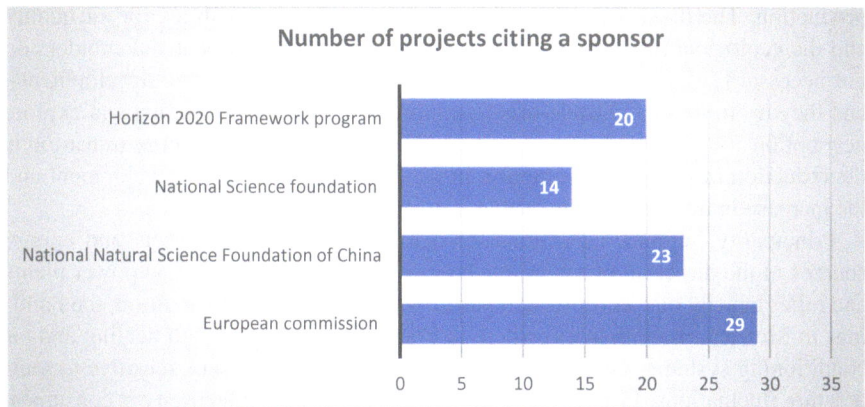

Fig. 2 Smart cities and energy studies sponsors. (*Source*: Scopus – Authors calculation. The ranking is based on the sponsors' occurrence frequency pooled from 2006 to 2021. The sponsors are generated from the list of sponsors referenced on the Scopus platform)

European Commission (29 projects), followed by the National Natural Science Foundation of China (23 projects), the National Science Foundation, and the Horizon 2020 Framework program (20 projects) and the National Science Foundation, i.e., an independent agency of the United States government, intended to financially support theoretical scientific research (14 projects).

As we can see, European institutions devote resources and funds to developing and exploring sustainable energy consumption related to smart cities. In the following, we will present an example of a study resulting from another funding source but in line with the European trend. In fact, the specific case of the Principality of Monaco provides exciting insights into how a city state country that clearly adopted a sustainability-oriented public policy invested in new behavioral strategies to reach its commitment to reducing greenhouse gas emissions in the built environment.

3.2   *Monaco, the Smart Energy City of Southern Europe*

The Principality of Monaco is a very attractive territory and provides important employment opportunities in Southern Europe. The population was 38,300 residents in 2018 for 52,000 salaried jobs. Seventy-six percent of private-sector employees live in France, and 14% live in Principality. A large proportion of its local residents are financially well endowed and live in apartments in tower blocks that were built mostly in the 1970s.

In addition, Monaco hosts and organizes numerous cultural, sporting, and professional events. These specificities have an impact on the environmental and energy balance of the Principality due to transport, energy requirements, and waste production. The downside of this dynamic is negative externalities for air quality and the ecological preservation of the marine environment. Local stakeholders do not necessarily see a conflict between economic and demographic developments, and they try to find local trade-offs to minimize the ecological burden and explore new options toward the energy transition. The challenge of the energy transition is the reduction of energy consumption in the context of economic development and the increase in the share of renewable energy.

Principality implements actions for defining consumption targets and energy sources in the direction of renewable energy. Monaco has only a few power plants and relies heavily on electricity and gas imports from France. In addition, the buildings in Monaco, more than elsewhere in France, are equipped with heating and air conditioning systems. Electricity consumption is, therefore, very sensitive to temperature fluctuations. Principality imports almost all of the electricity it consumes. Although the regional energy supply has been strengthened in recent years, the local government has tried to promote the creation and development of local capabilities and produce local energy. Fossil fuels, such as natural gas and fuel oil, are also imported. The balance is better for thermal energy thanks to the Fontvieille network and sea water heat pumps, which slightly reduce energy dependence.

Monaco's average electricity consumption per inhabitant tends to be below the average of its neighbor France. However, this is mainly because Monaco residents spend only part of the year in Monaco rather than being more energy-saving conscious, which makes comparison difficult. Ninety percent of Monaco's electrical energy is supplied by France and includes a high percentage of renewable electricity (75% for Monaco compared to only 20% for the whole of France). This promotes more careful use of energy and more attention to the environment.

3.3 The White Energy Book on the Energy Transition

In 2008, the local government started actions in the energy transition by implementing a climate-energy policy and a program of actions, the *Climate Energy Plan*, supported by the Department of the Environment. Monaco is also committed to the European Energy labeling process award to extend the *Climate Energy Plan* to promote continuous improvement for energy reduction. In this sense, Monaco was labeled an exemplar city of energy transition in 2014. In addition, the Principality has been engaged within the international community by ratifying the Framework Convention United Nations successively on Climate Change in 1994, the Kyoto Protocol in 2006, and the Paris Agreement in 2016. As an extension of these actions, the signature of the Paris Agreement in 2016 represents a turning point with new, very ambitious goals.

Three primary sources of greenhouse gases in Principality are the consumption of fossil energy for heating (31% of GHG), energy recovery from household waste (30%), and fuel consumption for transportation (31%). The 2017 *White Book on the Energy Transition* describes the aim of reducing greenhouse gas emissions by 80%

(compared to 1990 levels) by 2050 and achieving carbon neutrality in the long run. Actions implemented for reaching these very ambitious goals are the following:

1. Sea water heat pumps. Sea water heat pumps are thermodynamic systems that recover heat energy from sea depths to satisfy heat needs. They are sources of renewable energy. Principality is a forerunner in this area. The first sea water heat pump was installed in the 1960s. Today, Monaco has more than 70 units, along with a public heating network.
2. The e-bikes. With 15 stations, nearly 100 electrically assisted bicycles, and more than 500 users, the service continues to develop in Principality. This is the most visible action and one of the most cited for reducing the energy burden in transport.
3. Support for the acquisition of hybrids and electric vehicles. The Prince's Government grants aid finance for the purchase of new e-mobility registered in Monaco. Those grants are provided to households as well as companies.
4. Retrofitting. The local government is involved in a retrofitting program to reduce buildings' greenhouse gas emissions and ensure that all new buildings conform to environmental standards.
5. Behavioral tools and interventions. Monaco tries to promote large environmental awareness programs and new cultural values around energy consumption and devotes significant resources to behavioral tools. The Smartlook project, in partnership with the SMEG and CNRS/Université Côte d'Azur, detailed below, is illustrative of this new trend.

4   The Smartlook Experiment

Smartlook is a novel way of targeting energy transition with the experimentation of new behavioral tools to reduce electrical consumption and with an enrollment of volunteers to exemplify opportunities for changing energy behaviors (Lazaric & Toumi, 2022).

4.1   Context of the Smartlook Field Experiment in Monaco

The experiment took place in the Principality of Monaco from December 2018 to May 2019 with the support of the main local energy provider (SMEG). The sample included 77 households from a diverse range of buildings with different heating systems and, more importantly, dwellings that were not part of any current retrofitting program.

All participants were volunteers and completed the first questionnaire prior to the beginning of the experiment to grasp information on sociodemographics, ecological concerns, commitment, electricity use, heating system, and curtailment behaviors.

*The New Ecological Paradigm (NEP) scale is used as **a unidimensional measure of environmental attitudes***. It was developed to measure the overall relationship between humans and the environment. A high NEP score is associated with high ecocentric orientation (Stern et al., 1995)

At the end of the experiment, a second questionnaire was sent to capture changes in household composition or energy use. After completing the first questionnaire, households were randomly assigned to one of the experimental treatments or to the control group following the Harrison and List (2004) description of a framed field experiment.

More precisely, control group households received an email informing them they were simply a part of an experiment aimed at gathering information on Monegasque households' energy transition. Households in the three treatment groups received twice-monthly emails containing instructions with a reminder of their electricity use reduction goal and/or a set of boosts. More specifically, Treatment 1 ($n = 16$) set an ambitious electricity consumption reduction goal compared to the previous 6 months' usage (25%) and received boosts on electricity savings. Treatment 2 ($n = 17$) received a modest (15%) electricity consumption reduction goal compared to the previous 6 months of usage and a set of boosts. Treatment 3 ($n = 21$) provided only boosts (advice) about how to reduce their electricity consumption, and finally, the control group ($= 23$) received neither a goal nor boosts.

4.2   The Smartlook Project's Main Results

Table 1 presents the evolution of average electricity consumption across the four treatments.[1] From Table 1, we observe that the boost and modest goal treatment (T2) consumed the least electricity, followed by the boost-only treatment (T3) (Table 2).

The percentage of variation in electricity consumption presented in column (c) shows a similar trend of increased average electricity consumption during the experiment for all treatments due to the winter months. While the highest consumption is observed in the control group with a 31% increase in consumption compared to the average consumption of the whole sample, the T1 and T2 groups had the lowest increases at 12% and 7%, respectively. A Kruskal–Wallis (K-Wallis)[2] equality test of average monthly electricity consumption among treatments confirms that average electricity consumption during the 6 months of the observation period differed significantly across treatments (p value = 0.0001). Additionally, pairwise

---

[1] The statistics are based on average electricity consumption seasonally adjusted data provided by the SMEG.

[2] We rely on the K-Wallis and WMW tests as a nonparametric test to compare two or more independent samples of equal or different sizes. It is an extension of the MWM U test which is used to compare two groups.

Table 1 Evolution of electricity consumption before and during the field experiment

Treatment	Average electricity consumption per household during the pretreatment period (a) (kWh)	Average electricity consumption per household during the treatment period (b) (kWh)	Difference (%) (c)	Diff-in-Diff (%) (d)	P value (e)
Boost & ambitious goal (T1)	329.22	369.11	12%	−19%	0.0778*
Boost & modest goal (T2)	236.34	252.47	7%	−24%	0.0177**
Boost only (T3)	295.00	343.49	16%	−15%	0.1237
Control (CG)	314.96	412.67	**31%**	–	–
Average of the panel	296.71	352.82	18.91%		

Note: Column (a) is the average electricity consumption by treatment in kWh during the 6 months before the start of the experiment, June 2018 to November 2018. Column (b) is the average electricity consumption by treatment during the 6 months of the experiment from December 2018 to May 2019. Column (c) is based on treatment and shows the difference in average electricity consumption between the two periods, i.e., during the experiment period minus the average consumption in the 6 months before the experiment in percentage. Column (d) shows the percentage variation (between the periods and with respect to the control group) in the percentage of variation in the control group (CG). Column (e) presents the results of a t test of the difference between the average consumption of the treated groups compared to the control group
***$p > 0.01$; **$p < 0.05$; *$p < 0.10$

Table 2 Average electricity consumption per household during the treatment and pretreatment period

Average electricity consumption per household during the pretreatment period (a)	Control (CG)- Boost & modest goal (T2) = 314.96 − 236.34 = 78.62
Average electricity consumption per household during the treatment period (b)	Control (CG)- Boost & modest goal (T2) = 412.67 − 252.47 = 160.2

comparison by treatment based on a Wilcoxon-Mann–Whitney (WMW) test shows that average electricity consumption over the period of the experiment differed significantly across some treatments ($p = 0.0001$), although the consumption of the pair T3-CG shows no differences during the first 2 months of the experiment ($p = 0.495$).

The Smartlook field experiment explored the effectiveness of boosts and goals for driving potential electricity reductions. In fact, in line with the work of Belaïd and Garcia (2016) and Belaïd and Joumni (2020), an urgent need to increase electricity use transparency through the provision of information and education has been observed. Thus, tools to reduce the possible behavioral barriers by bringing new opportunities to adopt conservative energy behaviors may be effective and very robust in promoting new behaviors. Consequently, the combination of goal setting and boost seems efficient for transforming stated environmental concerns into concrete actions. In fact, the results of the study show that modest goals combined with specific information can translate concern for the environment into green behavior. A goal ranging from 15% to 25% reduction in energy use is efficient for households already concerned about the environment and committed to greener behaviors.

Combining goals (ambitious or modest) with boosts might significantly reduce electricity use (see Table 1). In fact, providing boosts reinforces households' knowledge and creates motivation toward sustainable electricity consumption and objective achievement (Martela, 2015). Concerning the type of goal, and in line with Harding and Hsiaw (2014) findings on the need to set realistic goals, it appears that the combination of modest (realistic) goals and boosts produces better results than a more ambitious goal and boosts. This suggests that households would prefer a long-term process of incremental learning to reach long-lasting efficiency in energy reduction.

Concerning boost only (treatment T3), although no electricity reduction has been observed, we could notice a significant impact on the specific environmentally concerned household with a high NEP scale score. This result can be interpreted as a profile-dependent result illustrating the complexity of the link between individual concerns and energy consumption (Nauges & Wheeler, 2017). Moreover, it confirms the need for thinking of nonmonetary incentives as context and target dependent. In fact, not only must the goal be realistic and associated with boosts, but it must also be sent to the right citizens who will be sensitive to it. Another interesting result, in line with the work of Pullinger (2014) and Shove et al. (2020), highlights that individuals with more time (retirees) and NGO members are more likely to have the resources and motivation to change their electricity use behavior. Moreover, higher education and greater environmental commitment are good predictors of such actions.

When focusing on low-concern households, it appears that none of the boost and goal settings were efficient in reducing electricity consumption. However, in the former household category, we observed a significant impact on education and some curtailment behaviors. These results contribute to the discussion on the difference between curtailment and energy-efficient behaviors. In fact, as argued by Nauges and Wheeler (2017), for some citizens with low environmental concern or low intrinsic motivation, monetary tools, and other behavioral interventions are required to encourage sustainable behavior and, more importantly, conditions for learning to play a detrimental role in maintaining changes over time. Indeed, Smartlook results outline the important issue of the translation of environmental values and concern for concrete behaviors with respect to the available dwelling materials and local conditions (Welsch & Kühling, 2009; Woersdorfer & Kaus, 2011; Babutsidze & Chai, 2018).

5 Conclusions

Reducing GHG emissions to prevent potentially catastrophic global climate change necessitates an essential reduction in energy consumption in the built environment. It is largely acknowledged that cities may contribute significantly to efforts to alleviate climate change. On the other hand, there is a remarkable consensus about the critical role that behavioral change may play in reducing residential energy consumption. However, the debate is still evolving concerning the optimal mechanism and effective instrument to promote individuals' actions on energy efficiency. Building on the emerging concept of "smart energy cities," this chapter examines options for reducing energy consumption in buildings. More precisely, it focuses on the role of behavioral change in reducing electricity demand in the residential sector. First, we discussed the leading role that behavioral economics may play in reducing the carbon footprint of residential energy demand. Then, we provide a brief overview of smart city research and initiatives in Europe and Monaco. Finally, we build upon a field experiment conducted in the Principality of Monaco to explore the complementarity of different behavioral interventions and their impact on residential electricity demand. Accordingly, we used three treatments: (Treatment 1) reduction goal combined with information; (Treatment 2) modest electricity reduction goal combined with information; and (Treatment 3) only information.

This research constitutes a step toward a more accurate evaluation of the behavior-driven energy demand reduction in the residential sector. In addition, the study offers various exciting results. Boosts appear to be a novel and promising instrument that involves a few prerequisites before it can be implemented. Goal setting is a standard and effective strategy that has a more substantial impact when implemented in combination with other instruments. The empirical findings illustrate the impact of boosts and goal setting on energy savings, as well as their complementarity and effectiveness in motivating individuals to reduce their energy use. This result suggests that behavioral change programs using multiple intervention methods save more energy than those using fewer intervention options. In addition, targeting more agents who are more sensitive or responsive to these kinds of interventions will significantly reduce the built environment's carbon footprint.

Given the increasing global recognition of the crucial role of behavioral change in achieving climate goals, the empirical results have several implications for the policymaking process and policy intervention. The results call for developing more proactive behavioral change programs using innovative instruments to curb residential energy consumption and related GHG emissions. In parallel, the discussion provides specific guidelines for harnessing empirical behavioral studies to improve the effectiveness of behavioral intervention programs. Endorsing these guidelines will be valuable in designing and executing successful energy efficiency programs.

References

Asmare, F., Jaraitė, J., & Kažukauskas, A. (2021). The effect of descriptive information provision on electricity consumption: Experimental evidence from Lithuania. *Energy Economics, 104*, 105687.

Babutsidze, Z., & Chai, A. (2018). Look at me saving the planet! The imitation of visible green behavior and its impact on the climate value-action gap. *Ecological Economics, 146*, 290–303.

Bakaloglou, S., & Belaïd, F. (2022). The role of uncertainty in shaping individual preferences for residential energy renovation decisions. *Energy Journal, 46*(4), 127–158.

Bakaloglou, S., & Charlier, D. (2019). Energy consumption in the french residential sector: How much do individual preferences matter? *The Energy Journal, 40*(3), 77–99.

Belaïd, F. (2016). Understanding the spectrum of domestic energy consumption: Empirical evidence from France. *Energy Policy, 92*, 220–233.

Belaïd, F. (2017). Untangling the complexity of the direct and indirect determinants of the residential energy consumption in France: Quantitative analysis using a structural equation modeling approach. *Energy Policy, 110*, 246–256.

Belaïd, F. (2022a). Mapping and understanding the drivers of fuel poverty in emerging economies: The case of Egypt and Jordan. *Energy Policy, 162*, 112775.

Belaïd, F. (2022b). Implications of poorly designed climate policy on energy poverty: Global reflections on the current surge in energy prices. *Energy Research & Social Science, 162*, 102790.

Belaïd, F. (2024). Decarbonizing the residential sector: How prominent is household energy-saving behavior in decision making? *The Energy Journal, 45*(1), 125–147.

Belaïd, F., & Garcia, T. (2016). Understanding the spectrum of residential energy-saving behaviors: French evidence using disaggregated data. *Energy Economics, 57*, 204–214.

Belaïd, F., & Joumni, H. (2020). Behavioral attitudes toward energy saving: Empirical evidence from France. *Energy Policy, 140*, 111406.

Belaïd, F., & Rault, C. (2021). Energy expenditure in Egypt: Empirical evidence based on a quantile regression approach. *Environmental Modeling & Assessment, 26*(4), 511–528.

Belaïd, F., Youssef, A. B., & Omrani, N. (2020). Investigating the factors shaping residential energy consumption patterns in France: Evidence form quantile regression. *The European Journal of Comparative Economics, 17*(1), 127–151.

Belaïd, F., Rault, C., & Massié, C. (2021). A life-cycle theory analysis of French household electricity demand. *Journal of Evolutionary Economics*, 1–30.

Harding, M., & Hsiaw, A. (2014). Goal setting and energy conservation. *Journal of Economic Behavior & Organization, 107*, 209–227.

Harrison, G. W., & List, J. A. (2004). Field experiments. *Journal of Economic Literature, 42*(4), 1009–1055.

Hertwig, R., & Grüne-Yanoff, T. (2017). Nudging and boosting: Steering or empowering good decisions. *Perspectives on Psychological Science, 12*(6), 973–986.

Lazaric, N., & Toumi, M. (2022). Reducing consumption of electricity: A field experiment in Monaco with boosts and goal setting. *Ecological Economics, 191*, 107231.

Maréchal, K. (2010). Not irrational but habitual: The importance of "behavioral lock-in" in energy consumption. *Ecological Economics, 69*(5), 1104–1114.

Martela, M. (2015). Fallible inquiry with ethical ends-in-view: A pragmatist philosophy of science for organizational research. *Organization Studies, 36*(4), 537–563.

Moran, D., Wood, R., Hertwich, E., Mattson, K., Rodriguez, J. F., Schanes, K., & Barrett, J. (2020). Quantifying the potential for consumer-oriented policy to reduce European and foreign carbon emissions. *Climate Policy, 20*(sup1), S28–S38.

Nauges, C., & Wheeler, S. A. (2017). The complex relationship between households' climate change concerns and their water and energy mitigation behavior. *Ecological Economics, 141*, 87–94.

Pira, S. (2021). The social issues of smart home: A review of four European cities' experiences. *European Journal of Futures Research, 9*(1), 1–15.

Pullinger, M. (2014). Working time reduction policy in a sustainable economy: Criteria and options for its design. *Ecological Economics, 103*, 11–19.

Shove, E., Trentmann, F., & Wilk, R. (2020). *Time, consumption and everyday life: Practice, materiality and culture.* Routledge.

Sovacool, B. K., Newell, P., Carley, S., & Fanzo, J. (2022). Equity, technological innovation and sustainable behavior in a low-carbon future. *Nature Human Behavior, 6*, 326–337.

Stern, P. C., Dietz, T., & Guagnano, G. A. (1995). The new ecological paradigm in social-psychological context. *Environment and Behavior, 27*(6), 723–743.

Thaler, R. H., & Sunstein, C. R. (2008). *Nudge: improving decisions about health, Wealth, and Happiness, 6*, 14–38.

Thornbush, M., & Golubchikov, O. (2021). Smart energy cities: The evolution of the city-energy-sustainability nexus. *Environmental Development, 39*, 100626.

Übelmesser, L., Klingert, S., & Becker, C. (2020). Comparing smart cities concepts. In *2020 IEEE international conference on pervasive computing and communications workshops (PerCom Workshops)* (pp. 1–6). IEEE.

United Nations. (2019). *Department of Economic and Social Affairs, Population Division. World Urbanization Prospects: The 2018 Revision (ST/ESA/SER.A/420).* United Nations. https://population.un.org/wup/Publications/Files/WUP2018-Report.pdf

United Nations. (2022). *Sustainable development.* https://sdgs.un.org/goals

Welsch, H., & Kühling, J. (2009). Determinants of pro-environmental consumption: The role of reference groups and routine behavior. *Ecological Economics, 69*(1), 166–176.

Woersdorfer, J. C., & Kaus, W. (2011). Will non owners follow pioneer consumers in the adoption of solar thermal systems? Empirical evidence for northwestern Germany. *Ecological Economics, 70*(12), 2282–2291.

Open Access This chapter is licensed under the terms of the Creative Commons Attribution 4.0 International License (http://creativecommons.org/licenses/by/4.0/), which permits use, sharing, adaptation, distribution and reproduction in any medium or format, as long as you give appropriate credit to the original author(s) and the source, provide a link to the Creative Commons license and indicate if changes were made.

The images or other third party material in this chapter are included in the chapter's Creative Commons license, unless indicated otherwise in a credit line to the material. If material is not included in the chapter's Creative Commons license and your intended use is not permitted by statutory regulation or exceeds the permitted use, you will need to obtain permission directly from the copyright holder.

Back to the Future: Tapping into Ancient Knowledge Toward Human-Centered Sustainable Smart Cities

Anna Laura Petrucci

Abstract Terms such as resilient, smart, and sustainable are often used as synonyms in defining future cities. While pointing out specific features and objectives, the implementation of each one of the concepts cannot exist if not implying the other two. A city cannot be resilient if not based on a wide range of sustainability concepts and principles of smart growth and social engagement. It cannot be smart without considering long-term perspectives and adaptability; it cannot, finally, be sustainable if not observing resilience and smart approaches to design and infrastructures. What is worth highlighting here, is how, on smart cities, academic and technical literature mainly focuses on technology, while omitting the traditional urban requirements of sustainability, resilience, and quality of life. Smart cities are cities first; are social built environments, where advanced technologies provide smart data collection and delivery to offer their inhabitants a better quality of life, and before that, offering livable physical solutions enabling social and facilitating interaction. In a metaphor, public space is the hardware, where technological software can be applied; public space is the platform where data exchange is made possible.

Keywords Cities · Resilience · Sustainability · People · Planet · Prosperity · Ancient knowledge · Holistic design · Complexity management

A. L. Petrucci (✉)
Universita' La Sapienza, Roma, Italy

King Saud University, Riyadh, Saudi Arabia

© The Author(s) 2024
F. Belaïd, A. Arora (eds.), *Smart Cities*, Studies in Energy, Resource and Environmental Economics, https://doi.org/10.1007/978-3-031-35664-3_7

1 Introduction: Cities and Their Smart Components

Cities are made of people, places, and built environments, and the unique network-ing these are creating together is commonly known as *genius loci* or the *gene of the place*. While the traditional Latin means the guardian of the place, in the contem-porary theory of architecture, C. Norberg-Schulz defined it as an "existential space" or "live experiential space" (Norberg-Schulz, 1971). In his view, the exis-tential task of any new architecture should be to encourage the city to become the place, i.e., to reveal all its potential meanings and expectations. It is a dual relation of a person with his/her living environment, manifested by orientation in the space and psychological identification (Čepaitienė, 2015). The *genius loci* reaffirm and renew itself as a site-specific practice of people acting and living in a certain place, according to their time; it's the eternal here and now, or what Le Corbusier defined as an "eternal present." Citizens, implicitly if not physically, are city-builders, are urban makers, each contributing through knowledge, experience, or social pres-ence. Each territory can be seen as "a palimpsest, where different generations have written in, and corrected, deleted, and added" (Curti et al., 1992), and our planet as an immense archive in progress, showing us intentions, projects, and concrete actions that, by stratifying, have given rise to a cumulative selection. Even sponta-neous constructions still had been put in place through (unwritten) regulations, through a modus operandi dictated by the cultural, technical, and social customs of those who imagined and invented new futures. Moreover, cities are sedimentation of practices, collective experiences, and a process of social interaction within poli-cies that seek to endow them with coherence and consistency; according to the plans, they show the form of the power structure and its continuous mutations to confer identity upon the different social groups. For centuries, city planning has always been an open process that encompasses different scales and is never reduced to the mere ordering of unbuilt space. The process reconsiders old and new urban materials and continuously upgrades architectural language until new vocabular-ies, grammar, and syntax are created to express new spatial conceptions. The archi-tecture of the city is an image of a society connected through a public realm, repeating the same parameter of urban porosity over time. It densifies while the urban meshes are pierceable through special impalpable, social, economic, and political transfers (Secchi, 2013). To be resilient, a city must build up a wide range system of sustainable measures and apply principles of smart growth and social engagement; it cannot be smart without considering long-term perspectives and adaptability; it cannot finally be sustainable if resilience and smart approaches to design and infrastructures are not observed.

2 People Planet Prosperity

The Italian G20 highlighted the importance of "People, Planet, and Prosperity" as parts of an integrated system (T20 Italy, 2021). Human-centered sustainable and smart cities might be considered under the same conceptual grid: cities are meant for people, are consistently interacting with the planet, and are major catalysts for prosperity.

People are the first beneficiary of the urban process, and cities are historically conceived as dense connectors of human relationships and logistic facilitators through infrastructures and opportunities. Cities are a network of opportunities, a system of differences starting from shared identity, memories, and resources; they are "ports of land" in an extended international economic model. Urban systems are not just a spatial concentration of uses or material sedimentation of activities but also a network of interdependencies between multiple social actors, which exchange relationships based on competition and cooperation relationships and medium- to long-term agreements. Cities, as part of reticular projects, must also respond to a social demand that is not indifferent to the location but, on the contrary, requires spaces for one's own social and cultural growth. The act of planning includes exploring the future while constructing "scenarios" by questioning and verifying alternative situations throughout the project (Viganò et al., 2010). In the past, cities were facing a technological deficit, and rationality guided the distribution of the settlements in the territory. Much of the landscapes we admire today result from this set of prohibitions and precautions due to avalanches, floods, unstable soils, generated shades, most fertile lands, available materials, and others. On the other hand, the modern world lives in a technological surplus and is guided by more general and abstract criteria of rationality while abandoning many traditional precautions (Secchi, 2013). The failure of the 1970s and the rational approach to architecture have shown how urban planning has strong, specific responsibilities in managing social and economic opportunities and inequalities. Cities were traditionally the space for social and cultural integration, a safe place protected from the violence of nature and humankind, a producer of new identities, and a privileged seat of every technical, scientific, cultural, and institutional innovation, but becoming today, the place where inequalities stand out, especially in large metropolitan areas, where urban projects often favor strategies of distinction and exclusion.

Only recently has urban design returned to nature-based solutions to provide sustainable management of ecosystems to tackle the different environmental challenges faced by an unresponsive approach to urbanism. Different planning approaches have been undertaken worldwide to explore the potential and limitations of different approaches, promote the participation and adaptive capacity of vulnerable groups, and facilitate overall urban resilience. Most of these solutions can reconnect the population with nature, mitigate air pollution, improve thermal comfort in cities, reduce the effect of urban heat islands, and manage stormwater runoff, among many other benefits to the environment. Nature-based solutions, modern urban ecology, traditional urban systems, and resilience have a commonality: to learn from observing nature and devising a suitable remedy while creating a win-win situation for humans and other living creatures to live in harmony.

Traditional urbanism was directly related to natural or manufactured environmental infrastructures within a mechanism of circular economies. Prosperity was the output of a complex equilibrium among short-, middle-, and long-term strategies. Prosperous cities were built around the public space as the founding principle, which today is defined as placemaking. The community (software), through tangible and intangible actions, implemented the space (hardware) co-creating stories and visions for the city's future. On a bigger scale, cities and villages are attractors of a whole regional area, destinations and catalysts of further regeneration all around, and connecting groups of people exploring a territory without apparent spatial-temporal linearity (Petrucci, 2021). The multistakeholder narrative enhances community-based participation; currently, engaging the community in PPPP (participatory public private projects) projects could effectively relieve the economic burden and provide constructive support in decision-making for national and local governments. The institutions could subside, strategize, and direct an inducted growth and development without risking urban financial struggles and activating self-regenerative and self-funded communities, spaces, and cities. It can be defined as a top-down bottom-up approach (Petrucci et al., 2022).

The "People, Planet, Prosperity" motto of the Italian G20 becomes a circular concept and a matrix and recalls traditional cities as sustainable communities based on collective effort and intelligence around public space. Therefore, smart cities need to generate well-designed dense urbanism, including old and new traditions, making the urban space a social connector where communities interact and exchange data and experience.

3   Tapping into Ancient Practices

If the European urban shape from the nineteenth century became a sort of ultimate achievement for new cities, it is due to its harmonious elegance, mixed-use, and amusing definition of a vibrant public space. Nevertheless, the European landscape was urbanized only after the demographic revolution of the Middle Age and maintained a strong interconnection with the countryside, where life developed in the rhythm of seasonal cycles (Astengo, 1966). When the industrial revolution broke into the metropolis, the ancient rhythm was replaced with the mathematical time of the clock, while small towns remained much longer in the agricultural economy and conservative mentality: a collective life where anonymity is not allowed. The longer the industrial process, the greater the inherent urban qualities of porosity, flexibility, resilience, and sustainability of historic cities were kept due to implicit, slow, collective urban planning of public spaces working in a cascade, or a circuit, and distributing the flows of collective space in increasing or decreasing intensity (Secchi, 2011).

This model survived centuries of socioeconomic, political, and cultural changes, responding to the changes with an efficient and robust structure. Decoding this DNA of best practices could potentially give contemporary times a deeper understanding of designing an effective and resilient spatial network, achieving what we call today "inclusivity" or social and economic resilience. Moreover, the same structure of medieval European villages can be identified in traditional settlements worldwide; the same kind of spatial strategy was applied in Tonkinese, Indian, Native American, Hinuit, Zulu, Arab, and Slavic settlements. The starting point for traditional planning was always the gathering space, the small-scale clustering of a few families, creating a sort of neighborhood unit, bigger or smaller depending on the social structure of the family and the community. All ancient villages were generated as subunits around a space or a focal point. Moreover, indigenous architecture and urbanism show an understanding of shared values while "responding to the preexisting local place," in direct opposition to imposing order from the outside. (Trancik, 1986) ensures that space responds to the surrounding context, is designed for people to use, and reveals a deep knowledge of social, economic, and environmental livability. Indigenous societies all had embedded very simple and recurring urban principles, arising into a level of universality, and mainly based on the relationship between humans (microcosm) and the environment (macrocosm), along with its laws and institutions, to be incorporated into the urban design and building processes. A holistic planning process was the key to ancient knowledge, where the city-makers (the inhabitants) were guided by spiritual and wide well-educated leaders in finding the best solution for a harmonious and prosperous life. Architects were also spiritual leaders. They were skilled in integrating different disciplines and able to master a wide multifactor complexity; they were priests, astronomers, doctors, and psychologists along with their being architects. A wide range of knowledge allowed the management of a full-spectrum holistic approach, where the interaction between the three essential phenomena of existence, space, and time was also considered. The planning and building process was deeply rooted in the way of life and the moral values of the community and connected with universal forces and laws. In the Roman Empire, the Pontifex Maximus; in the Ancient Chinese Empire, the Mandarins; and in the Arab tribal system, the Sheikhs were spiritual and mundane leaders who could download and share both philosophical and technical wisdom during the city foundation. As previously mentioned, industrial revolution labeling and specialization and the following rational approach to urbanism, based on fast and practical solutions instead of long-lasting solutions, cut off traditional knowledge. Just a few of the traditional practices are still alive, although not mainstreamed: The Indian Vastu Shatra and the Chinese Feng Shui.

The traditional Indian discipline of planning, the Vastu Shastra, is a compendium of literary text on architecture that dwells on topics such as urban design, building typology, materials, measurement systems, orientation, and building components. Its principles include scaling from the micro to the macro and vice versa, aiming at achieving a balance among functionality, bioclimatic design, and religious and cultural beliefs. The ancient Indian ideology-based sciences have been established for 6000 years: "Vastu is the art of living in harmony with the land, such that one derives the greatest benefits and prosperity from being in perfect equilibrium with Nature" (Acharya, 1981). Its metaphysical and philosophical nature is a highly sophisticated method of creating a living space that is a miniature replica of the cosmos and has been rooted since ancient times, having significant importance in the Indian way of living. Indian people relied on Vastu Shastra for centuries to design cities and settlements and build homes, temples, and palaces. Historically, it has greatly contributed to the environmental design and town planning of many cities that still flourish. Several texts are found all over India in the mid-twelve century, recalling today's urban guidelines, manuals, and codes and incorporating cultural, regional, and geographical contexts (Das & Rampuria, 2015). The content of the texts is comprehensive while intersecting with allied disciplines such as astrology, Ayurveda (medicine), and numerology and becoming part of an integral worldview. The texts were written by, and for, a variety of professionals (priests, architects, masons, patrons, and connoisseurs), and the specialism of the author or the audience is also reflected in their content.

Similarly, the ancient Feng Shui design code, also known as Chinese geomancy, was practiced by the noble cast of Mandarins, priests before architects, having the knowledge to read, interpret, and implement the astronomical, astrological, architectural, cosmological, geographical, and topographical dimensions to generate the most prosperous site-specific residences and cities. It was originally used as a practice to protect the ancestors' tombs, as the root and source of the power of reigning families, and later implemented in other structures. The literal meaning of Feng Shui is the wind's ability to flow within and around buildings and settlements; the wind is showing the path of Qi, the "cosmic current" or energy-improving wealth, happiness, and long life, essentially based on the main principles of void and polarity, built among the elements of force. The Void (as an urban space, as well as an inner courtyard) was always the focal point in collecting and directing the divine energy. The polarity was the expression of yin and yang theory: one part creating an exertion and one receiving the exertion and the balance between the two. The five elements or forces (wu xing) of metal, earth, fire, water, and wood are considered the forces essential to human life. As per Vastu, urbanism applies the same principles of traditional medicine, looking at the city as a physical body and aligning a city, site, building, or object with yin-yang force fields.

From a socioeconomic perspective, Feng Shui encourages an autarkic mode of production, prudent consumption, and a strong self-resumable function concerning the local ecological system so that a balance is achieved between the production, goods consumption, and exploitation of nature. Energy consumption through balance to address a balance between exploitation and consumption of renewable and nonrenewable energy is improved. Solar, wind, and geothermal energy both as warming and cooling strategies, use of local materials and waste, landscape, local climate, and biodiversity management of water resources (Al-Sadkhan, 2018). Its fractal conception of the universe allows, similar to Vastu, an application on different scales, from the micro up to the macro scale of cities and territories. Feng Shui applies to urban city planning, landscape architecture, and building design based on the idea that nature and humans should be in harmony.

The concept of void, the center for both concentrating and dissipating urban tension, is common to all cultures and later re-established in medieval villages as a civic center. Traditional Arab towns were also designed and built around the space, while mixed uses and further residential settlements were generated by volumetric addition around the space beside the mosque, the collective space for the whole community. Therefore, a mosque served the purpose of offering prayers and informing different contexts used by the community: formative, recreational, security, ritual, and directional, all complementary and related to each other. As a flexible and multifaceted complex of facilities, mosques serve as multifunctional and adaptive public spaces in cities. Connected by a complex system of irregular streets stemming from the main mosque, the settlement is organized into clusters. Urban space moved from public to private through unwritten but well-defined thresholds into clustered areas. Streets were not traced in grids or axes, as traditionally happened in Europe and later in the United States, but instead generated from juxtaposed buildings and their growing extension of units when the family gets bigger (Petrucci, 2022). Same as it used to be in all ancient civilizations until the Middle Ages, narrow and irregular tissue created a natural shadow and cross ventilation of the public space. Passive environmental solutions and the best use for local materials were taught from generation to generation based on ancient knowledge. Moreover, Arab cities surely can be considered one of the most resilient and sustainable constructs within extremely difficult environmental conditions; their system must be trusted as an incredible potential of traditional environmental construction strategies and implemented through the most modern technologies. The circular economy, integrated natural environment, multilayered indoor-outdoor exchange, cross ventilation, water management systems, use of natural materials, and building features, such as double skins, wind towers, solar chimneys, environmental courtyards, thermal mass, and earth pipes, are all sustainable solutions first developed in traditional Arab cities.

In the modern age, ancient disciplines were never incorporated into the curriculum of architecture schools and the entire infrastructure—the planning authorities, design and building processes, and the provision of materials—caters to an industry that is entirely separate from traditional practice. Nevertheless, the traditionally related micro and macro spaces do not fit dimensionally anymore; how can the sense of narrow urban streets within a six-lane urban highway be replayed? Urban planning regulations are not laid down with traditional design typologies. Designing a house on a standard rectangular plot within a modern city development does not allow the incorporation of a traditional central courtyard. Moreover, restrictive zoning and land use do not allow open spaces to adjust for multiple mixed uses as was happening in the past (Sachdev, 2011).

The question might be here: How did we go off track? Have we been lost on translation? And how did it happen?

We saw how the starting point for traditional planning is always the small-scale clustering of a few families, creating a sort of "neighborhood unit," a model shared by medieval European villages so as in traditional settlements worldwide: from south to north, from east to west. Depending on the geographical location, these are called *comunanze*, *zadrugá*, or *kraal* and always express the same principle of a shared community of life, work, and goods. The community is organized around a void; the void makes the family first and—on a bigger scale—the community. By the end of the nineteenth century, the first industrial revolution provoked the collapse of traditional urban infrastructures, making the central areas very hard to live in; the first Garden Cities were established, the urban expansion across the peripheries, and the "sprawl" happened. Cities moved increasingly toward gathered communities from a multifunctional and well-integrated urban mix of subsidiary communities reunited by the well-formed urban space, generating different kinds of urban ghettos: luxury-like compounds or pockets of economic and social poverty-like slums. Supported by the rational culture of modernism, the ideas of the city as an organism were replaced by classical critical tools such as proportion, number, regularity, and order, and cities became rigid and shapeless, and a crisis of mobility or denial of generalized accessibility to every place for every individual or social group, generating evident "spatial" injustices (Soja, 2010). That is how the interconnections among people, the planet, and prosperity were lost. The phenomenon has reached its apical point in recent decades, and the increase in individual autonomy pushes forward to single houses and gathered communities, increasing the gap among social groups in terms of accessibility to culture, education, and capital. This is how communities moved from universal values to globalism.

Many large cities and metropolises in recent years have returned to the urban project trying to contribute to their reduction by putting a new focus on environmental and mobility issues as appropriate and relevant ways to search for social justice and environmental problems, all topics requiring practical answers. In the process of returning to the traditional urban form, cities have rediscovered the importance of isotropy, porosity, permeability, connectivity, and accessibility and the physical and social barriers that had been created for the full use of the city itself. If we want, the 15 minutes' city, the slow city is an answer to the radical change in the physiognomy of demand (Secchi, 1999).

4   Looking Up to the Future, the Need for Upscaling

Contemporary cities are growing fast and are extremely dense and demanding in terms of services and requirements, larger-scale buildings, and infrastructures. Urban growth is forced into a faster track and is no longer following a natural path, generating greed of upscaling, instead of following the natural times or fitting the well-being of their inhabitants. Cities did not have the time to build up through time and to grow organically; instead, intensive real estate took over in most cases the planning while neglecting shared social goals. A layered use of the space requires a more complex and sophisticated design approach and language, a more fluid form-based system of urban coding and planning, not considering these as static normative pictures, but as an ongoing process to be detailed and declined within time and space. It requires a design system that places more constraints on the designers to make the solution an exercise in striking a balance among the various demands of context and use. Such a system would need a multidisciplinary approach to design a community rather than just the material enclosures for living, requiring a much bigger complexity and capacity for envisioning scenarios due to their upscaling and enhanced technological and comfort, and spatial requirements. Observing ancient architectural codes, and their ability to move from micro to macro, from object design into cosmology, and bringing any aspect of the city to a higher sense of unity, might be the answer. Looking at the spatial organizational principles, the traditional urban cultures ensured that the design of physical spaces responded to the people's cultural values and way of life, and these are still up-to-date and now part of the global manifesto of sustainable planning expressed by SDGs, the New Urban Agenda 2030, so as in all the quality assessment of urban space and green architecture, and could be outlined by the following principles:

(a) Coexistence of systems and relative wholeness: the components of the physical environment can be treated as complete entities in their context, individual cores, and part of a continuum wherein each center is defined by several other subcenters. A system of spatial networks that reminds the sacred geometry of Mandalas and establishes an active urban landscape.

(b) Individuality within a group, or unity within diversity: the system of concentric wholes, centers, and subcenters, implying independent identities and diverse positioning in terms of specialization and mixed use. Strategic positioning of districts within the city and cities nation- and worldwide as traveling, residential, and business destinations.

(c) Coexistence of extremes and celebration of junctures: multipolar systems reach opposition and juxtaposed spaces and functions; it requires intercommunicating units to generate over time a continuum balance of all the energies enhancing mutual existence.

(d) Timelessness of space: The multilayering of time is usually naturally reflected in space adaptability, making cities resilient and able to adapt to upcoming changes, which might be translated as a multiscale analysis of a city constantly changing and adapting toward circular economy-based systems.

(e) Holistic vision, interconnection, and constant evolution: things eternally change, and there is a fundamental correlation between all events; cities are fluid beings, and there is no distinction between macro and micro levels.

(f) Multiqualitative approach to urbanism as physical space, sensory experience, and activity. The life of streets and urban areas is longer than the life of individual buildings, and so is how individuals perceive and remember city elements in city space through paths, nodes, edges, districts, and landmarks.

(g) Appropriate exploitation of natural resources, sensitive use of energy, recycling and reuse of materials, and consistent protection of the natural environment.

From the sociocultural perspective, promoting a positive individual contribution in terms of personal responsibility, respect, care, and sharing while involving family and the local community.

5   Conclusions: Building Through Collective Intelligence, the Indigenous Path

What is civilization? Civilization is a collective effort of culture, knowledge, intelligence, and sensitivity acting together for the common good. It brings a fundamental morality into urban planning: when it betrays or neglects its social purpose in some way, it is doomed to failure. Contemporary cities are demonstrating how, when just interests or aestheticizing, technologies, or economies prevail on the fundamental respect for the social structure, the result cannot be other than a setback. The same happens when the planning is just a theoretical supra-structure and does not engage in solving the complex essence of the site on a physical, social, and economic level. Therefore, how can urban planning be brought back to the locally established civilization path to make cities more resilient, sustainable, and smart?

Several of the traditional principles are applied today under different labeling. The involvement of local know-how, for example, is defined today as participatory- or codesign or placemaking. It is how the community can benefit from the projects even before it starts while manifesting their collective intelligence into adaptable and inclusive visions, site-specific, dynamic, and transdisciplinary solutions, and activating collaborative scenarios focusing on creating destinations and the best path connecting those destinations. Following the ITC metaphor, the interface "public space" needs hardware to get manifested, which is the physical design of the space; this engages, guides, and allows citizens to make an experience of the public realm (Ellery et al., 2021) Short-term, community-based projects became a powerful and adaptable new tool for urban activists, planners, and policy-makers seeking to drive lasting improvements in their cities and beyond. These quick, often low-cost, and creative projects are the essence of tactical urbanism (Lydon & Garcia, 2015) and offer bottom-up support to traditional top-down decision-making. The key here is a simultaneous multidisciplinary layering and comprehensive vision of the public space and the whole urban structure by integrating urban design, landscape design, transport and sustainability strategies, and architecture to analyze and engage the layers of place, where different scales of interventions are analyzed and compared by constant up- and downscaling. This process would simulate and compress in time the collective work in building cities in the past, engaging people as city makers.

Here are some of the detected strategic tools to achieve this ambitious and necessary goal toward sustainable and smart future cities:

(a) Fractal principles and macromicro zooming. Applying a methodology independently from the scale, up- and downscaling during the process of moving up and down from micro to macro and vice versa, to assess the coherence and implementation of the integral system.
(b) Multipolarity matrix system for managing extremes. A matrix as the footprint of the place, multilayered through the recovery-restructuring of voids and/or abandoned areas, with more fluid reconsiderations on mobility and urban accessibility, and inducing urban redestination as a tool to increase the speed of urban transformation.
(c) Nature-based solutions. Rediscover a high and wide sense of Nature. Research reference to multiple disciplines and cosmology toward an extended environmental analysis. By incrementing passive and design-based environmental solutions to reduce the carbon footprint and enhance the well-being of citizens.
(d) Holistic approach for fluid master planning integrating strategies of place-making and flexibility of the planning to move from top-down to bottom-up solutions, such as short-, middle-, and long-term goals in terms of social and economic responses. Using the master plan as a facilitator for the generative process will result in an urban design plan.
(e) Comprehensive multidisciplinary design to be implemented technically, socially, and economically, and emphasizing walkability, connectivity, mixed-use and diversity, mixed housing, quality architecture and urban design, traditional neighborhood structure, increased density, green transportation, sustainability, and quality of life.
(f) Human-centered design includes a process toward cultural identity aiming to increase the sense of belonging and genius loci. Several cultural and design movements have been calling for action, such as New Urbanism, the Community Voice Method, and generative and form-based codes.
(g) Learning by doing the process is a good practice in cultural and professional training: knowledge was shared among generations with a sense of proportions and sustainable building construction and economy. It is crucial to activating a culture of continuous learning based on the experience of the past and the transgenerational transmission of values and wisdom.

Finally, the leading management toward sustainable and smart cities requires holistic and comprehensive knowledge; the same kind of wide experience in practice and philosophical thinking, the same deep knowledge of communities and their *genius loci*; the awareness of ancient builders, and the skills to reinforce those within an integrated sustainability plan, social, economic, and environmental. While in the past, solutions were embedded into construction, currently solutions are found through highly specialized fields, not cross-fertilizing each other into a complex holistic urban system. Building and managing a sustainable and smart city requires the individual and political will to be beneficial to the whole community as a healthy and wealthy organism, facilitating its growth and constant innovation. Pieces of information are all around and embedded in our ancient cities and the laws of nature within and around us. Informed design, decision-making, and implementation can now span a much wider multicriteria evaluation thanks to traditional and digital technologies, data collection and management, and the interexchange of information within a worldwide network. Going back and forth from past to future means having at the core integrity and unity of man and the environment in terms of organic interaction, resources, generation, and quality of space. Working on eco-cyclicity, as the coordination of the rhythms of the environment and humans, and phenomenology taking into account the specific situation of place-time, social conditions, characteristics of individuals, and shared culture.

References

Acharya, P. K. (1981). *Manasara series on Vastusashtra and Silpasashtras*.

Al-Sadkhan, A. (2018). The application of FengShui in the design of contemporary architecture and its environment. *Iraqi Journal of Architecture and Planning, 9*(1). https://doi.org/10.36041/iqjap.v9i1.172

Astengo, G. I nuclei sociali e l'urbanistica, lectures series at IUAV, 1966.

Čepaitienė, R. (2015). *Lithuania genius loci as a "nameless value" of natural and built heritage*. How to assess Built Heritage? Assumptions, Methodologies, Examples of Heritage Assessment Systems (ed by *B. Szmygin*). Florence-Lublin.

Curti, F., et al. (1992). *Gerarchie e reti di città: tendenze e politiche*. Franco Angeli.

Das, P., & Rampuria, P. (2015). Thinking spatial networks today: The "Vastu Shastra' Shastra' way. In *Proceedings of the 10th international space syntax symposium*. Bartlett School of Architecture.

Ellery, P. J., et al. (2021). Toward a theoretical understanding of placemaking. *International Journal of Community Well-Being, 4*, 55–76. Springer.

Lydon, M., Garcia, A. (2015). Inspirations and antecedents of Tactical Urbanism. In *Tactical Urbanism*. Island Press.

Norberg-Schulz, C. (1971). *Existence, space and architecture*. Paperback.

Petrucci, A. L. (2021). *Policy Report No 41. 21-09-26 How can cultural heritage serve human-centered smart cities in the GCC*. Konrad Adenauer Stiftung.

Petrucci, A. L. (2022). *Policy Report No 61 – Collective intelligence and cities' cities' smart growth in GCC*. Konrad Adenauer Stiftung.

Petrucci, A. L., et al. (2022). *Policy report no 57 – Resilience, growth, and GCC smart cities*. Konrad Adenauer Stiftung.

Sachdev, V. (2011). Paradigms for design: The Vastu Vidya Codes of India. In S. Marshall (Ed.), *Urban coding and planning*. Routledge.

Secchi, B. (1999). *Fisiognomica della domanda, in A. Clementi (a cura di), Infrastrutture e progetti di territorio*. Fratelli Palombi Editori, Roma.

Secchi, B. (2011). *First urban planning lesson*. Laterza.

Secchi, B. (2013). *La citta' dei ricchi, la citta' dei poveri*. Laterza.

Soja, E. W. (2010). *Seeking spatial justice*. Paperback.

T20 Italy. (2021). https://www.t20italy.org/

Trancik, R. (1986). *Finding lost space; theories of urban design*. Van Nostrand Reinhold Company.

Viganò, P., et al. (2010). Extreme City. Climate change and the transformation of the waterscapes. Università Iuav di Venezia.

Open Access This chapter is licensed under the terms of the Creative Commons Attribution 4.0 International License (http://creativecommons.org/licenses/by/4.0/), which permits use, sharing, adaptation, distribution and reproduction in any medium or format, as long as you give appropriate credit to the original author(s) and the source, provide a link to the Creative Commons license and indicate if changes were made.

The images or other third party material in this chapter are included in the chapter's Creative Commons license, unless indicated otherwise in a credit line to the material. If material is not included in the chapter's Creative Commons license and your intended use is not permitted by statutory regulation or exceeds the permitted use, you will need to obtain permission directly from the copyright holder.

Environmental Retrofitting, Fighting Urban Heat Island Toward NEZ Sustainable Smart Cities

Anna Laura Petrucci

Abstract The discovery of the physical phenomenon of heat island dates back to 1833 when Luke Howard undertakes on the air temperature in London and its surroundings. His research showed how, already at that time, winds get stopped and pulled up by the intensive urbanization, reducing the quality of the outdoor environment while turning cold into warmer areas within the urban settlements (Mills 2008). The last centuries confirmed how intense urbanization can make the temperature rise to several degrees, activating a vicious circle where car and energy use become more necessary, and co-cause for further temperature arise. No doubt, cities are—literally—heat islands if compared to their surroundings. The crescent environmental stress is the main challenge while targeting quality of life in today's urbanization, in order to enable the public space to welcome citizens and encourage their outdoor activities. It requires a consistent commitment to the built environment and the awareness of the main role of public space as an interactive platform for a sustainable and human-centered smart city. As matter of fact, it is the public space where social life shapes and grows. The public space was often defined as the smart city's interface. So, smart cities need to generate and maintain a welcoming, healthy, livable, vibrant public space to have a reason to be. The scope of smart cities is to create a space first, the infrastructure to strengthen the connections among people and between people to the place. (Petrucci 2022) Cities worldwide must respond to a growing and diverse population, ever-shifting economic conditions, new technologies, and a changing climate. The task becomes especially challenging in extreme climate environments, such as in the Far East, in the African continent, or in the GCCs, which are also the countries where the most extensive urbanizations are taking place. Both intense urbanization growth and extreme weather conditions make here mandatory, more than everywhere else, an integrated strategy to achieve livable and sustainable cities through the fight against the urban heat island.

Keywords Urban heat island · Environmental retrofitting · sustainability · NEZ · Quality of life · Densification · Urban growth · Urban regeneration · PPP · PPPP

A. L. Petrucci (✉)
Universita' La Sapienza, Roma, Italy

King Saud University, Riyadh, Saudi Arabia

© The Author(s) 2024
F. Belaïd, A. Arora (eds.), *Smart Cities*, Studies in Energy, Resource and Environmental Economics, https://doi.org/10.1007/978-3-031-35664-3_8

111

1   Intro: Heat Island and Urban Livability

The discovery of the physical phenomenon of heat islands dates back to 1833 when Luke Howard studied the air temperature in London and its surroundings. His research showed how, already at that time, winds were stopped and pulled up by intensive urbanization, reducing the quality of the outdoor environment while turning cold into warmer areas within urban settlements (Mills, 2008). The last centuries have confirmed how intense urbanization can make the temperature rise to several degrees, activating a vicious circle where car and energy use become more necessary and causing further temperature to rise. Undoubtedly, cities are—literally—heat islands if compared to their surroundings. Crescent environmental stress is the main challenge when targeting quality of life in today's urbanization to enable the public space to welcome citizens and encourage their outdoor activities. It requires a consistent commitment to the built environment and awareness of the main role of public space as an interactive platform for a sustainable and human-centered smart city. In fact, it is the public space where social life shapes and grows. The public space was often defined as the smart city's interface. Therefore, smart cities need to generate and maintain a welcoming, healthy, livable, vibrant public space to have a reason to be. The scope of smart cities is to create a space first and the infrastructure to strengthen the connections among people and between people and the place (Petrucci, 2022). Cities worldwide must respond to a growing and diverse population, ever-shifting economic conditions, new technologies, and a changing climate. The task becomes especially challenging in extreme climate environments, such as in the Far East, in the African continent, or in the GCCs, which are also the countries where the most extensive urbanizations are taking place. Both intense urbanization growth and extreme weather conditions make here mandatory, more than everywhere else, an integrated strategy to achieve livable and sustainable cities through the fight against the urban heat island.

2   Factors to the Urban Heat Island

The formation of urban heat islands (UHIs) in urban areas depends on numerous factors, and the assessment of the effect of each on temperature, both real and perceived, requires complex exponential calculations and numerical models. The complexity of the equation tells us that the issue can only be approached as a comprehensive process that takes into consideration the multiplicity of interactions among all factors. This is probably also the reason why the issue has not been properly addressed thus far, showing very few successful case studies worldwide. As demonstrated by Howard two centuries ago, the first cause of UHIs is urbanization itself at the density of the built masses. The increased density of human-made structures, surface materials drier than their surroundings and radiation of sensible heat, such as waste heat from vehicles and buildings, requires a complete rethinking of urbanism as a cooperative system and considering it as the only tool available to address and possibly solve the island heat effect; greater urban density means an increase in the anthropogenic sources of heat. Thus, the city's massiveness is the first cause of urban heat islands, and its growing congestion requires addressing the problem with drastic solutions to reverse it and apply CO_2 negative solutions.

Heat is given off by buildings, transport systems, and industry by dense surface materials in the ground and buildings, which massing limits the wind flow and reduces its potential cooling effect, while the sealed soil facilitates fast evaporation. These last points alone are extremely important if considering that wind velocity can double over open water and wet surfaces. Sun exposure and direct and diffuse radiation are also important factors, where diffuse radiation is directly related to air pollution. Albedo, as the measure of absorbed heat during the day by each urban surface and its heat emission overnight, plays a crucial role in generating UHIs. This means that streets, sidewalks, roofs, and buildings' walls release the accumulated heat and contribute to the formation of urban heat island days and nights. Even color can play an important role here, considering that black surfaces exposed to the sun can become hotter than white surfaces, on the order of 7–21 °C.

The abovementioned factors can be defined as passive factors to albedo rise. To the already extreme complexity in measuring the interaction among passive factors, the active factors to albedo rise must be added, such as the presence of heat waves related to district heating, traffic and industrial activity, and the high use of air conditioning. Peaks in temperature within the districts create peaks in heat islands through both passive and active factors, and the temperature increase makes it difficult for any outdoor activities, requiring more intensive use of cars and more intensive use of air conditioning. That is how the vicious circle is activated and further implemented.

To break the vicious circle of UHI generation and reverse it, it is essential to analyze the urban structure as a holistic body where passive buildings and a more extensive passive urbanism are aligned with the reduction of resource use through compactness. Unless the whole circle is taken into consideration and addressed through informed policies by the decision-makers, any solution will not lead closer to a Near Zero Energy (NEZ) city, only partially reducing the urban temperature and not offering outdoor livability in extremely hot climates.

3   Buildings' Contribution to the Urban Heat Island

Considering the extreme case of southern countries, which are experiencing one-third of the years at a temperature above 40 °C, buildings are completely dependent on air conditioning to protect inhabitants from heat stress. In the UAE, for example, buildings consume more than 80% of the total electrical generation, where the cooling systems are responsible for approximately 70% of the building's peak electrical load despite the government of Dubai initiating several efforts to improve building efficiency and move toward a more sustainable city (Biggart, 2021). The heat discarded by air conditioners in the air, on the other hand, contributes to dramatically increasing the street temperature. It can be locally measured in a range of temperature improvement from 1 to 4 °C, depending on the type of implant (Tremeac et al., 2012). This extra warmth gets piled on top of the urban heat island effect coming from other factors. A city of one million people can be as much as 3 °C hotter than the area immediately around it due to extra heat generated by air conditioning, and in fact, downtown buildings would require 25% more air conditioning than the same buildings in a rural area (Hyde, 2007). This means that the actual use of air conditioning generates a temperature rise in the districts and an exponential increase in energy consumption while offering a lower benefit to the inhabitants. It is truly a vicious cycle, a vicious circle of increasing street temperature and, therefore, increasing the air cooling demand. Furthermore, it has been demonstrated that in hot and dry cities, air conditioning use at night plays an important role in contributing to the nocturnal urban heat island and increasing with its cooling demands; having urban heat during the night means not allowing the natural cooling of the city due to a lack of thermal gap, making natural ventilation systems less efficient. Several studies have shown that the implementation of a comprehensive sustainable mitigation strategy for air conditioning use would achieve several objectives: a successful reduction in the urban heat island temperature by an average of 2 °C and a reduction in electricity consumption on a city scale while mitigating the urban climate and enhancing the environmental quality. The economic impact would also be significant: an estimation made for the Phoenix metropolitan area shows that successfully reducing the urban heat island temperature would bring at least 1200–1300 MWh of direct energy savings per day alone (Olson, 2014). The solution is not simple and requires some creative thinking, including research about reusing the heat and water wasted from HVAC (Salamanca et al., 2014).

Energy-efficient buildings require an integrated design approach from the very beginning of the design process and good integration of mechanical and electrical services with passive systems to avoid extra costs. Studies on energy consumption conducted in Europe found that a glass-to-wall ratio of less than 20% gives the minimum life-cycle cost of buildings (IEA, 2010). In the Middle East, it has been tested how building orientation and thermal insulation can save up to 20% in residential buildings and how the use of appropriate window glazing and orientation in high-rise office buildings saves up to 55% energy. Moreover, integrating natural ventilation can reduce energy consumption by up to 30% in villas to up to 79% in high-rise office buildings (Friess, et al., 2017). Long-term forecasts worldwide show that the heating energy demand is projected to decrease by 34% by 2100, while the cooling demand is estimated to increase by 72% over the same period, making air conditioning a necessity to maintain acceptable indoor comfort levels in wider geographical regions. The resulting electric load attributable to HVAC equipment accounts for 40% of the total annual average electrical load and up to 60% of the summer peak load (Krarti et al., 2018). The implementation of evaporative systems displays significant energy savings, and using mixed-mode ventilation in conjunction with a cooling tower for slab cooling makes night use of HVAC no longer needed, resulting in energy savings from 55% to 73% over the global variable air volume and up to 25% over the simple daytime mixed-mode system (Ezzeldin et al.,

2013). The above results indicate that natural ventilation is an effective passive cooling strategy that should receive more attention to reduce the environmental footprint. Passive cooling is a large part of the vernacular architecture and the urban solutions of ancient civilizations; it requires expanding the design process as part of a broader sustainability agenda, including identity scouting of the so-called "cool vernacular" by linking the themes of social progress, technological and industrial transformation within a discussion of global and local trends, climate types, solution sets, and relevant low-resource utilization technologies. This type of holistic assessment of the environmental impacts of buildings includes not only the form and fabric of the building but also a wider set of parameters, such as the lifestyle of the occupant (Brophy et al., 2011). New solutions in more efficient air conditioning or electrical systems, such as controls and building energy management systems, are an important part of the ongoing research; however, these should be considered only after a comprehensive evaluation of passive solutions to eliminate energy leaking from the building. Building plans and forms emerge in a complex process where functional, technical, and aesthetic considerations all contribute to a synthesis. Wind, solar availability and direction, shelter and exposure, air quality, and noise conditions will inform the relationship of the building to its external environment and affect the form and design of the envelope. Bioclimatic heating, cooling, daylighting, and energy strategies should mesh at an early stage with the architect's other priorities. The support of governments through incentives and other strategies is crucial, together with a strategic plan of developing an upgrade of the national industry to start a self-production of high-end and sustainable building construction elements.

Natural ventilation strategies and indirect lighting can be implemented, arriving at fully integrated building systems through high technology envelopes, sensors, and smart devices to support energy management. Finally, buildings can be upgraded through active systems for energy production by renewable sources: geothermal, natural air precooling/preheating, thermal mass, heat exchange systems, and photovoltaic. Governments must control this kind of strategic approach to avoid the propaganda effect of running into photovoltaic systems and any other manifesto approach without first respecting the right protocol for saving energy leakage. In contrast, it could have a boomerang effect on investments, sustainability, and social awareness, discouraging the public from obtaining the scarcest results despite a large investment (Petrucci et al., 2022). Measurement and assessment tools also play a significant role in defining the quality and not just the quantity in balancing inputs and outputs within the complexity of a holistic approach looking at the building as a part of the interconnected system, a fractal approach to the overall planning, and an objective within the city-planning and building regulatory framework, allowing policy changes to facilitate the process in both new and existing districts.

In the process of designing sustainable solutions for new buildings, the approach to the existing real estate offers a major challenge. Nevertheless, a massive intervention in existing real estate is essential to contribute to reducing the UHI effect and requires a combined short- to mid- to long-term approach strategy. Implementing the 4Rs of the Circular Carbon Economy Reduce, Reuse, Recycle, and Remove existing real estate addresses the existing gaps in the urban challenge toward Near

Zero Emission and CCE Index (KAPSARC, 2020), which is true, especially in the assessment of the construction value chain. The reducing, reusing, and recycling approach to new and existing buildings must include the implementation of passive environmental solutions as the starting point in eliminating the source of excessive heat in the buildings and any cause of energy consumption. Otherwise, the implementation of the technological solution will only generate extra costs without an effective reduction in the environmental level. Passive solutions for existing buildings are the main part of their environmental retrofitting and can be applied visibly or invisibly. In the first case, these contribute to a reshaping of the building and eventually enhancing its volume or livable spaces; in the second case, the building gets an extra coating of insulating materials to reduce the heat transfer from indoor/outdoor and a replacement of windows and doors. Those actions would better follow a synergic plan and eventually concur to a completely new appeal from the building by encapsulating it into a new façade or adding extra volumes as an immediate return for the investment. In sustainable architecture, the link between building performance and the design of the envelope is critical while balancing more expensive building solutions and the improved balance between heat gain and heat loss. A life cycle cost analysis should be used to evaluate the contribution of ventilation openings, thermal functions of thermal mass, acoustic and energy protection, orientation and functional reasons of glazed elements self-shading, screening, and integration of photovoltaic technology. Once the causes of energy leakage in buildings are removed, the minimization of energy consumption comes from the Rs of recycling and reuse: starting from graywater recycling and optimizing materials and moving forward, the question links the construction of the built environment with a wider industrial and economic strategy. The prevailing economic model is linear and translates into raw materials being mined to manufacture components that are subsequently used and ultimately end as waste at the end of their lifecycle. The demand for raw materials is predicted to double by 2050. The reuse is based on the fact that not necessarily changing means that everything existing deserves to be trashed or cannot be repurposed through an incentive of locally based business in handcrafting. This might be the case with furniture but also with doors and windows. Recycling requires a bigger scale program and an industrial plan toward systemic recycling of construction materials, making urban communities a leading model in developing circular economy models.

4   A Holistic and Local-Based Approach to Fight the Urban Heat Island

Field research and data analysis essentially demonstrate the urban paradox: urban density is at the same time a cause and solution to sustainable living. Cities are economic powerhouses, resource consumption centers, and significant producers of greenhouse gas emissions. On the one hand, compact cities are the most sustainable living model in terms of soil and energy consumption, and density makes cities catalysts for economic and social opportunities; on the other hand, urban density is the first cause of generating urban heat islands.

It means we need density, and in the meanwhile, we urgently need to rethink the way we densify. Instead of recurring to partial solutions, it is crucial to work on synergic strategies, addressing consistent remedies to the urban form, as density's generator, and the result of the complex interaction of interdependent pressures and influences: climatic, economic, social, and political, strategic, aesthetic, technical, and regulatory. Energy efficiency is part of an integrated search for sustainable development recognizing the local, regional, and global impact of cities on people and the environment. Planning decisions have a pervasive and long-lasting impact on social cohesion and the quality of life of the individual and the collective, locally and globally. A wider comprehensive approach to cities is needed, organically including a multiplicity of aspects concurring to it at the different scales of intervention; a comprehensive strategy from macro to micro scale, involving planning, urban design, architecture, and all their subdisciplines.

UHI is mainly researched and treated by urban planners and misses the importance of retrofitting the existing built environment to achieve a reduction in energy consumption and CO_2 emissions. The targeting of livable and walkable districts, as a matter of fact, requires a comprehensive approach, including macro- and microscale interventions that cooperate to minimize energetic consumption and the related CO_2 footprint for new and existing built environments. There are several critical relationships: among the buildings themselves; among the buildings and the topography of the site; and the overall harmony of behavior between buildings, vegetation, and natural and artificial landforms. When internal and external spaces are designed with bioclimatic aims, buildings and the space surrounding them react together to regulate the internal and external environment to enhance and protect the site, local ecosystems, and biodiversity. Aspects such as land use, density, transportation, green space, water and waste, energy, microclimate, site selection, site planning, building form, urban fabric, orientation, and facades are all components of the same macro project. Guidelines for an integrated effort of environmental requirements such as energy performance, heating, cooling, ventilation, lighting, indoor air quality, energy production elements, and the opaque transparent ratio of elevations must facilitate a holistic approach. It can be achieved in new constructions through the already consolidated and shared environmental labels and protocols such as SDGs and New Urban Agenda, LEED, BREAMS, and others, and—in the case of the existing districts—it can be further enhanced into a wider sustainable strategy, including a retrofitting of the built environment. Moreover, it contributes to a circular economy system that involves paradigm shifts in how urban environments are designed and built.

5 Conclusions and Recommendations

Density is a crucial factor in sustainability and UHI management, and cities are expanding faster than the urban population, with an urbanization growth rate of 1.8 versus the 1.2 rates of population growth. This phenomenon has generated urban sprawl, soil consumption, exploitation of natural resources and raw materials, and increasing production of CO_2. Peripheral districts load the environment and public administration with up to 10–100 higher costs in infrastructures than urban retrofitting. The extension of urbanized areas not only increases the environmental, construction, and management costs but also generates an expansion of the urban heat island, moving it from an urban to a regional level. If not properly addressed, the increase in urbanization up to 90% of the world population expected for the year 2050 will make UHIs a phenomenon not only limited to downtowns but also extended to peripheral and regional urbanized or semiurbanized areas.

Developing a dynamic urban growth strategy is the first step to densifying the city and achieving the expanded targets while also keeping the city from additional urban sprawl infringing on sensitive environmental ecosystems that have been there for thousands of years. The challenge of densification currently would be to curb urban sprawl. It would limit further land consumption and seal and generate controlled sprawl, with regular reviews of the defined borders according to expansion needs. We need smart growing cities based on multicentered and polycentric models targeting density, accessibility, and connectivity, planning along with the public transport system, and generating a strategic distribution of facilities and attractors (for citizens and investors) within the city. Implementation policies are required to encourage infill and urban regeneration, making the housing stock newly available within the city while preserving the historical flair of the existing heritage. New urban codes must acknowledge urban sprawl and restructuring the urban texture in terms of economic growth, sustainability, and quality of life. Being linked to measurable development outcomes, such as the enhancement of quality of life, the improvement of local economic opportunities, and growth with the reduction city's carbon footprint, are crucial. It implicates the recognition of the importance of a form-based approach to urbanization and a systemic approach from macro to micro scale. Moreover, density upgrades require a conjunct effort between the public, providing the legal infrastructure and proper incentives, and private investors through public private projects (PPPs) to finance costly infill development and urban regeneration projects. Private-sector investment can be a good source of capital if the local government can ensure that private investments meet public needs. Moving further from PPP to the wider collaborative form of Participatory Public–Private Projects, governments can induce added value and capture the opportunity to build up urban culture and a sense of belonging from developers and stakeholder engagement to ensure effective locally based strategies toward a circular carbon economy and maximize the benefit from environmental resources targeting the NEZ. It would further increase small and medium businesses and activate a circular economy process, strengthening the collectivity and linking it with stakeholders and decision-makers. An urban approach including retrofitting strategies indirectly contributes to the other main parameters of NEZ urbanism.

The fight against UHIs brings up disruptive considerations in the design and management of cities and requires a reassessment of planning policies. To win the challenge, cities must develop tailored and locally based strategies for regulating the built environment through multicriteria evaluation systems facilitating real changes in public and private behaviors and moving their future users toward active and collective awareness and actions. To embrace the challenge, here is a list of recommendations, surely not exhaustive, that can possibly be further implemented:

1. Urban interventions to be handled at each phase of their planning as a complex system considering deep interactions between macro and micro scales within a comprehensive sustainable system around its circular economy.
2. Cities to apply the first R-Reduce as the first step toward Near to Zero Emission (NEZ) and Circular Carbon Economy (CCE) Index toward a holistic approach to managing emissions across energy systems and economies and to achieving carbon circularity.
3. Cities and their governments to move forward to the next 3R-Recycle Reuse Remove with integrated industrial and strategic planning on construction materials and their comprehensive life cycle management.
4. Cities and their governments should include air conditioning as a part of the climate challenge, as well as its possible solution, as a major contributor to greenhouse gas emissions, requiring more effective and widely used methods for passively cooling buildings.
5. Cities and their government to endorse and implement informed research on the so-called "cool vernacular" targeting a revival of indigenous knowledge in human-centered urbanism and architecture.
6. Environmental concepts to be expanded include macro- and microclimatic conditions to influence bioclimatic strategies as a greater synthesis between building elements and local climate conditions.
7. Nature-based solutions should be adopted as the main intervention in urban design and architecture for passive environmental strategies and implementing technologies at the final stage of the project.
8. Passive and active cooling/heating strategies to be coded generate a range of opportunities to improve the effectiveness of design outcomes.
9. Public space and smart public transport must be enhanced as mutual factors to reduce urban heat island effects and generate vibrant livable cities.
10. Cities' reforestation should be encouraged throughout the city to contribute to human well-being and health and to environmental refurbishment.
11. Displacement economization should be made, allowing the diminution of particular vehicles and the reduction of gaseous emissions of pollutants.
12. Public and private sectors are co-creators, along with the people, in public–private projects and participatory public–private projects to activate a long-lasting urban culture and awareness.

References

AA.VV. CCE guide overview, KAPSARC 2020. https://www.cceguide.org/

Biggart, M., et al. (2021). Modelling spatiotemporal variations of the canopy layer urban heat island in Beijing at the neighbourhood scale. *Atmospheric Chemistry and Physics (ACP), 21*, 13687–13711. https://doi.org/10.5194/acp-21-13687-2021

Brophy, V., et al. (2011). Green Vitruvius, principles and practice of sustainable architectural design. *Earth*.

Ezzeldin, S., et al. (2013). The potential for office buildings with mixed-mode ventilation and low energy cooling systems in arid climates. *Energy and Buildings, 65*, 368–381.

Friess, W., et al. (2017). A review of passive envelope measures for improved building energy efficiency in the UAE. *Renewable and Sustainable Energy Reviews.* www.elsevier.com/locate/rser

Hyde, R. (Ed.). (2007). *Bioclimatic Housing: Innovative Designs for Warm Climates* (1st ed.). Routledge.

International Energy Agency. (2010). *World Energy Outlook.* www.iea.org/reports/world-energy-outlook-2010

Krarti, M., et al. (2018). Review analysis of economic and environmental benefits of improving energy efficiency for UAE building stock. *Renewable and Sustainable Energy Reviews, 82*(Part 1), 14–24.

Mills, G. (2008). Luke Howard and The climate of London. *Weather, 63*(6), 153.

Olson, R. (2014). *Excess heat from air conditioners causes higher nighttime temperatures.* Arizona State University, ASU News.

Petrucci, A. L. (2022, January). Policy Report No. 62 –Placemaking, a stress relief tool for deliverables in GCC smart cities.

Petrucci, A. L., et al. (2022, January). *Policy Report No. 57 –Resilience, growth, and GCC smart.* Konrad Adenauer Stiftung.

Salamanca, F., et al. (2014). Antropogenic heating of the urban environment due to air conditioning. *Journal of Geophysical Research Atmospheres, 119*(10), 5949.

Tremeac, B., et al. (2012). Influence of air conditioning management on heat Island in Paris air street temperatures. *Energy, 95*, Elsevier, 102.

Open Access This chapter is licensed under the terms of the Creative Commons Attribution 4.0 International License (http://creativecommons.org/licenses/by/4.0/), which permits use, sharing, adaptation, distribution and reproduction in any medium or format, as long as you give appropriate credit to the original author(s) and the source, provide a link to the Creative Commons license and indicate if changes were made.

The images or other third party material in this chapter are included in the chapter's Creative Commons license, unless indicated otherwise in a credit line to the material. If material is not included in the chapter's Creative Commons license and your intended use is not permitted by statutory regulation or exceeds the permitted use, you will need to obtain permission directly from the copyright holder.

A Comprehensive Smart System for the Social Housing Sector

Isam Shahrour

Abstract This chapter presents a comprehensive smart system for the social housing sector which considers technical, social, and environmental issues. The chapter is composed of four sections. The first section discusses the challenges of the social housing sector with an emphasis on the social and environmental dimensions. The second section presents the research methodology, including an analysis of the expectations of the tenants and the social housing manager and the specifications for the design of the comprehensive smart system. The third section describes the architecture of the smart system, including the stakeholders' communication channels, the monitoring system, and the smart services. The last section shows an application of smart system to a renovated social housing residence. The chapter shows that the comprehensive smart system should go beyond the smart building concept by extending this concept to the construction of a smart community and the involvement of this community in the improvement of the social housing environment.

Keywords Social housing · Tenants · Smart · Community · Energy performance · Sensors · Platform · Comprehensive

1 Introduction

The social housing sector constitutes a significant issue for several European countries. It accounts for 30% of the rental housing in the Netherlands, 18% in the United Kingdom, and 17% in France. The importance of this sector goes beyond the high housing sharing ratio because it generally combines a challenging social context and poor construction performance. Since this sector concerns mainly

I. Shahrour (✉)
Laboratoire de Génie Civil et géo-Environnement, Université de Lille, Lille, France
e-mail: Isam.Shahrour@univ-lille.fr

© The Author(s) 2024
F. Belaïd, A. Arora (eds.), *Smart Cities*, Studies in Energy, Resource and Environmental Economics, https://doi.org/10.1007/978-3-031-35664-3_9

low-income families, tenants are exposed to social and economic difficulties such as low income, unemployment, and lower access, compared to other populations, to essential services such as education, health, culture, and sports.

Some scholars have focused on analyzing the poor energy performance of the social housing sector and its consequences for tenants' quality of life and housing precarity (Esmaeilimoakher et al., 2017; Filippidou et al., 2016; Elsharkawy & Rutherford, 2015; Juan et al., 2018). They showed an urgent need for the renovation of the aged social housing sector for both energy savings and occupants' quality of life. Other scholars have analyzed the impact of social building renovation on energy savings (Elsharkawy & Rutherford, 2015; Enertech, 2018; Kavgic et al., 2012). The results showed that the renovation program did not result in the expected energy savings. They attributed this disappointing result to an insufficiency in the monitoring program and occupants' involvement in energy management. Kavgic et al. (2012) showed that building characteristics are not the most relevant factors in energy savings because of the dominant role of resident behavior in energy consumption. Belaïd and Garcia (2016) explored the main factors influencing residential energy-saving behaviors in France. They highlight the impact of five main attributes incentivizing energy-saving behaviors: energy price, household income, education level, age of head of household, and dwelling energy performance. A questionnaire to the residents of an energy-saving program in Nottingham showed that the achievement of high energy performance requires occupants' involvement and policy interventions (Elsharkawy & Rutherford, 2015).

A recent report about the portrait of social housing tenants in France indicated the following features (USH, 2019): approximately 40% of tenants live alone, 20% are single-parent families, 30% are over 60 years old, and 30% live below the poverty line. In addition, approximately 70% of the social housing buildings in France were built before 1990, which means that most of this sector does not respect today's building standards. Approximately 66% of these buildings are energy intensive, which means high energy consumption and too many expenses for low-income tenants.

In a recent study, Freund et al. (2022) showed the following well-being needs of social housing tenants in Australia: paying unexpected bills, feeling sad or anxious, feelings of anger or frustration, and memory or concentration problems.

The portrait of the social housing sector shows that this sector is facing significant and explosive challenges because of the critical social and construction contexts. To address these challenges, national and local governments launched several initiatives and strategies to renovate the social housing stock. These strategies are, of course, necessary (Jensen et al., 2022; Bal et al., 2021). Nevertheless, considering the social context, they should be carried out within a comprehensive strategy that combines social development, tenants' participation, and innovative services with guarantees that tenants can have easy access to these services (Shahrour & Xie, 2021; Jnat et al., 2020; Duvier et al., 2018). This chapter presents how the smart city concept could help create this environment. The chapter first presents the research methodology, including establishing the specifications for the smart system. Then, it describes the architecture of this system and, finally, its application to a renovated social housing residence. The main contribution of this chapter consists of the design of a comprehensive smart system for the social housing sector, which goes beyond the smart building concept. The design is based on analyzing the expectations of the tenants and managers of social housing. The capacity of the system is illustrated through its application to a renovated social housing residence.

2 Research Methodology

2.1 Overview

This research was conducted through cooperation with social housing managers with the objective of designing a comprehensive smart system for social housing management that considers technical, social, and environmental issues at both the apartment and complex levels. The research started by analyzing the process of managing the social housing complex and identifying the responsibility of the social housing manager. Concertation was then conducted with tenants to understand their expectations from the smart transformation of social housing. The combination of the manager's and tenants' expectations allowed us to determine the specifications for designing the comprehensive smart social housing system. Since this system is expected to offer a wide range of services, its construction requires time, resources, testing, and readjustment. For this reason, the first phase of its implementation concerned the evaluation of the impact of renovation on tenants' quality of life and energy consumption.

2.2 Manger's Expectation of the Smart Transformation of the Social Housing Complex

The responsibility of the social housing managers concerns providing services related to the maintenance of buildings and infrastructures, supply of drinking water and heating, lighting and cleaning of common areas, local waste collection, and access security. The manager is also responsible for tenants' information about maintenance works, water, energy consumption, and current shared expenses.

Generally, social housing managers work with many companies and service providers. Since each company has its communication channel and management system, the social housing manager has to deal with a multitude of unconnected systems that cause frequent manual interventions.

Maintenance work, particularly that requiring emergent interventions, seems to be the most critical. The challenge is how to capture, transmit, and efficiently organize urgent interventions. This challenge could be addressed through the development of a smart monitoring system that (i) ensures a real-time survey of the

Table 1 Tenants' expectations from the smart transformation of social housing

Expectation	Concerned stakeholders
Report complaints and follow up responding measures and actions	Social housing manager, tenants' community, services providers
Access to information about maintenance works, disorders, new services, etc.	Social housing managers, services providers, maintenance companies
Access to data about energy and water consumption, indoor comfort, and air quality	Social housing manager, services providers, occupants
Security and safety	Social housing managers, security companies, emergency services, police departments
Delivering services for occupants with specific needs (people with disabilities and older people)	Social housing manager, services providers, medical services, social services, tenants' community
Addressing energy precarity	Social housing manager, services providers, social services, tenants' community
Build a smart community	Social housing manager, tenants'community

critical equipment and spaces, (ii) transmits the collected data to concerned people, including the local building manager, the maintenance companies, and the tenants, (iii) tracks the technical interventions, (iv) informs concerned people about the intervention progress, and (v) establishes automatic reports about the technical interventions. In a large social housing complex, urgent interventions could be delayed by difficulties related to localizing and accessing the intervention area or requirement. Rapid localization could be provided by a smart indoor navigation system based on digital tags displayed in the main indoor sections and equipment. Building information modeling (BIM) offers extensive 3D graphic facilities to digitally identify buildings' components and equipment, including their properties and spatial localization and indoor navigation.

The manager is also concerned with establishing an efficient communication channel with the occupants to receive their complaints and observations and to follow up on the measures and responding actions.

Table 1 summarizes the expectations of the social housing manager from the smart transformation of the social housing sector.

2.3 Tenants' Expectations

Discussion with the social housing tenants showed several expectations, which could be classified into seven categories. The first concerns the communication channel. It is about how the occupants can easily report complaints and observations and follow up on the responding actions. The second concerns receiving information about the maintenance works, service interruption, and new services. The third category concerns access to real-time and historical data about indoor comfort, air quality, and energy and water consumption. The fourth category is related to

Table 2 Tenants' expectations from the smart transformation of social housing (Jnat, 2018)

Expectation	Concerned stakeholders
Report complaints and follow up responding measures and actions	Social housing manager, tenants' community, services providers
Access to information about maintenance works, disorders, new services, etc.	Social housing managers, services providers, maintenance companies
Access to data about energy and water consumption, indoor comfort, and air quality	Social housing manager, services providers, occupants
Security and safety	Social housing managers, security companies, emergency services, police departments
Delivering services for occupants with specific needs (people with disabilities and older people)	Social housing manager, services providers, medical services, social services, tenants' community
Addressing energy precarity	Social housing manager, services providers, social services, tenants' community
Build a smart community	Social housing manager, tenants' community

security, emphasizing secure access to the building and external common areas such as parking, green spaces, and children's playing areas.

The fifth category concerns delivering customized services to occupants with specific needs, such as persons with disabilities and older people. The question is, how could the manager help provide specific services? The sixth category concerns energy precarity, which harms low-income families. The tenants expect to benefit from measures to reduce energy consumption through construction renovation and replacing energy-intensive appliances. The last category concerns building a smart community to create a friendly social environment to enhance mutual aid, cultural and collective activities, services and products exchange, and break individual isolation and loneliness.

Table 2 summarizes the tenants' expectations from the smart transformation of social housing.

3 Design of the Comprehensive Smart System

3.1 Communication Channels

The architecture of the comprehensive smart system is illustrated in Fig. 1 (Shahrour & Xie, 2021). A smart platform ensures communication with social housing stakeholders, including social housing managers, tenants, maintenance companies, service providers, social services, and emergency services. Each stakeholder can communicate with the platform using a mobile application or a web service. Stakeholders receive information on the mobile application or through email or messages. Each stakeholder has secure identification access. Tenants have a personal profile.

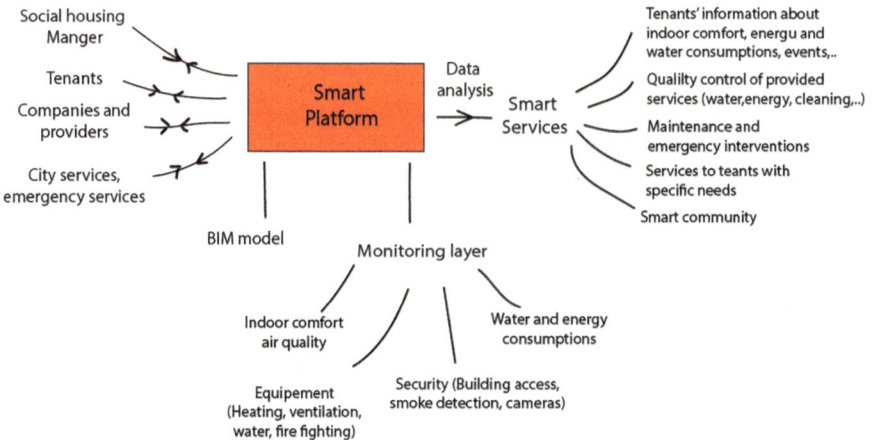

Fig. 1 Architecture of the comprehensive smart platform for social housing (Lagsaiar et al., 2021)

The platform is connected to a BIM model of the social housing complex, which includes a spatial description of the complex and the properties of its components. It provides tools for the 3D visualization of the complex and its components. It also provides graphics for dynamic data such as indoor comfort, air quality, and energy consumption.

3.2 Data Collection

The social housing complex is monitored by smart sensors that measure and transmit real-time data to the platform (Lagsaiar et al., 2021). Smart monitoring covers a multitude of parameters, such as indoor comfort, air quality, energy and water consumption, and security parameters, including building access, smoke detection, movement detection, and survey cameras. In addition, some equipment, such as heating and ventilation systems, are monitored to track their functioning. They are also equipped with actuators that permit their online or automatic control.

Social housing stakeholders, particularly tenants, can transmit their observations through mobile applications. For example, when they observe an anomaly, they can take a photo, record a voice message, and send their comments via the mobile application.

For maintenance work or emergency interventions, professionals can access information and data about the equipment through the mobile application. They use the mobile application to upload the intervention reports, which are then automatically transferred to concerned people.

3.3   Data Analysis: From Data to Smart Services

The platform operates data analysis to turn the collected data into smart services. Figure 1 shows the smart services provided by the system. They include (i) information of tenants about indoor comfort and energy and water consumption; (ii) control of the quality of provided services such as water and energy supply, cleaning, indoor and outdoor lighting, and waste collection; (iii) maintenance of buildings and emergency interventions; (iv) services for people with specific needs, particularly people with disabilities and older people; and (v) building a smart community with an emphasis on mutual aid, cultural activity, services exchange, and breaking individual isolation and loneliness.

4   Application to a Renovated Social Housing Residence

4.1   Objectives

This section presents the application of the proposed system to renovated social housing residents in northern France. This application concerned the first stage of implementing the comprehensive smart system with emphasis on the impact of the renovation on tenants' quality of life and energy consumption.

4.2   Description of the Social Housing Residence and the Monitoring System

The social housing residence is approximately 50 years old. It was completely renovated (Jnat et al., 2020). It uses a central heating system with a manual regulation system. Despite the recent renovation, occupants complained about the high energy expenses. To understand the causes of these increased expenses, some apartments were monitored by indoor comfort sensors, including temperature and humidity. Data were recorded at 30-minute time intervals. The outdoor temperature and humidity were obtained from the nearby weather station. The monitoring program covered the heating period from October to April.

4.3   Results of the Monitoring Program

Figure 2 illustrates the variations in indoor and outdoor temperatures. The outdoor temperature varied between −5 and 23 °C with an average value of 5.1 °C, while the indoor temperature varied between 19 and 29 °C with an average value of

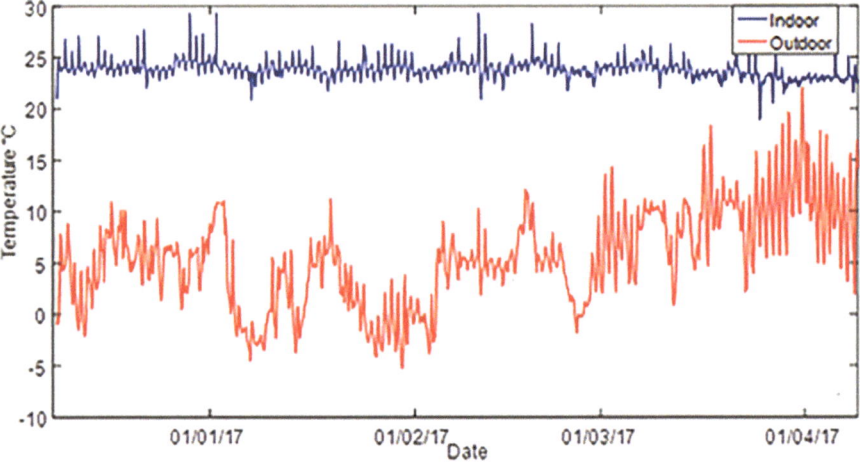

Fig. 2 Variation in the indoor and outdoor temperatures (Jnat et al., 2020)

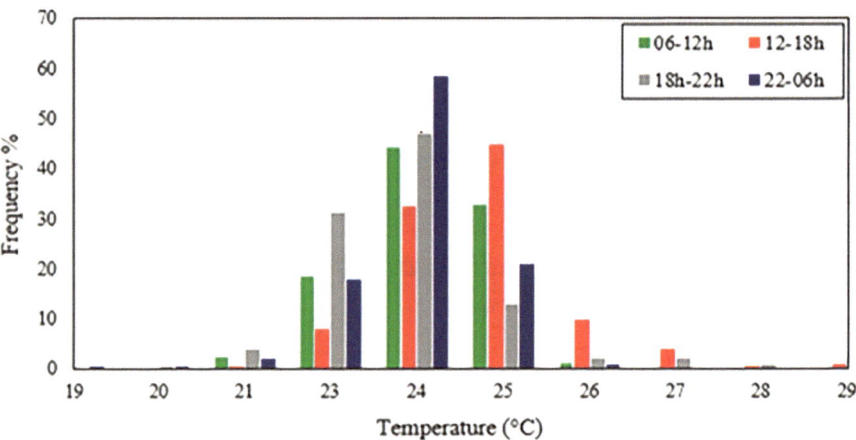

Fig. 3 Temperature distribution according to time slots during the heating period (A) (Jnat et al., 2020)

23.8 °C. This average exceeded the French regulation heating temperature by approximately 4.8 °C.

Figure 3 compares the distributions of the indoor temperature during four daily periods. The temperature distributions over the four periods are very close. The high-temperature values in the afternoon and the evening are related to tenant activity. According to French thermal regulations, the maximum indoor temperature during this time should not exceed 19 °C. During sleeping time, it should be limited to 16 °C. As shown in Fig. 4, indoor temperature highly exceeds the French thermal regulations values all over the day.

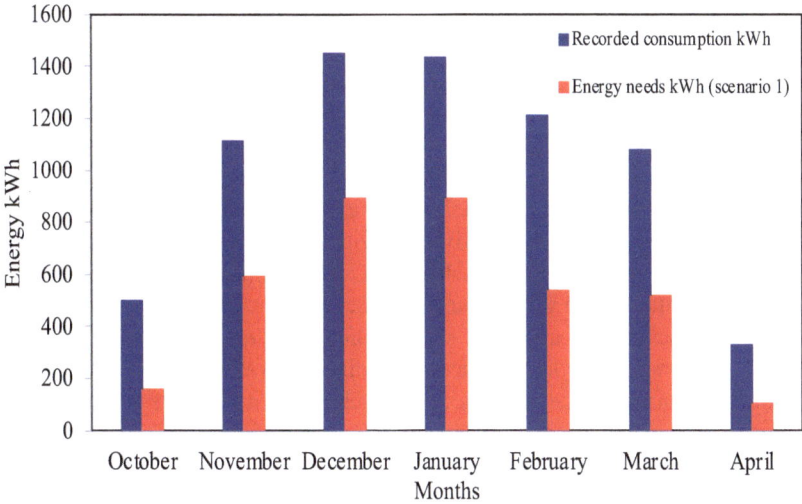

Fig. 4 Comparison between recorded consumption and estimated energy needs according to the French thermal regulations RT 2012 (Jnat et al., 2020)

Discussion with tenants showed that they were not aware of the excess indoor temperature and its consequences on their energy expenses. They attributed the high indoor temperature level to the manual regulation of the heating system and to the absence of a monitoring system that alerts them about the indoor temperature excess and its consequences on their expanses. Therefore, they were interested in estimating the impact of a friendly heating system on energy expenses.

To respond to the tenants' demand, the thermal software ArchiWIZARD (Jnat, 2018) was used to estimate the energy expanses according to the French thermal regulations RT 2012: an indoor temperature of 19 °C during the occupancy period, 16 °C in the case of a vacancy for a period less than 48 hours, and 8 °C for a vacancy exceeding 48 hours.

Figure 4 compares the recorded heating consumption and the estimated heating energy consumption according to the French thermal regulation RT2012. The monthly energy savings vary between 229 kWh in April and 675 kWh in February. Over the heating period, the overall heating energy savings equal 3420, which is approximately 45% of the heating consumption.

This example shows how the smart monitoring program could help identify the causes of high energy consumption in renovated social housing buildings, which combines poor monitoring and regulation of the energy system and the lack of information and awareness of tenants. The use of the proposed concept could help address these two issues.

5   Conclusion

This chapter presented the design of a comprehensive smart system for the social housing sector. The design is based on analyzing the expectations of the tenants and managers of social housing. This analysis showed that the comprehensive smart system should go beyond the smart building system. It should extend the conventional services of the smart building to social issues, including assistance to tenants with specific needs, cultural and sports activity, services exchange, and building a smart community. It also requires the establishment of an efficient communication channel between the tenants and the social housing manager.

The first phase of this approach was applied to a renovated social housing residence to understand the low impact of building renovation on energy expenses. The smart monitoring of the residence and exchange with tenants showed global indoor overheating due to poor management of the heating system and to the absence of a monitoring system and an efficient communication channel between tenants and the social housing manager. This example shows the importance of implementing the proposed comprehensive smart system in the social housing sector.

The concept proposed in this chapter could help policymakers improve the performance of both new and renovation social housing programs because it takes into consideration both technical and social issues. In addition, it reduces the current expenses of low-income tenants and facilitates their integration into the community as well as their access to cultural, sportive, and education services.

References

Bal, M., Stok, F. M., Van Hemel, C., & De Wit, J. B. F. (2021). Including social housing residents in the energy transition: A mixed-method case study on residents' beliefs, attitudes, and motivation toward sustainable energy use in a zero-energy building renovation in The Netherlands. *Frontiers in Sustainable Cities, 3*, 656781. https://doi.org/10.3389/frsc.2021.656781

Belaïd, F., & Garcia, T. (2016). Understanding the spectrum of residential energy saving behaviors: French evidence using disaggregated data. *Energy Economics, 57*, 204–214. https://doi.org/10.1016/j.eneco.2016.05.006

Duvier, C., Anand, P. B., & Oltean-Dumbrava, C. (2018). Data quality and governance in a UK social housing initiative: Implications for smart sustainable cities. *Sustainable Cities and Society, 39*, 358. https://doi.org/10.1016/j.scs.2018.02.015

Elsharkawy, H., & Rutherford, P. (2015). Retrofitting social housing in the UK: Home energy use and performance in a pre-Community Energy Saving Programme (CESP). *Energy and Buildings, 88*, 25–33. https://doi.org/10.1016/j.enbuild.2014.11.045

Enertech. (2018). Evaluation par mesure des performances énergétiques des 8 bâtiments construits dans le cadre du programme européen concerto. 2012. Available online: http://www.enertech.fr/pdf/66/zdb_rapport-synthese-vl.pdf. Accessed on 13 May 2022.

Esmaeilimoakher, P., Urmee, T., Pryor, T., & Baverstock, G. (2017). Influence of occupancy on building energy performance: A case study from social housing dwellings in Perth, Western Australia. *Renewable Energy and Environmental Sustainability, 2017*. https://doi.org/10.1051/rees/2017018

Filippidou, F., Nieboer, N., & Visscher, H. (2016). Energy efficiency measures implemented in the Dutch nonprofit housing sector. *Energy and Buildings, 132*, 107–116. https://doi.org/10.1016/j.enbuild.2016.05.095

Freund, M., Sanson-Fisher, R., Adamson, D., et al. (2022). The wellbeing needs of social housing tenants in Australia: An exploratory study. *BMC Public Health, 22*, 582. https://doi.org/10.1186/s12889-022-12977-5

Jensen, S. R., Gabel, C., Petersen, S., & Kirkegaard, P. H. (2022). Potentials for increasing resident wellbeing in energy renovation of multifamily social housing. *Indoor and Built Environment, 31*(3), 624–644. https://doi.org/10.1177/1420326X211039883

Jnat, K. (2018). *Smart Bâtiment: Analyze et optimization des dépenses d'énergie dans le logement social* [Doctoral dissertation]. Université de Lille, Lille, France, 13 Novembre 2018.

Jnat, K., Shahrour, I., & Zaoui, A. (2020). Impact of smart monitoring on energy savings in a social housing residence. *Buildings, 10*, 21. https://doi.org/10.3390/buildings10020021

Juan, A., Zabalza, I., Llera-Sastresa, E., Scarpellini, S., & Alcalde, A. (2018). Building energy assessment and computer simulation applied to social housing in Spain. *Buildings, 8*, 11. https://doi.org/10.3390/buildings8010011

Kavgic, M., Summerfield, A., Mumovic, D., Stevanovic, Z., Turanjanin, V., & Stevanović, Z. (2012). Characteristics of indoor temperatures over winter for Belgrade urban dwellings: Indications of thermal comfort and space heating energy demand. *Energy and Buildings, 47*, 506–514. https://doi.org/10.1016/j.enbuild.2011.12.027

Lagsaiar, L., Shahrour, I., Aljer, A., & Soulhi, A. (2021). Modular software architecture for local smart building servers. *Sensors, 21*, 5810. https://doi.org/10.3390/s21175810

Shahrour, I., & Xie, X. (2021). Role of internet of things (IoT) and crowdsourcing in smart city projects. *Smart Cities, 4*, 1276–1292. https://doi.org/10.3390/smartcities4040068

USH - Union Social pour l'Habitat. (2019, September 22–24). Les HLM en chiffres. 8ème congrès HLM, Paris. https://www.union-habitat.org/sites/default/files/articles/pdf/2019-09/ush_les_hlm_en_chiffres_2019.pdf. Accessed on 13 May 2022.

Open Access   This chapter is licensed under the terms of the Creative Commons Attribution 4.0 International License (http://creativecommons.org/licenses/by/4.0/), which permits use, sharing, adaptation, distribution and reproduction in any medium or format, as long as you give appropriate credit to the original author(s) and the source, provide a link to the Creative Commons license and indicate if changes were made.

The images or other third party material in this chapter are included in the chapter's Creative Commons license, unless indicated otherwise in a credit line to the material. If material is not included in the chapter's Creative Commons license and your intended use is not permitted by statutory regulation or exceeds the permitted use, you will need to obtain permission directly from the copyright holder.

Smart Green Planning for Urban Environments: The City Digital Twin of Imola

Mansoureh Gholami, Daniele Torreggiani, Alberto Barbaresi, and Patrizia Tassinari

Abstract Urban green spaces are significant in adjusting the urban microclimate. Street trees are the most influential type of urban vegetation in reducing heat stress. However, simulating trees' 3D models, wind flow, surface temperature, and radiation parameters in complex urban settings and producing high-resolution microclimate maps is often time-consuming and requires extensive computing processes. Therefore, efficient approaches are needed to visualize green scenarios for the future development of the cities. Smart green planning of Imola aims at developing a microclimate digital twin for the city that provides complementary and supportive roles in the collection and processing of micrometeorological data, automates microclimate modeling, and represents climatic interactions virtually. This chapter sets out to explore the smart green planning of Imola in two parts. The first part is focused on the potential and intentions of developing the urban microclimate digital twin for the city of Imola and its conceptual framework. The second part aims at testing and evaluating the applicability of the proposed microclimate digital twin by implementing it in the city of Imola. This digital twin can provide urban planners and policymakers with a precise and useful methodology for real-time simulation of the cooling effects of the trees and other green systems on urban-scale, pedestrian-level thermal comfort, and also a guarantee for the functionality of policies in different urban settings.

Keywords Digital twin · Smart green planning · Urban transformation · Microclimate · Real-time simulation · Imola

M. Gholami (✉) · D. Torreggiani · A. Barbaresi · P. Tassinari
Department of Agricultural and Food Sciences, University of Bologna, Bologna, Italy
e-mail: mansoureh.gholami2@unibo.it

© The Author(s) 2024

F. Belaïd, A. Arora (eds.), *Smart Cities*, Studies in Energy, Resource and Environmental Economics, https://doi.org/10.1007/978-3-031-35664-3_10

1   Introduction

The Mediterranean countries are mostly vulnerable regions to climate change, extreme weather events such as heatwaves, cold snaps, heavy rainfall or snowfall, ice or hailstorms, droughts, extratropical or tropical cyclones, storm surges, and tornadoes (Cramer et al., 2018).

Cities are notably vulnerable to these climate changes and must have a key role in making adaptation policies that require action at the local level. The main aim of these adaptation plans must be to prevent further deterioration of the climate crisis and to slow down the increase in global temperature, as it is considered in the Paris Agreement to reach net-zero greenhouse gas (GHG) by 2050. To accomplish this, it is necessary to keep up-to-date knowledge of the deteriorative impacts of climate change (Gholami, 2022). Because climate change is already expanding to some level, responding to climate change involves a two-pronged approach:

- Mitigation: Reducing emissions and stabilizing the levels of heat-trapping greenhouse gases in the atmosphere.
- Adaptation: Adapting to climate change that is already in the pipeline. The mechanism for climate adaptation is defined as adjusting to actual or possible expected scenarios for future climate (NASA, 2022).

These mechanisms are not limited to the urban public environment but also include building complexes and individual buildings.

Global energy demand is set to increase 4.6% in 2021, more than offsetting the 4% contraction in 2020 that occurred due to the COVID-19 pandemic (Energy Agency, 2021). Building energy consumption is responsible for over one-third of global final energy consumption and 38% of total global energy-related GHG emissions.

The EU climate and energy policy sets the following targets for 2030 (EU, 2021):

- At least 40% cuts in greenhouse gas emissions (from 1990 levels)
- At least 32% share of renewable energy
- At least 32.5% improvement in energy efficiency

Europe, however, has made itself committed to raising the 2030 GHG emission reduction target to at least 55% in comparison to 1990. Thus, important measures are needed to reduce energy consumption and GHG emissions in urban areas. Without consistently pursuing reduction policies in urban areas, Europe will miss these targets.

1.1   State of the Art

Italy is in a region that is specifically vulnerable to global climate change. Climate data in Italy show an increase in extreme temperatures, which puts this country at risk of natural hazards. Global climate change is expected to increase vulnerability to climate-related events in subsequent years. Urbanization contributes to a considerable increase in human activities (Yang et al., 2017). The high-level concentration of human activities causes a decrease in green urban spaces and a shift in the microclimate of urban areas (Gholami et al., 2020a). Climate change and urbanization are unavoidable. Howard in the 1800s (Mills, 2008) and later Landsberg in the early 1900s (Landsberg, 1981) have been considered among the pioneers who have studied urban climate and demonstrated the difference between urban microclimate and the microclimate in surrounding suburbs and rural areas. This variation can be found in air temperature, humidity, solar radiation, wind speed and direction, air pollution level, and amount of precipitation (Fig. 1). Urban Heat Island (UHI) has been counted as a synthetic factor in using temperature in microclimate variation, and it is defined as a heating effect when air temperature rises in a developed (urbanized) area in comparison to the surrounding rural area (Taha, 2004). Urban spaces cut out vegetation from the land area by the development of building construction, roads, roofs, and other urban infrastructure throughout open lands, and these urban surfaces hold the heat of shortwave radiation and longwave radiation more than vegetation (Gholami et al., 2018). Thus, UHIs are partially due to a lack of greenery and inappropriate urban design.

UHI is defined by calculating the difference between average and maximum air temperature in rural and urban areas (Khare et al., 2021). There are two types of UHIs, air and surface heat islands, each of which is caused by a variety of dominant physical processes and has different temporal and spatial patterns, but they are largely caused by the transformation of the surrounding landscape into corrugated, mostly manufactured, and less vegetation-covered surfaces. Air UHI is the effect in the canopy layer or boundary layer. The UHI in the urban canopy can be measured at meteorological stations; however, for the UHI in the boundary layer, more special installments are needed in aircraft or tall towers (CUHK, 2008). On the other hand, surface UHIs measure the difference in radiant temperature between urban and

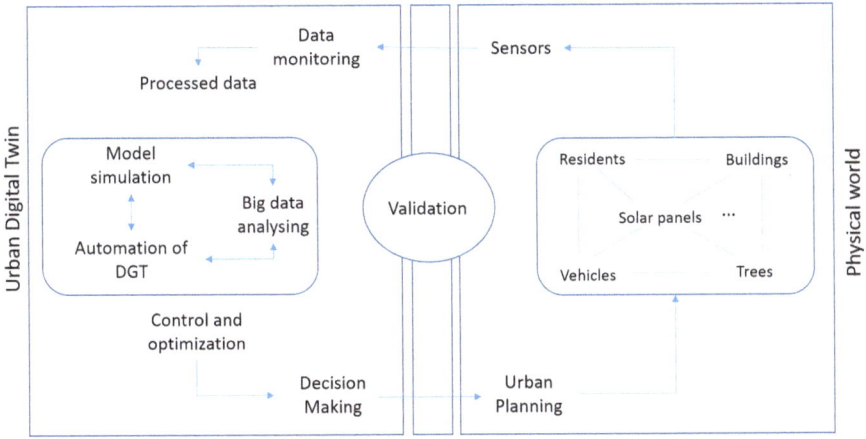

Fig. 1 Conceptual framework of an urban digital twin

suburban surfaces, and this UHI can be measured directly by satellite thermal temperature remote sensing data. In urban spaces, the mean radiant temperature depends on the building geometry, street layout green cover, and albedo of urban walls and ground (Wong et al., 2011).

Green spaces in an urban area can affect UHIs in various aspects of the surface energy balance (Gholami et al., 2020b). It can reduce radiation and air temperature and decelerate wind. Trees can decrease the mean radiant temperature and cause a reduction in the sky view factor (Khare et al., 2021). The mean radiant temperature (MRT) is the uniform temperature of an environment in which the radiant heat transfer from the human body is equal to the radiant heat transfer in the actual environment (Li, 2016). Assessing the effect of trees on heat stress proved that the mean radiant temperature can be decreased by 77% (Milošević et al., 2017). Wang and Akbari showed that the highest range of MRT reduction can be achieved by enlarging the tree canopy (Wang & Akbari, 2016). Zheng et al. examined the lawn and shaded ground with shrubs and trees through a CFD model and demonstrated that the lawn has the lowest surface temperature in summer. However, trees are more effective in reducing the Tmrt (Zheng et al., 2016). A comprehensive review of all the published studies in different climate zones (Santamouris, 2013) has demonstrated the importance of green roofs for the air temperature at the pedestrian level by reducing Tmrt by 0.3–3.0 K by the condition that it has been applied on the city scale. A test carried out in Singapore demonstrated that the cooling effects of a green roof can be considered if the height of the building is less than 10 m (Wong et al., 2003).

1.2   Conceptual Framework Planning for an Urban Microclimate Digital Twin

Several models and methods have been developed for the evaluation of microclimate, energy balance, and air quality. However, the complexity of urban environments has limited the functionality of urban models to simulate all the interconnectivity and interdependencies in urban systems. To overcome this limitation, multilevel modeling methodologies have been increasingly employed to perform several single simulations and then combine the output as a comprehensive urban model. Digital twins are among the well-known multilevel models.

Performing any digital twin is in three separate domains, namely, the real world, digital world, and intermediate space, which should work together in a real-time frame (Wright & Davidson, 2020). A digital twin is a virtual representation that serves as the real-time digital counterpart of a physical object or process. A city digital twin is a series of interconnected digital twins representing certain aspects of the functioning of urban environments (Austin et al., 2020). Every digital twin receives a continuous flow of data collected by sensors in real time, analyzes the data, and presents the outcome in various virtual models (Cioara et al., 2021). Since the data are collected in real time, the ability of the digital twin relies on processing data flows. Digital twins perform change predictions and scenario modeling for what-if questions (Qi & Tao, 2018).

A microclimate digital twin provides complementary and supportive roles in the collection and processing of micrometeorological data, automates microclimate modeling, and virtually represents climatic interactions. A digital twin on the first step needs an urban climatic mapping method. Urban climatic mapping integrates parameters in urban planning and climate into mapping. This means that it takes meteorological data, land use, building footprints, topography, and green spaces into consideration to analyze thermal loads and climate comfort into various categories in a spatial thermal impact map (CUHK, 2008). Urban characteristics can create variation in the horizontal and vertical wind profiles, and this effect alters the climate conditions in urban canopy layers (Oke, 1989). As the frictional drag on airflow increases, wind speed and orientation can massively change in built-up areas in comparison to rural areas. This difference is originally referred to as the roughness of a region that plays the main role in affecting the local wind flow. Roughness points out any obstruction in cities such as buildings from, traffic networks, tree canopy outlines, density, and even banner boards in microclimate examination (Yang & Chen, 2020). The impact of urban characteristics on microclimate becomes more complicated when the terrain is not flat. For example, wind profiles and urban ventilation can change according to the topography. Thus, it is crucial to develop climate models based on urban planning characteristics.

An urban microclimate digital twin has three performance steps: information gathering, analyzing processes, and postperformance and assessing resolutions. The main aim of a digital twin is to create a parallel world to a real world (Cioara et al., 2021). The postperformance step is required when the uncertainty in the measurements and modeling is not completed. In this case, a validation of the real-time databases can overcome the uncertainty. This microclimate digital twin runs offline 3D urban simulations and then applies a real-time microclimate (Fig. 2). The method also proves how the digital twin can be expanded to include a wide variety

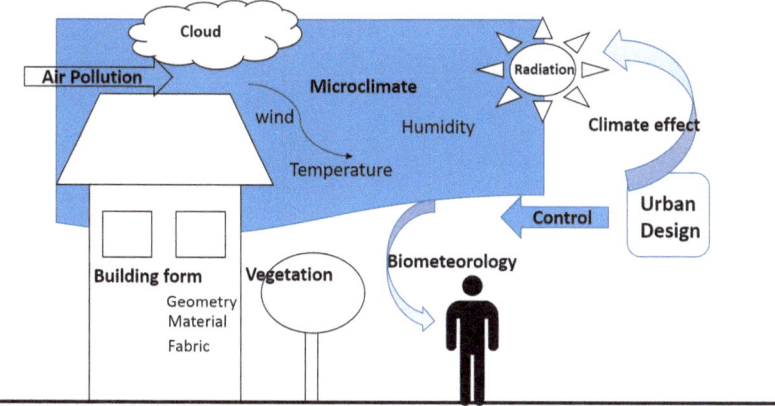

Fig. 2 Urban microclimate and involved parameters

of data resources and simulations on various scales and subjects. Moreover, a balance must also be created between urban management and urban engineering to maximize the benefits of trees in retrofitting strategies. The digital twin can provide urban planners and policymakers with a precise and useful methodology for simulating the effects of trees on urban-scale, pedestrian-level thermal comfort and help them guarantee the functionality of policies in different urban settings.

2 Methods

The main aim of this microclimate digital twin is to measure the mean radiation temperature and the Universal Thermal Climate Index (UTCI) based on a hybrid model. The Universal Thermal Climate Index (UTCI) was used for thermal comfort evaluation. The UTCI was introduced in 2011 by the International Society of Biometeorology (ISB) as a new thermal index for outdoor thermal comfort. It evaluates outdoor thermal conditions one-dimensionally in terms of air temperature, wind speed, humidity, and longwave and shortwave radiant heat fluxes.

The model is run through a Python 3.8.1 code that integrates three engines: EnergyPlus (U.S. Department of Energy, 2020), Rhino (McNeel, 2021), and OpenFOAM (Greenshields & Weller, 2022). Each parameter is modeled in different software. To ensure the accuracy of the modeling, the interconnectivity of the parameters is coded in a single comprehensive Python script, and input and output data are directly linked and validated in single steps and the ultimate results.

Figure 3 illustrates the interrelations and order of the steps.

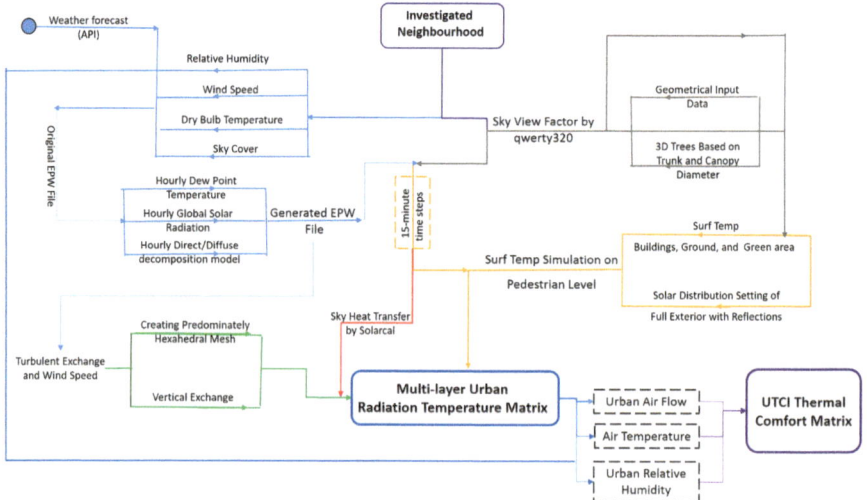

Fig. 3 Modeling methodology of the city digital twin

2.1   Automated Workflow Execution

One of the main challenges in developing the microclimate digital twin is a real-time weather data file for short-term forecasting. To accomplish this, a model is developed in Grasshopper that aims at regenerating the epw file based on forecasting data. The forecast weather data are acquired from API (OpenWeatherMap) in JSON format, which provides short-term forecasting weather data at various time scales. Four elements are required from the API file for the regeneration of the epw file: dry bulb temperature, relative humidity, wind speed, and sky cover. These elements were used to generate hourly dew point temperature, global solar radiation, and direct and diffuse decomposition models. Since Grasshopper cannot directly interact with EnergyPlus, the Honeybee tool was utilized to link open studio, energy plus, radiance, and daysim for an automated process to rewrite an epw file based on the existing epw file. The weather elements were replaced through the automated process in hourly format. This weather file will be used in the next steps for simulations.

3   Case Study

3.1   Smart Green Planning of Imola: Urban Microclimate Digital Twin of the City of Imola

The biosystems engineering group of the University of Bologna has initiated the project of developing an urban microclimate digital twin for the city of Imola to improve thermal comfort, livability, and sustainability in this city. In this phase of the project, developed within the "health, safety, and green systems" Ph.D. program of the district of Imola of the University of Bologna, the multidisciplinary team of the project has proposed a climate-sensitive digital twin that simulates possible green scenarios with the purpose of providing outdoor climate comfort.

Prof. Patrizia Tassinari, chair of the biosystems engineering group of the University of Bologna and coordinator of the Ph.D. program, mentions that the project will help urban planners and policymakers understand the impact of greener urban design and infrastructure on outdoor thermal comfort for the residents of the city of Imola.

Marco Panieri, the mayor of the city of Imola, brings up that this "digital twin" is the result of a scientific and technological study that shows how the Imola territory aims at cutting-edge goals, focusing on the knowledge and expertise developed within the university and continuing strengthening the connection among the university, the municipality, and the other local stakeholders, in the first place, the Cassa di risparmio di Imola Foundation, and Conami.

3.2 Objectives for Microclimate Digital Twin of the City of Imola

Imola is selected for the investigation of new greening policies in urban spaces. The outcome of this digital twin enables data scientists and infrastructure planners to validate and optimize the impact of new infrastructure before investing and deploying capital equipment (Gholami et al., 2022). By using historical databases, this tool helps planners simulate the impact of data-driven goals before they are implemented and helps operators monitor and maintain smart city services. This digital twin enables city planners to confidently and efficiently transform their energy systems by:

1. Identifying and optimizing spatial maps of existing, proposed, and planned infrastructure enhancements before investing in equipment;
2. Creating scenarios based on combined models to optimize service to the community; and
3. Predicting and effectively managing the green parameters in a given neighborhood, such as planting new trees in the city or managing rooftop green systems.

3.3 Urban and Climate Context of Imola

This chapter bases the study on the city of Imola, Italy. Italy is ranked fifth among European countries, with a population of 60 million in 2020. The district of Imola is a city in the Metropolitan City of Bologna, located on the river Santerno, in the Emilia-Romagna region of northern Italy (44° 21′ 11.0088″ N, 11° 42′ 52.9992″ E) (Fig. 4). It covers an area of 204 km² and has a population of approximately 70,000 inhabitants. According to Koppen's climate classification, Imola is in the zone of the humid subtropical climate and has a relatively continental and four-season climate. In Imola, the summers are warm and mostly clear, and the winters are very cold and partly cloudy. Over the year, the temperature typically varies from −0.5 to 31.11 °C and is rarely below −5 °C or above 35.5 °C. The city receives approximately 2230 hours of sunlight per year. July has an average maximum temperature of 30 ° C. The coldest month is January with an average minimum temperature of −1 ° C. The wettest month is November with 95 mm of rainfall. The sunniest month is August, with approximately 11 hours of sunshine. The driest month is July, with 43 mm of precipitation (Figs. 5, 6, 7, and 8). The selected neighborhood itself has an area of 900 m × 500 m, 9% of which is covered by street tree canopies. The neighborhood consists of detached two- to three-story single buildings with a dense pattern of tree planting in some parts (Fig. 9).

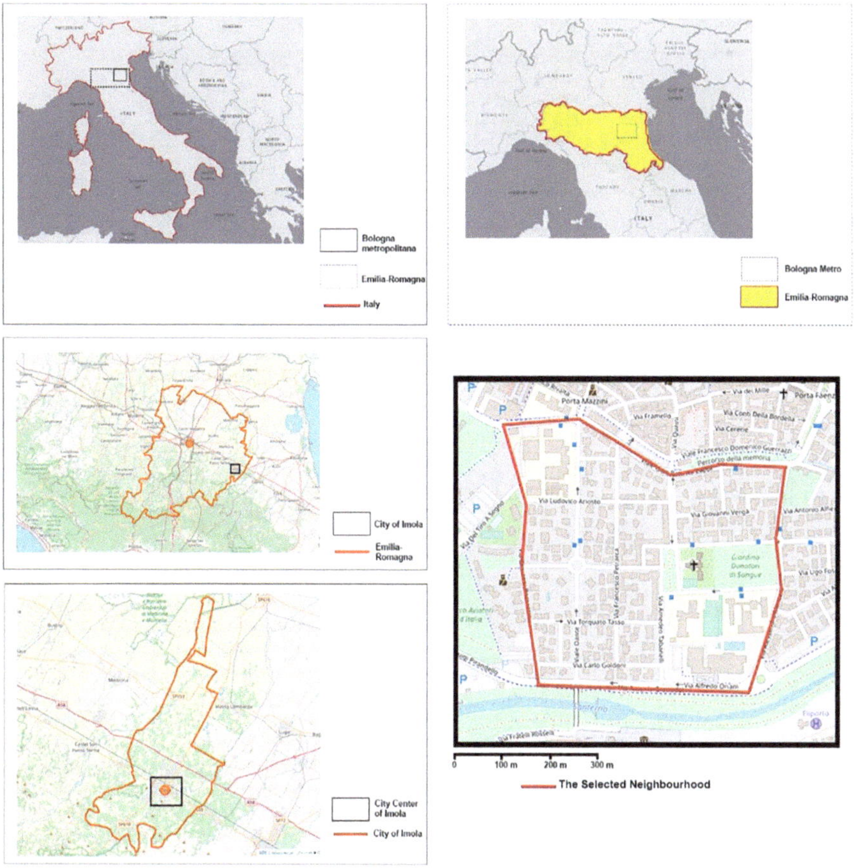

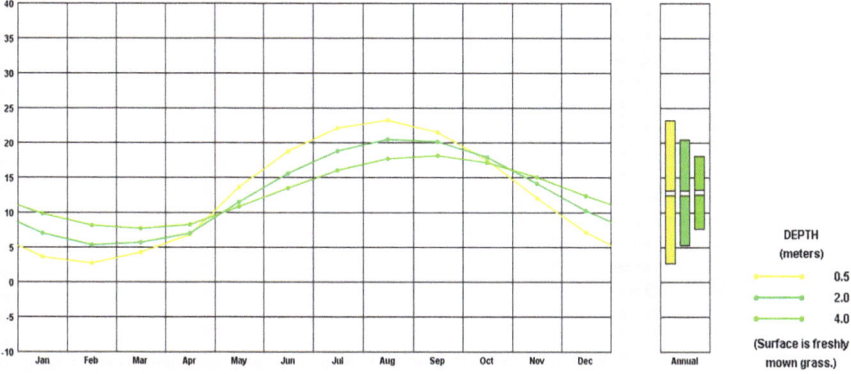

Fig. 4 Aerial view of the study area in Imola, Italy

Fig. 5 Monthly average ground temperature in Imola

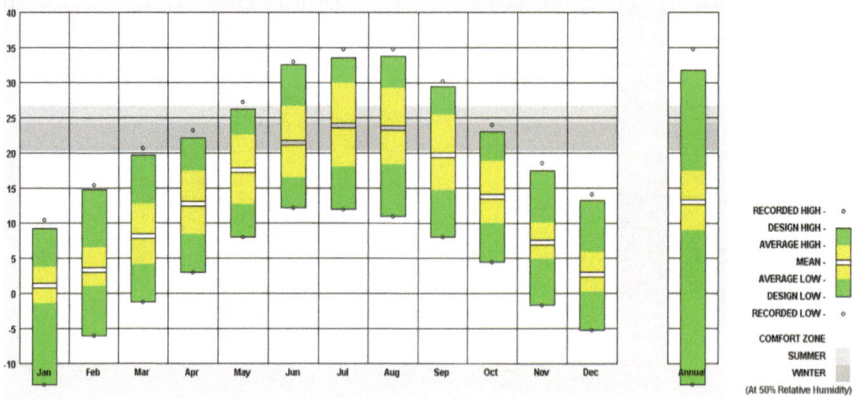

Fig. 6 Monthly average temperature in Imola

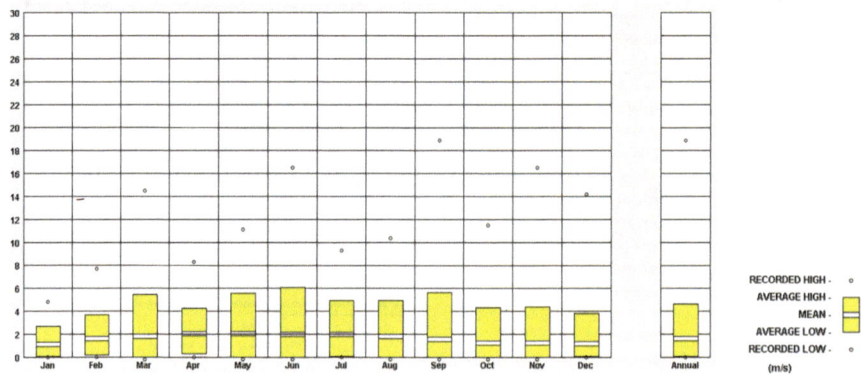

Fig. 7 Monthly average wind velocity in Imola

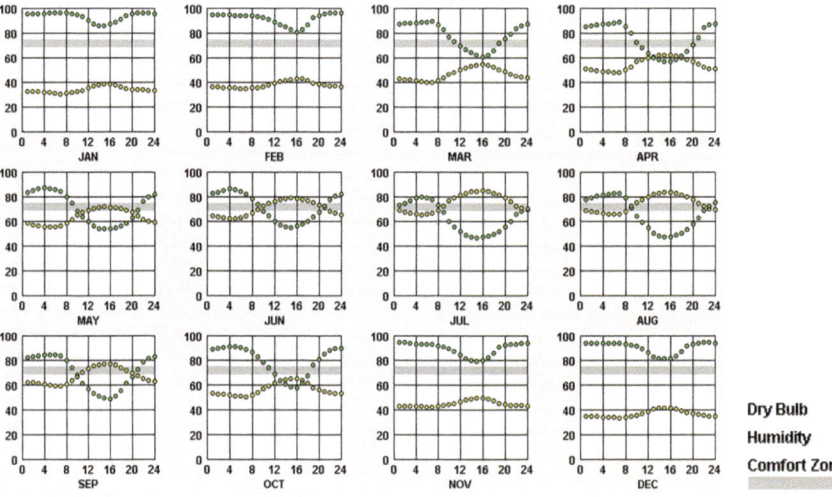

Fig. 8 Monthly dry bulb and humidity in the city of Imola

Fig. 9 Bird view of Imola, Italy (http://mappegis.regione.emilia-romagna.it/moka/download/sig/FOTO_IBC_HD/05DB_03_036.png; http://mappegis.regione.emilia-romagna.it/moka/download/sig/FOTO_IBC_HD/05DB_07_073.png

4 Results

Four different urban blocks inside the neighborhood are selected: (a) low-rise blocks with high canopy cover (LR-HCC), (b) mid-rise blocks with medium canopy cover (MR-MCC), (c) mid-rise blocks with low canopy cover (MR-LCC), and (d) high-rise blocks with high canopy cover (HR-HCC). After selecting the blocks and feeding the required data into the model, simulations are developed in the digital twin to generate a series of high-resolution mean radiant temperature (MRT) and Universal Thermal Climate Index (UTCI) maps at the pedestrian level. Moreover, a data matrix was created based on the hourly simulation of the hottest week of the year, August 3–9 (the study period). The maps and data were generated using a detailed 3D model of the geometric properties of buildings and canopies as well as several digital surface models (DSMs). As Fig. 3 illustrates, the process took into consideration the local microclimate, hourly turbulent exchange, surface temperature, and sky heat exchange. Therefore, the generated maps are highly accurate and clear for investigating tree parameters as well as optimal adaptation strategies in urban retrofitting.

4.1 Mean Radiant Temperature at the Street Level

The highest level of heat stress in the urban blocks was recorded at LR-HCC, which is a church in a designed landscape in a community-scale park. During the hottest day of the year, the mean radiant temperature exceeded 65 °C (Fig. 10). The canopy cover of the site is scarce due to the large distance between trees. Contrary to this site, the archetype HR-HCC has been shaded due to the high density of the canopies and buildings, and this shaded area has made this site the coolest part of the neighborhood during the hottest days of the year. The MRT in this block on the hottest day of the year was 45.5 °C, which is 20° lower than that in the archetype LR-LCC. As Fig. 11 shows, the archetype MR-LCC is a good example of the building shade effect in Imola. The dense fabric of the site helps create cool and comfortable pedestrian spaces.

4.2 The Universal Thermal Climate Index (UTCI) at the Street Level

This measurement system employs the correlations observed in human adaptive outdoor thermal behavior to predict the clothing of subjects of average age, height, and weight (Jendritzky et al., 2014). Heat stress occurs when the body fails to keep the internal temperature status at a certain level. Table 1 shows the categorization of thermal stress based on UTCI ranges. There are different types of UTCI, including heat and cold stress. In this chapter, we only consider heat stress.

The peak of the temperature on the 9th of August created a large gap in the level of UTCI in different urban blocks in Imola, but as the temperature decreased, this gap also decreased. Heat stress, on the other hand, was at an all-time high in Imola. The archetypal LR-HCC suffered the greatest UTCI level, as shown in Fig. 12, with

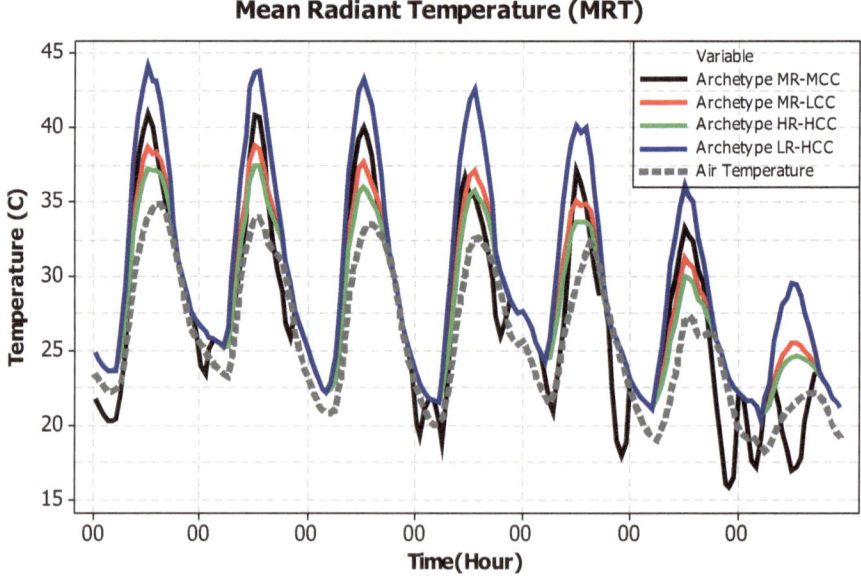

Fig. 10 Mean radiant temperature in the four archetypes in Imola

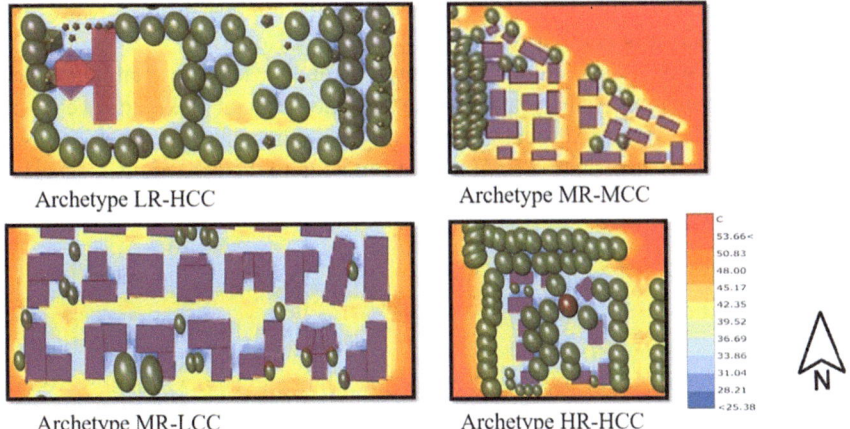

Archetype LR-HCC	Archetype MR-MCC
Archetype MR-LCC	Archetype HR-HCC

Fig. 11 Spatial variations in MRT at the microscale in the city of Imola

Table 1 The UTCI assessment scale: the UTCI categorized in terms of thermal stress

UTCI(C) range	Stress category
Above +46	Extreme heat stress
+38 to +46	Very strong heat stress
+32 to +38	Strong heat stress
+26 to +32	Moderate heat stress
+9 to +26	No thermal stress

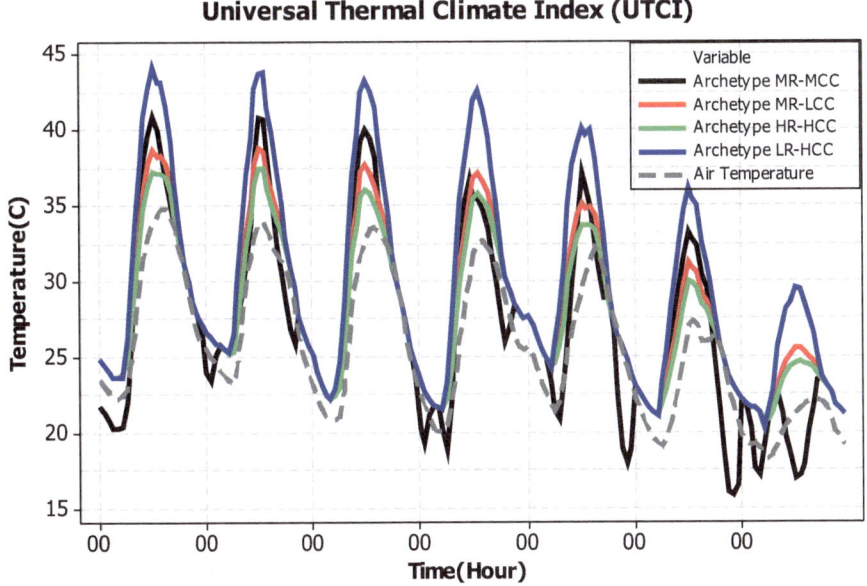

Fig. 12 Universal thermal climate index (UTCI) in the city of Bologna (above) and Imola

a temperature of 43.4 °C, indicating extremely significant heat stress. Meanwhile, the HR-HCC had less heat stress, with a mean UTCI of 37.5 °C at midday on August 4. It is worth mentioning that sunny parcels bordered by buildings were 1–2 °C warmer than wide sunny spaces. The reflection of solar and longwave radiation onto parcels close to vertical surfaces might be the cause. Similarly, shaded parcels next to trees had a 0.5–1.5 °C cooler temperature than shaded parcels close to buildings.

5   Discussion

The proposed model in this chapter employs Ladybug and Honeybee; however, to date, various simulation engines have been presented in different contexts. Ladybug and Honeybee are capable of analyzing complex 3D models. The model is based on surface temperature and generated sky view factor by the Rhino platform. Envimet is one of the most-employed engines over the last decade for the calculation of Tmrt. It uses 3D models based on the equation of Bruse. Ground evaporation and the transpiration of urban vegetation are also calculated. Rayman is another engine that simplifies the environment and calculates the MRT based on Stephen Boltzman's radiation law. Any model can be challenged by the complexity of urban environments, and the validation of the model and input data is necessary.

5.1   Performance Validation of the Digital Twin of Imola

The model was validated through field measurements in the city of Tempe, USA, on June 9th, 2018. The field measurements were conducted in a clear sky in the city for model validation. Four sample locations were selected for collecting human-bio meteorological data on June 9th, 2018. Sensors measure georeferenced 6-directional longwave (L) and shortwave (K) radiation flux densities with three Hukseflux 4-Component Net Radiometers, Ta, Ts, v, and RH in 2-s intervals at pedestrian height. Every sample site is archived through hemispherical 180° photos taken at 1.1 m height with a Canon EOS 6D and Canon EF 8–15-mm f/4 Fisheye USM Ultra-Wide Zoom lens. Table 2 indicates the minimum and maximum and average mean radiant temperature measured and modeled from 8 am to 8 pm. The results have been calibrated based on these four points in the area.

5.2   Application of the Cooling Scenario

The study used a strategy scenario that involves putting trees in specified spots as well as increasing the number of trees in public spaces to evaluate the model's performance in analyzing the impacts of increased tree coverage on communities.

Table 2 Measurements and model outputs of mean radiant temperature

Point code		Min Tmrt	Max Tmrt	Average Tmrt	Standard deviation
1	Measurement	30.4800	69.4017	43.8754	13.66891
	Model	37.2510	44.5320	39.7083	4.871698
2	Measurement	32.0426	40.3306	35.2940	3.229757
	Model	38.9000	43.3000	40.0000	3.579011
3	Measurement	24.8197	61.8556	48.6251	20.65881
	Model	25.0000	72.0000	55.6667	26.57693
4	Measurement	31.6427	48.7733	40.5803	8.589577
	Model	26.1000	33.9000	30.2000	3.915354

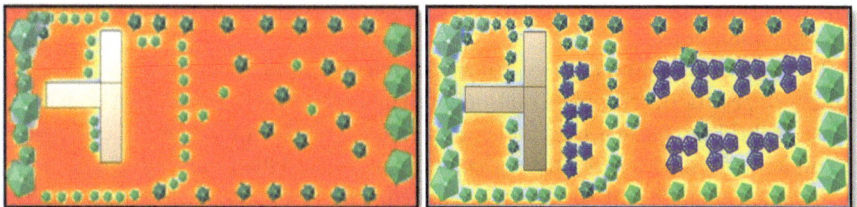

Fig. 13 Spatial variations in the UTCI of the cooling scenario for archetype LR-LCC

Numerous experiments have examined how shade caused by urban surfaces influences the spatial variance of radiant temperature (Thorsson et al., 2011). On a small scale, the number of trees and the spatial location of extra trees are both essential factors in lowering the MRT. To decide the priority of parcels to receive new trees, the first stage in putting trees in optimum areas is to create an accurate map of each block's climate comfort based on the UTCI (Fig. 13).

In examining the situations, two findings from previous research were taken into account: First, trees are more beneficial at lowering MRT in clusters than individually (Streiling & Matzarakis, 2003); and second, trees are more effective at reducing MRT in clusters than they are individually (Streiling & Matzarakis, 2003). The tree addition cooling scenario was put up in the block with the highest MRT in the modeled neighborhood in Imola. At each site, 10 clusters of trees with a canopy width of 7 m and a height of 10 m were placed (additional trees are highlighted in blue in Fig. 8). The present layout of the sites was not modified in any capacity, ensuring that the findings are applicable to the creation of appropriate retrofitting solutions. During the hottest hours of the year, the cooling strategy can reduce the radiant temperature by 9 °C and increase neighborhood climate comfort by up to 35%.

6 Conclusion

Today's urbanization has resulted in worldwide environmental and economic challenges that have a detrimental influence on cities and, as a result, urban planning procedures. Urban planning and design have been deeply involved in reducing carbon emissions in recent decades. One method of reaching low-carbon towns is through the upgrading technique of restructuring urban trees. Street trees are now more than ever valuable for our urban environments given how extreme heat in cities has reached concerning levels in recent decades. Overall, the number of trees must increase to prevent this effect, and urban design principles are essential for optimizing the distribution and placement of trees. Digital twins are one of the most recent technological innovations in urban planning. The goal of this study is to provide a digital twin concept that includes a real-time climate model in complicated 3D city simulations.

The suggested digital twin for the city of Imola simplifies complicated ideas and critical stages for developing a real-time urban microclimate model, and it incorporates a hybrid model that uses a unique method for MRT modeling. The output of the digital twin is visualized through thermal maps, and the future scenarios for the smart green planning of Imola are evaluated. The results are validated through a measured dataset, and the reliability of the real-time microclimate forecasting is indicated. The proposed microclimate digital twin in Imola can act as a platform to enhance green infrastructure and engage the community. A powerful community engagement tool can educate citizens and enhance the awareness of locals by coupling real city conditions with a digital twin platform. It is an incentive program for changing the materials of the buildings or planting trees on the streets.

We believe that this digital twin platform can be used to expand the possibilities for predictive maintenance, infrastructure expansion, and clean energy and green planning. Creating a digital twin library of existing and proposed infrastructure elements enables stakeholders to make action plans, transform cities and improve the quality of life for locals in urban spaces for a more sustainable world.

References

Austin, M., Delgoshaei, P., Coelho, M., & Heidarinejad, M. (2020). Architecting smart city digital twins: Combined semantic model and machine learning approach. *Journal of Management in Engineering, 36*, 04020026. https://doi.org/10.1061/(asce)me.1943-5479.0000774

Cioara, T., Anghel, I., Antal, M., Salomie, I., Antal, C., & Ioan, A. G. (2021). *An overview of digital twins application domains in smart energy grid.* undefined.

Cramer, W., Guiot, J., Fader, M., Garrabou, J., Gattuso, J. P., Iglesias, A., Lange, M. A., Lionello, P., Llasat, M. C., Paz, S., Peñuelas, J., Snoussi, M., Toreti, A., Tsimplis, M. N., & Xoplaki, E. (2018). Climate change and interconnected risks to sustainable development in the Mediterranean. *Nature Climate Change, 811*(8), 972–980. https://doi.org/10.1038/s41558-018-0299-2

CUHK. (2008). Urban climatic map and standards for wind environment-feasibility study FINAL REPORT C U H K, Working paper 1A: Draft Urban Climatic Analysis Map. Hong Kong.

Energy Agency, I. (2021). *Review 2021 Assessing the effects of economic recoveries on global energy demand and CO₂ emissions in 2021 Global Energy.*

Gholami, M. (2022). *Analyzing urban green adaptation opportunities: concepts, approaches, & strategies for existing neighborhoods.* http://amsdottorato.unibo.it/10052/

Gholami, M., Mofidi Shemirani, M., & Fayaz, R. (2018). A modelling methodology for a solar energy-efficient neighbourhood. *Smart and Sustainable Built Environment, 7,* 117. https://doi.org/10.1108/SASBE-10-2017-0044

Gholami, M., Barbaresi, A., Torreggiani, D., & Tassinari, P. (2020a). Upscaling of spatial energy planning, phases, methods, and techniques: A systematic review through meta-analysis. *Renewable and Sustainable Energy Reviews, 132,* 110036. https://doi.org/10.1016/j.rser.2020.110036

Gholami, M., Barbaresi, A., Tassinari, P., Bovo, M., & Torreggiani, D. (2020b). A comparison of energy and thermal performance of rooftop greenhouses and green roofs in Mediterranean climate: A hygrothermal assessment in WUFI. *Energies, 13,* 2030. https://doi.org/10.3390/en13082030

Gholami, M., Torreggiani, D., Tassinari, P., & Barbaresi, A. (2022). Developing a 3D City Digital Twin: Enhancing Walkability through a Green Pedestrian Network (GPN) in the City of Imola, Italy. *Land 2022, 11*(11), 1917. https://doi.org/10.3390/LAND11111917

Greenshields, C., & Weller, H. (2022). *Notes on computational fluid dynamics: General principles.* CFD Direct Ltd..

Khare, V. R., Vajpai, A., & Gupta, D. (2021). A big picture of urban heat Island mitigation strategies and recommendation for India. *Urban Climate, 37,* 100845. https://doi.org/10.1016/J.UCLIM.2021.100845

Landsberg, H. E. (1981). The urban climate. *International Journal of Geophysics, 28,* 1–277.

Li, H. (2016). Impacts of pavement strategies on human thermal comfort. In *Pavement materials for heat island mitigation* (pp. 281–306). https://doi.org/10.1016/B978-0-12-803476-7.00013-1

McNeel, R., & and others. (2021). *Rhinoceros 3D, Version 7.0.* Robert McNeel & Associates.

Mills, G. (2008). Luke Howard and the climate of London. *Weather, 63,* 153–157. https://doi.org/10.1002/WEA.195

Milošević, D. D., Bajšanski, I. V., & Savić, S. M. (2017). Influence of changing trees locations on thermal comfort on street parking lot and footways. *Urban Forestry & Urban Greening, 23,* 113–124. https://doi.org/10.1016/J.UFUG.2017.03.011

NASA. (2022). Mitigation and adaptation | Solutions – Climate change: Vital signs of the planet [WWW Document]. URL https://climate.nasa.gov/solutions/adaptation-mitigation/. Accessed 31 Jan 2022.

Oke, T. R. (1989). The micrometeorology of the urban forest. *Philosophical Transactions of the Royal Society B: Biological Sciences, 324,* 335–349. https://doi.org/10.1098/rstb.1989.0051

Qi, Q., & Tao, F. (2018). Digital twin and big data towards smart manufacturing and industry 4.0: 360 degree comparison. *IEEE Access, 6,* 3585–3593. https://doi.org/10.1109/ACCESS.2018.2793265

Santamouris, M. (2013). Using cool pavements as a mitigation strategy to fight urban heat Island—A review of the actual developments. *Renewable and Sustainable Energy Reviews, 26,* 224–240. https://doi.org/10.1016/J.RSER.2013.05.047

Taha, H. (2004). Heat islands and energy. In *Encyclopedia of energy* (pp. 133–143). https://doi.org/10.1016/B0-12-176480-X/00394-6

Thorsson, S., Lindberg, F., Björklund, J., Holmer, B., & Rayner, D. (2011). Potential changes in outdoor thermal comfort conditions in Gothenburg, Sweden due to climate change: The influence of urban geometry. *International Journal of Climatology, 31,* 324–335. https://doi.org/10.1002/joc.2231

U.S. Department of Energy. (2020). EnergyPlus | EnergyPlus. U.S. Dep. Energy.

Wang, Y., & Akbari, H. (2016). The effects of street tree planting on Urban Heat Island mitigation in Montreal. *Sustainable Cities and Society, 27*, 122–128. https://doi.org/10.1016/J.SCS.2016.04.013

Wong, N. H., Chen, Y., Ong, C. L., & Sia, A. (2003). Investigation of thermal benefits of rooftop garden in the tropical environment. *Building and Environment, 38*, 261–270. https://doi.org/10.1016/S0360-1323(02)00066-5

Wong, N. H., Jusuf, S. K., & Tan, C. L. (2011). Integrated urban microclimate assessment method as a sustainable urban development and urban design tool. *Landscape and Urban Planning, 100*, 386–389. https://doi.org/10.1016/J.LANDURBPLAN.2011.02.012

Wright, L., & Davidson, S. (2020). How to tell the difference between a model and a digital twin. *Advanced Modeling and Simulation in Engineering Sciences, 7*, 13. https://doi.org/10.1186/s40323-020-00147-4

Yang, F., & Chen, L. (2020). *High-rise urban form and microclimate* (The urban book series). https://doi.org/10.1007/978-981-15-1714-3

Yang, X., Leung, L. R., Zhao, N., Zhao, C., Qian, Y., Hu, K., Liu, X., & Chen, B. (2017). Contribution of urbanization to the increase of extreme heat events in an urban agglomeration in East China. https://doi.org/10.1002/2017GL074084

Zheng, S., Zhao, L., & Li, Q. (2016). Numerical simulation of the impact of different vegetation species on the outdoor thermal environment. *Urban Forestry & Urban Greening, 18*, 138–150. https://doi.org/10.1016/J.UFUG.2016.05.008

Open Access This chapter is licensed under the terms of the Creative Commons Attribution 4.0 International License (http://creativecommons.org/licenses/by/4.0/), which permits use, sharing, adaptation, distribution and reproduction in any medium or format, as long as you give appropriate credit to the original author(s) and the source, provide a link to the Creative Commons license and indicate if changes were made.

The images or other third party material in this chapter are included in the chapter's Creative Commons license, unless indicated otherwise in a credit line to the material. If material is not included in the chapter's Creative Commons license and your intended use is not permitted by statutory regulation or exceeds the permitted use, you will need to obtain permission directly from the copyright holder.

MUST-B: A Multiagent Model to Address the Future Challenges of Sustainable Urban Development

Seghir Zerguini and Nathalie Gaussier

Abstract This chapter presents the MUST-B model based on a systemic modeling that aims to address some of the future challenges of the smart cities. MUST-B stands for "Integrated Modeling of Land-Use – Transport for application in the Bordeaux agglomeration" and is funded by the Region Nouvelle Aquitaine. MUST-B is based on systemic land-use/transport modeling about how the land and property markets operate, and the interdependent factors for selecting the locations of households and employment. MUST-B is an agent-oriented model which simulates household and job location choices. It is based on an auction mechanism that models competition between agents in the real estate market (existing property holdings, including residential, industrial, and tertiary) and in the land property market (from buildable land reserves) over a given timeframe. This auction procedure is based on maximizing the utility provided to the agent by a given location: housing for a household and an area for business activity for employment premises. Utility is a function of several characteristics relating to the space and premises occupied, such as accessibility, surface area, energy quality of the building, notoriety, agglomeration effects and taxes, or property prices, the latter being endogenous. In this chapter, the mechanisms and functioning of the land property and real estate markets, which prevail in MUST-B, are presented. Methodological choices and behavioral guidelines for agents (households/workplaces) are also set out.

Keywords Choice of location · Households · Employment · Accessibility · Real estate prices · Sustainable urban development · LUTI model · Multiagent simulation · Systemic modeling

S. Zerguini (✉) · N. Gaussier
Université de Bordeaux, CNRS, BSE, UMR 6060, Pessac, France
e-mail: seghir.zerguini@u-bordeaux.fr

© The Author(s) 2024
F. Belaïd, A. Arora (eds.), *Smart Cities*, Studies in Energy, Resource and Environmental Economics, https://doi.org/10.1007/978-3-031-35664-3_11

1 Introduction

Several works have demonstrated that we are constantly pushing the limits of our daily travel, leading to urban sprawl (Orfeuil, 2000). In France, for example, the population living in urban areas is constantly increasing, estimated by the Banque Mondiale to be 80% of the total population today, compared with only 62% in 1960. At equal cost, with increasingly efficient means of transport, infrastructure, systems, and vehicles, it is possible to live further and further from the city center to seek environmental amenities or access a property. Urban sprawl produces an explosion in commuting, which results in increasing congestion of transport infrastructures, air pollution at peak hours, and a dependency on cars (Newman & Kenworthy, 1998).

Within this perspective, many authors question the fragility of populations living in peri-urban areas and whose transport budgets may be very sensitive to changes in energy costs (Crozet & Joly, 2004). At the same time, scarcity is what is generally observed: scarcity of space, energy, and time. This requires an understanding of residential locations and daily household movements from another perspective: that of travel conditions and planning urban space. Households are living in increasingly tight markets, with a constrained budget for housing, transportation, and heating. As an example, the average property prices in Bordeaux quadrupled between 1998 and today, pushing people further out toward the periphery. Controlling spatial urban growth is clearly becoming a requirement for sustainable development and public policies.

To limit greenhouse gas (GHG) emissions and fulfill France's international commitments to fight global warming, the government has implemented a series of plans (low-carbon strategy, multiyear energy programs, etc.) and laws such as the adoption of the Climate, Air, and Territorial Energy Plan. Thus, the law on energy transition for green growth requires all regions with more than 20,000 inhabitants to adopt a Climate, Air, and Territorial Energy Plan by December 31, 2018, at the latest, to be renewed every 6 years. The most critical issues regarding energy transition are found in urban areas, where most human activities (social, economic, etc.) are concentrated and consequently generate nearly two-thirds of GHG emissions. Indeed, urban areas raise the question of the most effective policy level at which global and local problems can be addressed with sustainable city policies. A wide part of the problem is felt at the local level (such as congestion and pollution), and many effects exist that have a global nature (such as global warming). Urban areas may be regarded as efficient starting points for sustainability policies because they operate locally with a global advantage (Camagni et al., 1998). Then, sustainable urban areas need to develop strategic tools of an urban sustainability policy such as smart cities.

On the other hand, local authorities have no tools at their disposal, which are capable of assessing the prospective effects of Climate, Air, and Territorial Energy Plans in their regions. While the current diagnosis of GHG emissions is undertaken in a detailed and precise manner per sector (mobility, building, etc.), future forecasts remain based on single-sector approaches that ignore intersectoral interactions

(Zerguini & Gaussier, 2020). For example, building a tramway line impacts not only GHG emissions linked to mobility (reduction of emissions through the mechanism wherein travel flows are transferred from cars to public transport) but also impacts GHG emissions linked to urban planning (through the mechanism wherein the urban area of the region is made more accessible by the tramway line).

The MUST-B model addresses this double issue of urban growth and sustainability (Bouanan et al., 2018). It aims to address the growing preoccupations of sustainable urban development, focusing specifically on prospective assessment in the field of transport and land use planning. Many studies have thus confronted the complexity of urban dynamics to provide models that integrate the interactions between transport and urbanization, known by the LUTI acronym, which stands for "Land Use – Transport Interaction." While the use of LUTI models has been compulsory in the United States for the past 30 years (the ISTEA Act, 1991 and TEA 21 Act, 1998), some local communities in France (Ile-de-France, Lyon, Lille, Grenoble, Besançon) have begun to explore urbanization scenarios in terms of investments and government policies.

The MUST-B model is designed to aid decision-making (helping to define and develop government policies) to guide the "trajectory of an area toward improved sustainability." Its ambition is to be the simulation model for actions and urban policies that work toward building the city of tomorrow, that is, a smart and sustainable city. According to (Giffinger, 2011), a smart city is a city that performs well in six characteristics (environment, economy, mobility, quality of life, habitat, governance) built on the "smart" combination of endowments and activities of self-decisive, independent, and conscious citizens. The city is a complex system characterized by human–human and human–environment microinteractions; the emergence of unexpected phenomena that arise from the behavior of interdependent units; nonlinear dynamics, which means that it is difficult to predict the output of the system from its inputs; and feedback loops. MUST-B is based on a systemic approach that allows us to take into account all these characteristics of the complexity and functioning of the city system.

It aims to enable the prospective assessment of the impact of its Climate, Air, and Territorial Energy Plan, particularly in the Bordeaux metropolitan area. This is conducted according to actions that can be implemented in the Energy–Climate policy. The MUST-B tool can provide interesting insight to support institutional decision-makers in the Bordeaux area and rise to the challenge of sustainable urbanization.

This chapter first exposes the state of the art of LUTI modeling and the positioning of the MUST-B model in relation to existing LUTI models. It then presents the principles underpinning the development of the MUST-B model and its implementation for a given future time prospect. The third section describes the theoretical and methodological choices made to model the utility function and location mechanisms of households and workplaces, as well as the systemic interactions between households and workplaces. The final part, which presents the implementation of MUST-B in the Bordeaux Urban Area (AUB), highlights the main indicators of the complexity of the urban phenomenon produced by the model.

2   MUST-B: Context and Positioning

There is a wealth of literature on the subject of LUTI models (Wegener, 2004) dating back to the 1950s in the United States, which study the interaction between transport and urban development. It aims to better "open up" the black box of transport – urbanization interactions with the aim of proposing applications in the context of large foreign and French agglomerations. It raises many questions, particularly about the representation and articulation of the systems that make up these models. Wegener (2004) thus identifies approximately 20 models that he compares using an articulated reading grid according to nine characteristics: (1) Their unified or composite structure, developed from hierarchically ordered subsystems, (2) The complete or partial integration of the transport system, (3) The theoretical foundations – auction-based models, expected utility theory, equilibrium, etc., (4) Modeling techniques according to their consideration of space and time, (5) Simulated dynamics, (6) Necessary data, (7) Parameter and validation exercises of the model, (8) Operationality, and (9) Model applicability.

In the MUST-B model, the mechanism of the location choice of the agents (households, firms) is based on the theory of maximizing the utility that this location will provide (Zerguini & Gaussier, 2019). Here, we are interested in the balance between the real estate and the land markets that emerges following the simulation of intra-agent competition (competition between households on residential supply and between workplaces on business premises supply) and interagent competition (competition between households and workplaces on buildable land stock). More generally, MUST-B is founded on a formalized set of principles taking into account the behavior of a large number of urban agents/actors and their mutual interactions: households, businesses, planners, developers, government policies, regulations, etc. To account for the complexity of the urban phenomenon, a multiagent simulation is used wherein most of the mechanisms that govern land use are endogenized, such as how property prices are defined, occupancy of the buildable land stock, and access to jobs and labor.

MUST-B proposes a novel approach in comparison to the numerous previous works on LUTI modeling for the following reasons:

- Its systemic articulation of the land and real estate markets (planner/developer/ occupier – households and workplaces) enables us to consider all interactions between the different urban actors using a countdown mechanism, according to which the land price is deducted from the other costs of an operation and the sale price of real estate.
- The collaboration of multidisciplinary researchers (in the fields of the economy, urban planning, geography, transport, IT, etc.) enables us to consider the specificities of the different disciplinary fields in relation to the urban phenomenon.
- The diversity of the project team (researchers and consultants) enables us to develop a tool that is compatible with the requirements of local and operational authorities to conduct research and provide advice.

- Its multiagent simulation enables us to better understand the complexity of the city system based on individual behaviors. Using computer power, multiagent simulation enables us to model collective behaviors that are not otherwise easily accessible through intuition or analytical calculation (Lemoy et al., 2011).
- It compares theoretical approaches with operational actors (developers, corporations, etc.), enabling us to validate the modeled mechanisms.
- It includes social housing, which represents a quarter of the housing stock in France. The modeling of the choice of location in social housing follows the same approach as for private housing (maximization of residential utility), and the price of social housing is indexed to that of private housing resulting from the auction mechanism.

3   Methodology

3.1   Architecture and Operation of the Model

Households compete with each other in real estate and workplaces to occupy buildable land stock.

MUST-B operates like a "four-stage" transport model, in that, for a given time prospect (Bonnel, 2001), the balance results from the confrontation between supply (transport networks) and demand (flow matrix). Thus, MUST-B can be considered a supply-and-demand model wherein spatial entities (land, housing, business premises) interact with social entities (households, institutions).

- Land occupation by households and workplaces in the different zones of an urban area at a given future time prospect is based on the following process (Fig. 1):

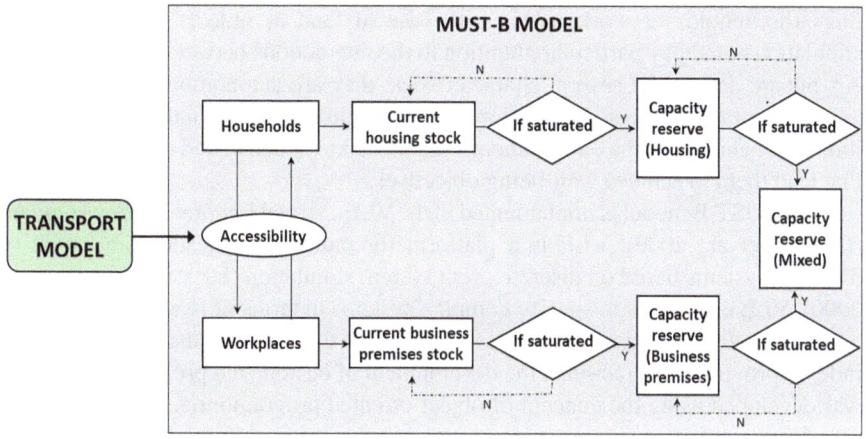

Fig. 1 Architecture and operating principles of MUST-B

- Exogenous demand in the future is defined by distinguishing between households (population) and workplaces (employment).
- At the beginning of the modeling process, agents are arbitrarily assigned to the respective current housing stock, respecting capacity constraints. For this, a procedure is developed in MUST-B, which enables an automatic preassignment of agents in the zones.
- Households will start by occupying the current housing stock. Once this is saturated, they will occupy the buildable land stock dedicated to housing according to the developer's profitability conditions. Deciding the location of workplaces follows the same process.
- When the dedicated buildable land stock is saturated, households and/or workplaces will be located in buildable land stock designated for mixed housing/business premises until this is saturated.

The MUST-B model is thus able to account for urban sprawl or tightness: apart from the current exogenous stock, supply becomes endogenous, capped by the buildable land stock available, under pressure from demand from agents (households and workplaces) and financial profitability for the developer (countdown mechanism presented in Sect. 3.3).

3.2 Formalization and Implementation of the MUST-B Model

Agent-based modeling is defined as a modeling and simulation technique that works on the level of microunits such as workplaces and households. Each microunit contains several attributes and follows a set of behavioral rules. This technique simulates the decision-making processes of individuals based on the heterogeneous attributes of agents and their interactions with the environment and other agents. The agent-based modeling approach has recently emerged and gained popularity within the scientific community regarding urban planning. These models use agent links (households or workplaces) – the use of land as objects of analysis and simulation – and pay particular attention to the interactions between these "agents." Agents are defined by several characteristics: they are autonomous, they share an environment through communication and interaction, and they make decisions that link their behavior to the environment. Agents make inductive and dynamic choices that lead them to achieve well-being objectives.

The MUST-B model is implemented in the VLE - Virtual Laboratory Environment (Quesnel et al., 2009). VLE is a platform for multimodeling and simulation of dynamic systems based on discrete event system simulation (DEVS) (Zeigler et al., 2000). VLE enables us to specify complex systems in terms of reactive objects and agents, simulate the system dynamics, and analyze the results of the simulation. The indexes provided also facilitate the development of customized programs. MUST-B was developed using the concept of object-oriented programming, in particular the C++ language.

3.3   The Notion of Equilibrium in MUST-B

MUST-B locates each randomly selected agent. The modeled territory is assumed to be that of a closed city, with a given future time perspective, in which the total number of agents (population, jobs) is fixed in advance. Thus, a large number of selections (several million) are conducted in a simulation to achieve equilibrium. The equilibrium is derived from the simulated dynamics of household and workplace location choices. Equilibrium is considered to be achieved when agents no longer improve the utility they can derive from a new location. This dynamic urban equilibrium, as opposed to the static equilibrium, which can be calculated in urban models, is similar to the Wardrop equilibrium (Wardrop, 1952; Corea & Stier-Moses, 2011) used in traffic flow allocation models. The Wardrop equilibrium is achieved in the allocation of traffic on a road network when no user can change itinerary without compromising their travel time.

In practice, in the MUST-B model, the aggregate utility of a given type of agent (household or workplace) converges toward a U^* level after a number I^* of iterations (Fig. 2). I^* corresponds to more than four million iterations for households, while workplaces require a higher number of iterations.

It can therefore be considered that after a certain number of iterations I^*, the aggregate utility of households will no longer increase and that no household can increase its utility by being in a given area without at least decreasing that of another. As with Pareto efficiency, there is nothing to say that all households are satisfied with their location. The same applies to business premises.

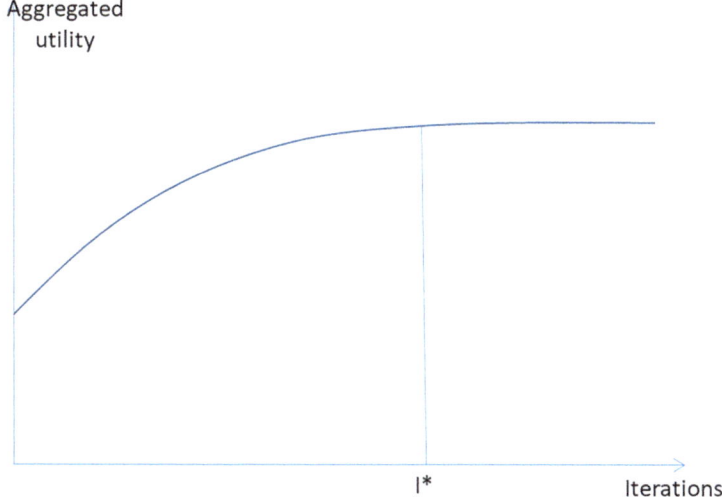

Fig. 2 Evolution of aggregated utility over iterations (MUST-B calculation)

3.4 Theoretical Principles and Modeling

3.4.1 Utility Functions

The proposed approach consists of developing a residential utility function that integrates household behaviors into their residential location choices. This enables households to be assigned to the different residential options according to an auction procedure. The residential utility function reflects the economic well-being that the household can derive from a given location and type of housing. It depends on several parameters relating to housing and space, such as the accessibility and reputation of the zone, the surface area of the housing, or the price of real estate.

The utility function of household h residing in zone z can be expressed as follows:

$$U_{h,z} = \alpha_{1_h} AC_z + \alpha_{2_h} NO_z + \alpha_{3_h} SA_h - EB_z \times SA_h - P_z^l \times SA_h \tag{1}$$

where:

- AC: Accessibility of the zone under consideration (reflecting access to employment)
- NO: Notoriety of the zone under consideration (reflecting the image and amenity of a zone that can be qualified by various parameters such as the atmosphere, style, specificity, and diversity of the businesses that are present)
- SA: Surface area of the desired housing
- EB: Energy bill per m² of the zone in question
- P: Price per m² of housing in the area under consideration (reflecting the energy costs related to the use of the housing such as heating, air conditioning, lighting, etc.)
- α_i: Parameters to be estimated according to the household's socioprofessional category (SPC)

To take the heterogeneity of households within the same category (size × SPC) into account, a statistical distribution is introduced within each category. This distribution concerns the α_i parameters of the residential utility function. The statistical distribution used is the normal distribution of the mean of the value of α_i, and the standard deviation is assumed to be 10% of the mean.

The proposed method for simulating workplace location choices is similar to that of households. It consists of developing a job location function that integrates the behavior of companies in the location choices of their workplaces and enables jobs to be assigned to the premises according to an auction procedure. This function reflects the utility that the company can derive from a given location and type of premises. This depends on several parameters, such as the accessibility and reputation (image) of the zone, the surface area of the premises, property prices, taxes, financial assistance, etc. In the same way, as for a household, the company will seek to acquire the premises that it considers the most useful for its business, taking the size of the workplace into account.

The utility function of workplace w (characterized by its size and its sector of activity) located in zone z can be expressed in this way:

$$U_{w,z} = \left(\lambda_{1_w} \text{AC}_z + \lambda_{2_w} \text{NO}_z + \lambda_{3_w} \text{RW}_z - \text{TD}_z \times \text{SA}_{wa} - P_z^w \times \text{SA}_{wa} \right) \times S_w \qquad (2)$$

where:

- AC: Accessibility of the zone under consideration (i.e., access to labor)
- NO: Notoriety of the zone under consideration (i.e., the image and specificity of a zone)
- SA_{ea}: Surface area of a job per type of business activity
- RW: Ratio of workplaces operating the same business activity as the premises under consideration out of all of the premises present in the zone (i.e., agglomeration effects)
- TD: Level of taxes and duties in the zone under consideration
- P: Price per m² of business premises in the zone under consideration
- S_w: Size of the workplace
- λ_i: Parameters to be estimated according to the business activity of the premises

3.4.2   Location Selection Mechanism

The mechanism for choosing the location of households is the same as that of workplaces: it is based on maximizing the utility of a location for the agent (household/workplace).

The assignment of agents to the different zones that make up the agglomeration is based on an auction mechanism for the acquisition of housing/business premises. The principle is that each agent will locate itself in a given zone, seeking to maximize their utility. The bid made by the agent (candidate wishing to move) is composed of the price of their current zone and the "monetarized" gain of the utility provided by their potential move.

Concretely, at iteration[1] n of the simulation, the bid that agent a will make to move into zone j depends on the price of housing in his home zone i at iteration n − 1 and on the difference in utilities between zones i and j at iteration n − 1. This is expressed as follows:

$$\pi_{j,n}^a = P_{i,n-1} + \varepsilon \left(U_{j,n-1}^a - U_{i,n-1}^a \right) \qquad (3)$$

---

[1]An iteration corresponds to the random selection of an agent subjected to the location selection process.

where ε, the amplitude of the auction, determines the utility gain transformed into a price added to the initial price in their current zone.

With the above mechanism (3), the simulated prices of the different zones will increase as they attract an ever-increasing number of agents. To simulate the inverse mechanism of price decline or stability, it is assumed that the agent can renounce moving if they obtain a reduction in the price of real estate in their home zone i. The auction that the agent makes to remain in their home zone is expressed as (4):

$$\pi_{i,n}^{a} = (1 - \beta) P_{i,n-1} \tag{4}$$

Agent a chooses to be located in the zone where they derive the most utility (Fig. 3). If $U\left(\pi_{j,n}^{a}\right) > U\left(\pi_{i,n}^{a}\right)$, the agent has chosen to be located in zone j; otherwise, they will remain in zone i.

In the case where the utility of going to zone j is greater than that obtained in zone i and zone j is already saturated (zone j has reached its total capacity), then agent a is relocated in any case to zone j, and it is the agent that derives the least utility in zone j, which is relocated to a randomly selected destination, zone k (Fig. 4).

At the end of the location process for the randomly chosen agent assigned to a zone, the zone's price is updated and calibrated with the auction made by the last agent. Thus, at each iteration, there is necessarily a zone that will be subjected to a price modification, upward or downward.

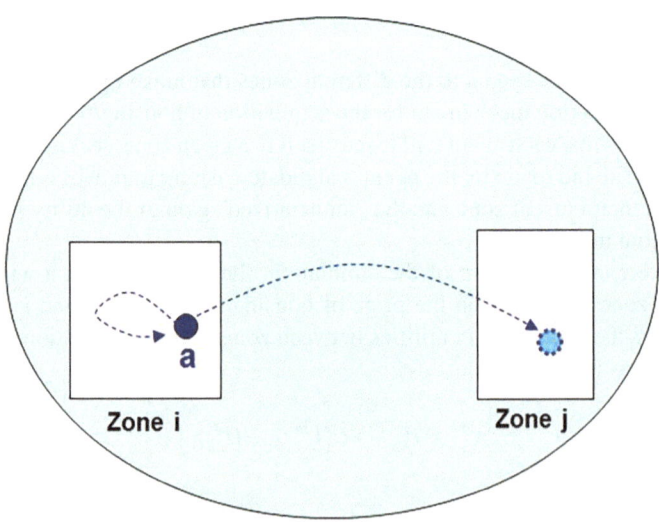

Fig. 3 Arbitrating the location of an agent. (Authors' graphic)

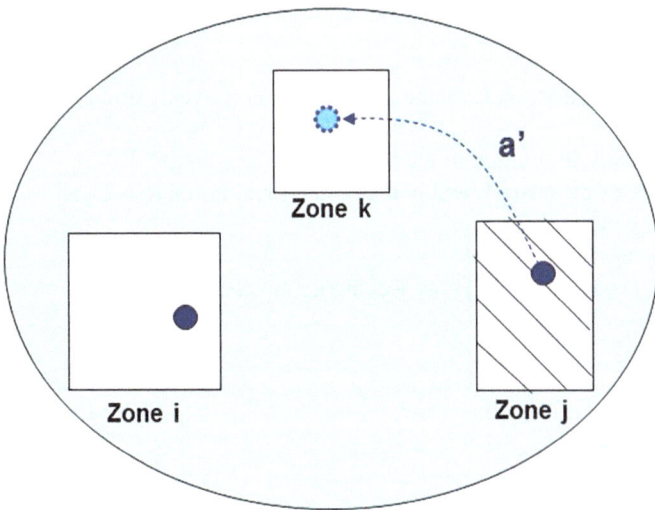

Fig. 4 In the event that zone *j* is saturated. (Authors' graphic)

Equilibrium is attained when, for a given type of agent, the level of utility is the same wherever they are located. The agent cannot improve its utility by changing the zone.

3.4.3 Procedure for the Endogenous Offer (Developer)

For a future time prospect, the MUST-B model locates homes and workplaces in a zone, having knowledge of their current respective land stocks. Once each of these current land stocks is saturated, they will begin allocating to the buildable land stock (land that can be built on in accordance with regulations). Consumption of the buildable land stock is activated by the countdown mechanism:, i.e., the price of land is deducted from all other costs of a real estate transaction. When developers wish to know the maximum price available to buy land and initiate a real estate program, they will deduct from the expected sales figure the costs of construction, financial costs, fees, taxes, and margin (Vilmin, 2015). The only item that ultimately determines the decision to invest and the profitability of the project is the land, given that the margin conditions the bank's ability to obtain loans and guarantees. The difference between expected revenues and expenses, therefore, corresponds to the maximum land charge that the developer can incur. The endogenous capacity mechanism will therefore be activated with the confrontation between property prices and land charges.

Decisional Investment Mechanism

Therefore, for 1 m²:

π: Price of the auction made by the agent at iteration n (in relation to the transferable surface)

C_L: Cost of land (in relation to the buildable surface area)

p: Weighting of the cost of land in the price of real estate ($p = C_L/\pi$)

C_C: Cost of construction

M: Developer's margin

C_P: Cost of production (land cost + construction cost + margin)

We obtain:

$$C_P = C_L + C_C + M \tag{5}$$

$$C_P = p\pi + C_C + xC_P \tag{6}$$

We can deduce that:

$$C_P = \frac{p\pi + C_C}{1 - x} \tag{7}$$

The developer's profitability condition is expressed as follows:

$$\pi > C_R \tag{8}$$

We finally obtain:

$$\pi > \frac{C_C}{1 - x - p} \tag{9}$$

As illustrated in Fig. 5, the mechanism for occupying the buildable land stock of a zone is conditioned by a double constraint: the saturation of the current land stock and the expected profitability for the developer to build in this zone (Eq. 9).

From Fig. 5, we can identify four configurations:

1. Between the beginning of the simulation and iteration I1, the agents are located in the current zone.
2. Between iteration I1 and iteration I2, the agents are located in the buildable land stock because, on the one hand, the current land stock under consideration is saturated and, on the other hand, the price of the property enables the developer to gain a profit (auction is higher than cost price).
3. Between iteration I2 and iteration I3, the zone is still considered saturated, but the buildable land stock is not mobilized because the developer's condition of profitability is not fulfilled within this interval.
4. Between iteration I3 and iteration I4, the agents are located in the buildable land stock as the profitability condition is fulfilled once again.
5. From iteration I4, the zone is considered definitively saturated (the capacities of the current zone and buildable land stock being full).

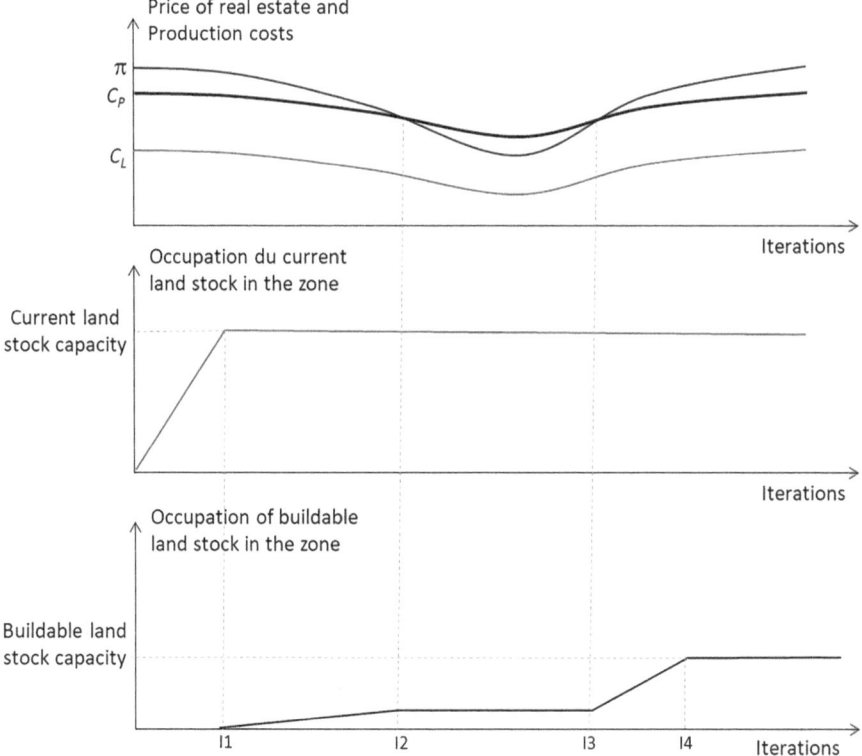

Fig. 5 Occupation of buildable land stock. (Authors graphic)

Process of Spatial Occupation of the Buildable Land Stock
Thus:

- IS: the inhabitable surface area of the real estate sought by the agent (real estate demand)
- BF: Building footprint
- GS: Ground surface area of the land used
- NF_{max}: Maximal number of floors allowed in the zone
- k_1: add-on factor ($k_1 > 1$) of the occupied surface area taking external walls of the building into account
- k_2: add-on factor ($k_2 > 1$) taking into account networks (road, water, sanitation, street lighting, etc.) and urban planning easements (view, right of way, etc.)

The building footprint satisfying real estate demand IS is expressed as:

$$BF = k_1 \times \frac{IS}{NF_{max}} \qquad (10)$$

Fig. 6 Process for the
occupation of buildable
land stock

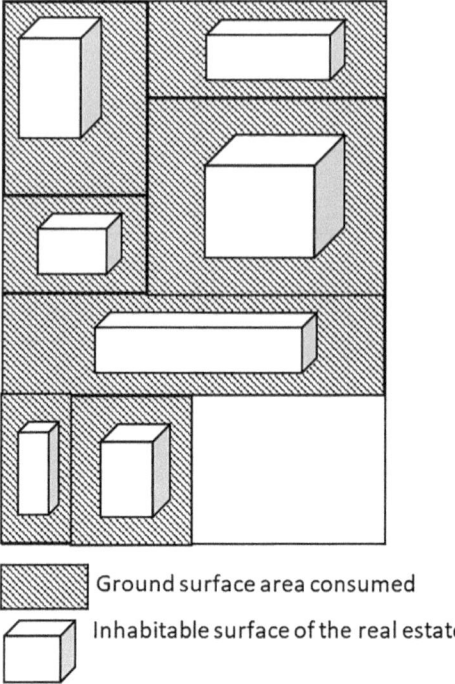

Ground surface area consumed

Inhabitable surface of the real estate

The ground surface area of land consumed by this real estate demand is equal to:

$$GS = k_2 \times BF \tag{11}$$

We finally obtain:

$$GS = \frac{k_1 \times k_2}{NF_{max}} \times IS \tag{12}$$

Each time an agent (household or workplace) is located in this zone, the ground
surface area consumed by this demand is subtracted from the buildable land stock
surface until the total consumption of the buildable land stock surface area of the
zone in question is attained (see Fig. 6).

3.4.4 Mechanism of Endogenous Accessibility

Households and premises interact via the accessibility variable (see Fig. 1). Indeed,
household accessibility must take into account not only transport supply but also the
location of workplaces (employment opportunities). On the other hand, the
accessibility of workplaces must take into account transport supply and household
location (labor force). There is then feedback between the two types of agents in the
system (Fig. 7).

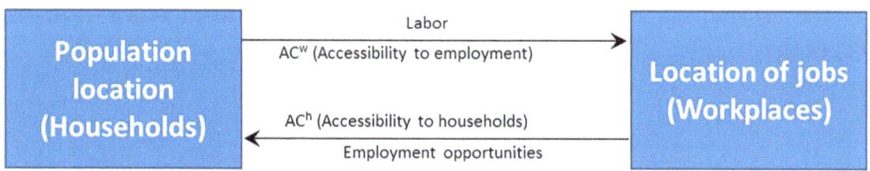

Fig. 7 Interactions between households and workplaces

The accessibility of a zone consists of two components: an exogenous component that reflects the performance of the transport network (speed and capacity) serving the zone in question and an endogenous component that reflects the spatial distribution of the population and jobs that evolve during the simulation.

At this stage of development of MUST-B, two simplifying hypotheses are made on how to take accessibility into account:

- The accessibility of households is reduced solely to access jobs, while there are other dimensions such as access to services, equipment, shops, schools, etc. These elements, which are present in the utility function of households, do not contribute to the calculation of endogenous accessibility.
- It is assumed that accessibility linked to the performance of transport networks is completely exogenous, whereas part of this component could be endogenized by taking transport network congestion resulting from the attractiveness of certain zones at certain times into account during the simulation.

The determination of accessibility for both types of agents (households and workplaces) is based on the following approach:

- Determination of the matrix of generalized costs of interzone travel for PV (private vehicle) and PT (public transport) modes of travel
- Determination of accessibility vectors for PV and PT modes of transport
- Aggregation of all modes into a single accessibility vector

Generalized Cost of Travel
The generalized cost of traveling by private vehicle (PV) between zones i and j is expressed as follows:

$$C_{ij}^{PV} = V_t \times tt_{ij}^{PV} + \left(CF + CK\right) \times d_{ij}^{PV} + C_j^{Park} \qquad (13)$$

where:

V_t: Value of time (€/h)
tt_{ij}^{PV} : Travel time via PV between zones i and j (h)
d_{ij}^{PV} : Distance covered in PV between zones i and j (km)
CF: Cost per kilometer of fuel (€/km)
CK: Cost per kilometer of vehicle use excluding fuel (€/km)
C_j^{Park} : Cost of parking in zone j (€)

The generalized cost of using public transport (PT) between zones i and j is expressed as

$$C_{ij}^{PT} = V_t \times tt_{ij}^{PT} + T^{PT} + \eta \times C_{ij}^{MC} \qquad (14)$$

where:

tt_{ij}^{PT} : Journey time on public transport between zones i and j (h)
T^{PT}: Tariff per trip on public transport
N_{ij}^{MC} : Number of mode changes during the trip
η: Parameter to be estimated

Accessibility per Mode of Transport
The accessibility provided for household h in zone i at iteration n, composed of an exogenous part (first term of the equation) and an endogenous part (second term of the equation), is expressed as:

$$AC_{i,n}^{h,PV} = \theta \sum_j e^{-C_{ij}^{PV}} + \mu \sum_j \sum_k \frac{s^k w_{j,n-1}^k}{\left(C_{ij}^{PV}\right)^2} \qquad (15)$$

$$AC_{i,n}^{h,PT} = \theta \sum_j e^{-C_{ij}^{PT}} + \mu \sum_j \sum_k \frac{s^k w_{j,n-1}^k}{\left(C_{ij}^{PT}\right)^2} \qquad (16)$$

The accessibility provided for workplaces w in zone i at iteration n, composed of an exogenous and an endogenous part, is expressed as:

$$AC_{i,n}^{w,PV} = \theta' \sum_j e^{-C_{ij}^{PV}} + \mu' \sum_j \sum_k \frac{s^l h_{j,n-1}^l}{\left(C_{ij}^{PV}\right)^2} \qquad (17)$$

$$AC_{i,n}^{w,PT} = \theta' \sum_j e^{-C_{ij}^{PT}} + \mu' \sum_j \sum_k \frac{s^l h_{j,n-1}^l}{\left(C_{ij}^{PT}\right)^2} \qquad (18)$$

where:

s^l: Size of household (number of people)
$h_{i,n-1}^l$: Number of t^l sized households in zone i at iteration $n-1$
t^k: Size of the workplace (number of jobs)
$w_{j,n-1}^k$: Number of workplaces of t^k size in zone j at iteration $n-1$
θ, θ', μ and μ': Factors to estimate

Aggregated Accessibility
To estimate the all-mode accessibility of a given zone, the accessibility of each mode is weighted by its modal share (MS) according to the following expressions:

$$AC_{i,n+1}^h = PM_{i,n}^{PV} \times AC_{i,n}^{h,PV} + PM_{i,n}^{PT} \times AC_{i,n}^{h,PT} \qquad (19)$$

$$AC_{i,n+1}^w = PM_{i,n}^{PV} \times AC_{i,n}^{w,PV} + PM_{i,n}^{PT} \times AC_{i,n}^{w,PT} \qquad (20)$$

Modal shares are estimated during the simulation using a sequential approach: generation, distribution, and modal choice. The generation and distribution steps are based on a gravity model and the location of households and workplaces. The modal choice step is based on a logit model and the generalized costs of the two competing modes (PV and PT).

4   MUST-B: Indicators of the Complexity of the Urban Phenomenon

The simulation of a governmental policy can be undertaken using several indicators that are determined in the MUST-B model.

4.1   Indicators Linked to Urban Planning

The occupation of the space following a simulation is characterized by defining the following for each zone:

- Number of accommodations/households/population
- Number of workplaces/jobs
- Social diversity and segregation
- Functional diversity
- Price of residential property
- Price of land property for business activities
- Price of land property

These indicators are output data calculated directly in the MUST-B model.

4.2   Sustainability Indicators

Following a simulation, the effects of an urban policy on sustainable development are estimated according to a definition for each zone in terms of:

- Energy consumption (home-work mobility, accommodation)
- GHG emissions (home-work mobility, accommodation)
- Artificialization of the land (built land/total surface of the zone ratio)

This last indicator can be directly calculated, whereas the two former indicators require prior calculations, such as the number of kilometers in question or the surface areas of the accommodation occupied according to the degree of emission and energy consumption.

5 Conclusion

The aim of the MUST-B model is to simulate the choice of location of households and firms. It is based on an auction mechanism that enables us, according to different timeframes, to model the competition between agents (households and firms) in the real estate market (existing property holdings: residential, tertiary, and industrial) and on the land market (reserves of buildable land stock).

The multidisciplinary approach to the design and development of MUST-B is a unique and defining characteristic wherein the collaboration of multidisciplinary researchers (from the fields of economics, urban planning, geography, transport, computer science, etc.) The specificities of the different disciplines could be considered in combination with the urban phenomenon, and the multiagent simulation enabled us to model the complexity of the city system from individual behaviors.

Today, MUST-B is an operational tool to support decision-making for urban development planning, integrating the effects of the different sectors of the city (mobility, accommodation, economic activities, etc.) on the climate and atmospheric pollution. Indeed, MUST-B is designed to assist decision-making (helping to define and develop government policies) to guide the trajectory of a region toward increased sustainability.

Simulation using MUST-B, therefore, enables us to assess certain effects of urban policies and actions that communities can undertake, as well as assist them in prioritizing these actions according to their financial resources and objectives and to articulate them in coherent local policies in favor of ecological transition. The MUST-B model can thus significantly contribute to identifying the conditions and levers for action that help to lead to a sustainable territory. For instance, MUST-B is a way to address the foresight exercise at the urban area scale: it makes it possible to test the spatial impact of the development of road, residential, and firm activities. It enlightens different aspects of urban sustainability, such as economic, social, and environmental criteria. It is a way to have a smart perspective on urban planning.

However, some limitations and paths of progression can be noted in the MUST-B model. The first concerns the calibration of the parameters and the validation of the model in relation to the real data and observations of the territory under consideration. For the model to be valid, it must reconstruct for each zone the location of households and jobs as well as the price of residential real estate and that of economic activities. However, the calibration of the model depends on the territories and on their capacity to keep information in the long term to reconstitute it within the framework of the simulations. The second progression of the model consists of integrating the evolution of the behavior of the agents in the mechanism of choice of location of new variables such as, for example, "connectivity to the Internet" in the utility function of households and companies. Indeed, since the crisis induced by the COVID-19 pandemic and the strengthening of telework, high-speed internet connections have become a decisive criterion in the attractiveness of territories for households and firms.

Acknowledgments The authors would like to thank all of the researchers involved in the MUST-B project for their remarks and observations throughout the discussions that took place: Anne Bretagnolle from Géographie-cités (Paris 1 University/Paris 7 University/CNRS), Sonia Guelton from Lab'Urba (Paris-Est University), Moez Kilani from LEM (University of Lille/CNRS), Nicolas Coulombel from LVMT (Paris-Est University), Laurent Guimas from Explain Consultancy and Ouassim Manout from ForCity.

We also thank Région Nouvelle Aquitaine for their co-financing of the STRATEGIE project and the MSHA for the financial management of the project and use of their premises.

References

Bonnel, P. (2001). *Prévision de la demande de transport*. Rapport HDR, Université Lumière Lyon 2, 409 p.

Bouanan, Y., Zerguini, S., & Gaussier, N. (2018). Agent-based modeling of urban land-use development: Modeling and simulating households and economic activities location choice. *International Journal of Service and Computing Oriented Manufacturing (IJSCOM), 3*(4), 253.

Camagni, R., Capello, R., & Nijkamp, P. (1998). Sustainable city policy: Economic, environmental, technological. *Ecological Economics, 24*(1), 103–118.

Corea, J. R., & Stier-Moses, N. E. (2011). Wardrop equilibria. In J. J. Cochran (Ed.), *Encyclopedia of operations research and management science* (12 p). Wiley.

Crozet, Y., & Joly, I. (2004). Budgets temps de transport: les sociétés tertiaires confrontées à la gestion paradoxale du "bien le plus rare". *les Cahiers Scientifiques du Transport, 45*, 27–48.

Giffinger, R. (2011). *European smart cities: The need for a place related understanding*. Department of Spatial Development, Infrastructure and Environmental Planning of Vienna University of Technology.

Lemoy, R., Raux, C., & Jensen, P. (2011). ILOT: un modèle multiagents de structuration sociale de la ville. In *Modéliser la ville – Formes urbaines et politiques de transports* (pp. 333–364). Economica.

Newman, P. W. G., & Kenworthy, J. R. (1998). *Overcoming automobile dependence*. Island Press.

Orfeuil, J. P. (2000). La mobilité locale: toujours plus loin et plus vite. In M. Bonnet & D. Desjeux (Eds.), *Les territoires de la mobilité*. PUF.

Quesnel, G., Duboz, R., & Ramat, É. (2009). The Virtual Laboratory Environment–an operational framework for multimodeling, simulation and analysis of complex dynamical systems. *Simulation Modelling Practice and Theory, 17*(4), 641–653.

Vilmin, T. (2015). *L'aménagement urbain. Acteurs et système* (141 p). Editions Parenthèses.

Wardrop, J. G. (1952). Some theoretical aspects of road traffic research. *Proceedings of the Institution of Civil Engineers, Part II, 1*, 325–378.

Wegener, M. (2004). Overview of land use transport models. In *Handbook of transport geography and spatial systems* (pp. 127–146). Emerald Group Publishing Limited.

Zeigler, B. P., Praehofer, H., & Kim, T. G. (2000). *Theory of modeling and simulation: Integrating discrete event and continuous complex dynamic systems*. Academic Press.

Zerguini, S., & Gaussier, N. (2019). MUST-B an agent-based model of urban land-use development: Simulating "households" and "firms" location choices in urban area of Bordeaux. In *45th Annual conference – Eastern Economic Association*, February 28 – March 3, 2019, New York, USA.

Zerguini, S., & Gaussier, N. (2020). MUST-B: Un modèle LUTI multidisciplinaire au service de la complexité du phénomène urbain. *Canadian Journal of Regional Science/Revue canadienne des sciences régionales, 43*(2), 50–59.

Open Access This chapter is licensed under the terms of the Creative Commons Attribution 4.0 International License (http://creativecommons.org/licenses/by/4.0/), which permits use, sharing, adaptation, distribution and reproduction in any medium or format, as long as you give appropriate credit to the original author(s) and the source, provide a link to the Creative Commons license and indicate if changes were made.

The images or other third party material in this chapter are included in the chapter's Creative Commons license, unless indicated otherwise in a credit line to the material. If material is not included in the chapter's Creative Commons license and your intended use is not permitted by statutory regulation or exceeds the permitted use, you will need to obtain permission directly from the copyright holder.

A Systematic Literature Review on Station Area Integrating Micromobility in Europe: A Twenty-First Century Transit-Oriented Development

Dylan Moinse

Abstract The increasing popularity of the bicycle, coupled with the emerging new micromobility solutions, such as personal electric micro-vehicles or sharing systems, calls for renewed attention to the smart urban and transport planning strategy advocated by the conventional Transit-Oriented Development (TOD) model. These personal and shared mobility devices constitute an opportunity to enhance accessibility to the public transport network, leading to a TOD vision revisited by the contribution of individual light modes. Given the relatively recent and extensive documentation related to micromobility and public transport integration, a systematic literature review was undertaken to reflect the state of research literature on the redefinition of the TOD perimeter by micromobility, with a focus on the European context. Nineteen of the 3955 articles recorded met the inclusion criteria specified in the methodology protocol. The analyzed papers clearly highlight TOD boundaries extended to about 3 km, suggesting the redistribution of variables toward residential and cycling-friendly areas beyond the first walking kilometer. This chapter uncovers gaps in existing academic literature, with the near absence of Eastern and Southern Europe case studies, innovative micromobility options such as private or shared electric bikes and scooters, impact assessment on neighborhoods, and the application of qualitative research methods.

Keywords Catchment area · Micromobility · Systematic literature review · Transit-oriented development

---

The original version of this chapter was revised. The correction to this chapter is available at https://doi.org/10.1007/978-3-031-35664-3_20

D. Moinse (✉)
LVMT, Université Gustave Eiffel, IFSTTAR, École des Ponts, Marne-la-Vallée, Paris, France
e-mail: dylan.moinse@univ-eiffel.fr

© The Author(s) 2024, Corrected Publication 2024
F. Belaïd, A. Arora (eds.), *Smart Cities*, Studies in Energy, Resource and Environmental Economics, https://doi.org/10.1007/978-3-031-35664-3_12

1 Introduction

1.1 Research Questions

The aim of this chapter is to reflect the state of research literature on the redefinition of the TOD perimeter through the contribution of micromobility. Which forms take the TOD model when looking at extended station areas? What distance ranges are covered by this modal synergy? Do TOD characteristics evolve? Are there similarities and distinctions according to local contexts?

To collect, analyze, and compare all the studies on this subject, a systematic literature review of peer-reviewed papers was conducted. Only English-written studies that address accessibility around European stations with the use of micromobility were included. This chapter presents the current state of knowledge by identifying international scientific articles that address the relevant geographical perimeter of station areas for micromobility use. The aim is to provide a spatial overview of these intermodal practices on a European TOD, although this work does not pretend to be exhaustive.

In this way, this chapter is organized as follows: Section 2 describes the methodological approach. Section 3 discusses the results. Section 4 considers avenues for future research on a TOD integrating micromobility.

1.2 Conceptualizing Renewed Smart Growth Regions

Since the 1990s, the urban and transport planning strategy of transit-oriented development (TOD) has become the dominant urban growth planning paradigm, particularly in the United States (Papa & Bertolini, 2015, p. 70). Closely connected with Smart Growth and New Urbanism movements, TOD, as formalized, promotes urban development along with public transport (PT) corridors as a tool and a target for mitigating uncontrolled urban sprawl and achieving more sustainable regions and smart transport networks. These urban approaches represent an invitation to revisit the suburban, car-oriented, and segregated way of life (Jamme et al., 2019, p. 411).

While TOD is not a recent concept, drawing inspiration from Garden Cities for the development around railway nodes, the challenge of adapting it to the auto-oriented metropolis is novel (Dittmar & Ohland, 2012, p. 5). European and Asian planners consider TOD to be an innovative instrument but not an invention (Jamme et al., 2019, p. 411), although the origin of the label is commonly attributed to Calthorpe (1993). This emerging vision of smart territories facing auto-oriented systems is seen as delivering multiple benefits, such as improving quality of life, creating attractive places, enhancing transit, walking, and cycling, while supporting polycentric regions and economic growth (Papa & Bertolini, 2015, p. 77). Internationally formulated TOD has the potential to fundamentally rethink multi-scale communities (Dittmar & Ohland, 2012, p. 19), even if most theories on how to design successful TOD projects have been developed in the United States (Pojani & Stead, 2015, p. 131).

The transit-oriented development (TOD) urban model is designed mainly to encourage the use of public transport and create a pedestrian-friendly urban environment to reduce vehicular traffic congestion (Nasri & Zhang, 2014, p. 172). TOD's fundamental principles are based on the formulation of the 3Ds (Density, Diversity, and Design) built environment hypothesis measured by Cervero and Kockelman (1997, p. 199) to reduce vehicle miles traveled (VMT) and encourage transit use.

Additional D-requirements also complete the list of factors coordinating and integrating planning and transportation issues, including the 6Ds with 3Ds variables supplemented by distance to transit, destination accessibility, and demand management (Ewing & Cervero, 2010, p. 274). As a critical dimension in the TOD equation, the "Distance to Transit" variable is crucial to measure the accessibility of a transit station and to make decisions on land development. "Destination accessibility" is based on the logic of providing greater mobility by moving people around the city more swiftly, while "demand management" can be considered actions that are implemented at specific sites or strategies that are implemented at an area-wide level (Ogra & Ndebele, 2014, p. 541).

One-half mile has become the accepted distance for gauging a transit station's catchment area in the United States (Guerra et al., 2012, p. 2), being a standard for planning TODs. The scope of TOD in most countries corresponds to a "pedestrian pocket" around a transit stop (Calthorpe, 1993, p. 44), with a radius between 400 and 800 m by walk or bike (Cervero et al., 2004, p. 238). The conventional TOD approach is based on the development of a dense commercial and employment "Primary Area" in proximity to a station (600 m), followed by residential zones with densities gradually decreasing (Calthorpe, 1993, pp. 43–87). However, this standardized rule is likely to evolve and change the model by revisiting the stations' service area.

1.3   An Extension of the Walking Bull's Eye

Academic works focusing on the revision of TOD principles seem to reflect two trends. The first one considers that the range of access and egress walking exceeds the established spatial limit of 500–800 m (Pojani & Stead, 2015, p. 133). The second is interested in mobility feeder solutions inviting the integration of first- and last-mile connections to public transport (Park et al., 2021, p. 38). Both approaches converge on a common critique of this distance threshold and suggest a broader view of TOD neighborhoods to include a more diverse share of communities. Calthorpe (1993, p. 60) refers originally to the surrounding environment as the "Secondary Area" (Fig. 1). According to him, lower density and auto-oriented urban systems characterize this 1.6-km adjacent perimeter (Banai, 1998; Ibraeva et al., 2020).

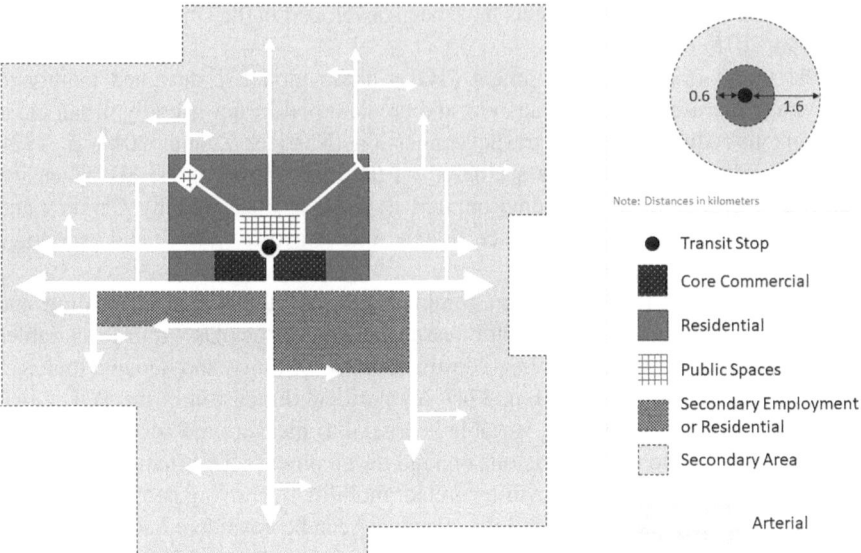

Note: Distances in kilometers

●   Transit Stop

■   Core Commercial

■   Residential

⊞   Public Spaces

     Secondary Employment
     or Residential

     Secondary Area

              Arterial

Fig. 1 Spatial configuration of TOD at the station level. (Redesigned from Calthorpe (1993, p. 60). Source: Author elaboration)

The first trend observed on the pedestrian side has been extensively studied in the scientific literature in connection with the idea of "bursting the [pedestrian TOD] bubble" (Canepa, 2007, p. 34). A variety of studies and reports have shown the increased reach of walking around high frequency hubs surrounded by attractive facilities, and the boundaries of a TOD district do not have to be confined to 1 km (Ker & Ginn, 2003, p. 79). Through a review of existing scientific works, L'Hostis (2016, p. 5) shows that the arbitrary idea of the 0.5 mile is not a relevant limit for getting to the stations and therefore requires extending the analysis beyond this boundary.

In contrast, the research body lacks, as far as the author knows, a discussion of the role of sustainable feeder modes in the redefinition of a TOD model. In fact, many studies dealing with the close links between cycling and transit are emerging but generally do not explicitly mention the urban planning vision of an extended TOD perimeter. Among the modes analyzed that require a new understanding of TOD are bikes (Lee et al., 2016, p. 979), buses, and personal rapid transit (Schneider, 1992, p. 151), as well as shared and digitized mobility solutions (Knowles et al., 2020, p. 7).

This trip chain perspective is leading to the development of emerging concepts, such as the "Extended-TOD" (E-TOD) and "Bicycle-based TOD" (B-TOD), which complete the guiding philosophy of TOD. The two derivatives of the urban model attempt to provide access to stations and to places further away than the conventional distance of walking by means of the development of a secondary mobility system offering intermediate capacity. E-TOD assumes a Feeder-Distributor-Circulator network (Schneider, 2012) similar to Personal Rapid Transit (PRT), whose deployment strategy has evolved (Schneider, 1992, p. 152). Part of the common thought of an extension of the TOD, B-TOD, focuses on the contribution of individual light modes and, in particular, of the bicycle. In combination with the re-evaluation of walking accessibility to and from a public transport station, secondary modes that complement the public transport network can also play a role in the TOD model.

Feeder services to mass transit are superimposed on the pedestrian bubble, significantly widening the catchment area of the stations and thereby increasing potential ridership. In addition, new mobility solutions are emerging to enhance the attractiveness of public transport systems and thus improve the accessibility of stations. The renewed popularity of the bicycle in Europe (Héran, 2015, p. 142), rising development of bike and e-scooter schemes, and the rapid adoption of emerging Personal Mobility Devices (PMDs), such as the standing scooter, play a positive role in public transport patronage (Kostrzewska & Macikowski, 2017, p. 2). This research focuses on this category of devices qualified as micromobility, with the aim of linking two approaches: the association of micromobility and public transport in connection with TOD areas.

2 Materials and Methods

2.1 Study Selection Procedure

Given the relatively recent and extensive existing documentation related to the integration of micromobility and public transport supporting the TOD, a systematic literature review (SLR) was undertaken to reflect recent trends toward new personal and shared mobility embodied in the micromobility system. A review earns the adjective "systematic" when the questions are clearly formulated, studies are relevantly identified, quality is appraised and finally, methodology is summarized. According to Transportation Research Board of the National Academies (2015, p. 2), the benefits of an (1) SLR are to uncover a solution to a problem; (2) identify concurrent or previous work on the same topic; (3) validate a particular method; (4) provide a focus for investigations; and (5) confirm that further research is needed.

The aim of this study is to determine the state of knowledge about the range of TOD catchment areas extended by the use of micromobility on a European scale to identify similarities, differences, and research gaps. In a reflexive way, the point is to apprehend a renewal or a redefinition of the TOD with regard to the emergence of microvehicles supplementing the accessibility stations. This main objective is based on four questions underlying the systematic literature review:

1. Which approaches and parameters are used to assess the accessibility of station neighborhoods regarding micromobility?
2. Do academic studies on the size of station areas address the principles of TOD?
3. Are the estimated micromobility ranges close to a common threshold across European cases?

The choice to focus on an enlarged vision of B-TOD to analyze the literature on the integration of private and shared microvehicles with transit is then explained by their increasing mode split in Europe, which raises interesting issues in the making of sustainable and smart territories. This choice is also in line with the design guidelines made by Calthorpe (1993, pp. 54–60) regarding the promotion of "walk-and-ride" and "bike-and-ride" with bicycle connections to transit stops rather than "park-and-ride." Indeed, the bicycle has the potential to expand the TOD boundaries from 4 to 25 times (Cottrell, 2007, p. 118) for less physical energy (Sebban, 2003, p. 50), with stations being connected to considerably more households, substantially increasing the potential number of users, services, and facilities (Jonkeren & Kager, 2021, p. 455). In parallel, this association is presented as the most efficient intermodal integration with walk-and-ride (Yang et al., 2013, p. 714), environmentally friendly (Cervero, 2001, p. 17), and fair (Cervero et al., 2013, p. 85), three objectives put forward by TOD.

A geographical lens is also applied to this research, with the deliberate selection of European case studies to exclude results from other parts of the world that might differ, as shown by the relative range of the walk. This geographic scope is motivated by the intention to supplement the systematic literature review on bike-and-ride published by Oeschger et al. (2020), which provides a solid and extensive overview of the subject. This chapter's contribution is therefore to update studies available since the publication of this scientific article, while giving a more specific insight into this region: among the 48 articles analyzed by the authors, 14 have a European country or city as a study area, representing one-third of the sample.

Methods for reviewing research systematically are still emerging, and there is much ongoing development and debate. This SLR method was developed following the thematic synthesis described Thomas and Harden (2008) and adapted by the Transportation Research Board of the National Academies (2015) in the mobility area, as systematic reviews differ between disciplines (Padeiro et al., 2019, p. 738). Thus, the review protocol selected is derived from TOD and micromobility SLR published by Bozzi and Aguilera (2021, p. 4); Neilson et al. (2019, p. 36); Oeschger et al. (2020, p. 4); Padeiro et al. (2019, p. 738); Pritchard (2018, p. 2); Şengül and Mostofi (2021, p. 2). The selection procedure under the SLR involves four main steps (Jain et al., 2020, p. 2544):

1. Keywords and omission criteria
2. Reading titles and omission criteria
3. Reading abstracts and metadata
4. Reading full papers and snowballing

2.2   Search Strategy and Data Sources

Relevant academic papers were identified with the Scopus™ and Web of Science™ (WoS) search databases using an expression based on three categories: Public Transport (1) + Micromobility (2) + TOD Perimeter (3), detailed in Table 1. The

Table 1 SEARCH A: Keyword phrase for the systematic search

Category	Keywords and Boolean operators
Public transport	ALL = ("Transit-Oriented Development" OR "Public Transport" OR "Transit" OR "Train" OR "Rail" OR "Metro" OR "Tram" OR "Bus" OR "Rail")
	AND
Micromobility	ALL = ("Micromobility" OR "Bicycle" OR "Bike" OR "Bike-and-Ride" OR "Cycling" OR "Scooter" OR "Device" OR "Transfer Mode" OR "Intermodal")
	AND
Geographical area	ALL = ("Catchment" OR "Isochrone" OR "Buffer" OR "Service Coverage" OR "Shed" OR "Station Area" OR "Access Distance" OR "Last-Mile" OR "Perimeter")

Source: Author elaboration

Boolean operators AND and OR were used to select only papers containing the three required keyword categories, depending on term variations. Each category's terms are synonyms and have been incorporated within the phrase, requiring at least one word from each category for the article to appear in the search. The choice of synonyms was made possible and enriched with the help of a preliminary reading period, which made it possible to highlight the most important keywords.

Consultation of these electronic databases relies on different filters, starting with the publication period that has been widened to cover all collections, with English as the written language, and by type of resource, defined as scientific articles, proceedings, and book chapters available on these two search portals. The online search was executed and saved on December 3, 2021 for both Scopus™ and WoS, resulting in 2920 entries.

In addition to these two academic search engines, search strategy was drawn from the top-rated academic journals according to their H-index, an author-level metric that measures both the productivity and citation impact of the publications. To this end, search B relies on the classification drawn up by the SCImago Journal Rank™ (SJR) website in 2020, using the top 10 journals in "Transport," on the one hand, and in "Urban Studies," on the other (Table 2). This second direct source of data reflects an approach adopted by Padeiro et al. (2019, p. 738), who manually searched issues through a selection of journals from among the top 40 rankings. The electronic database was collected on February 8, 2022 and has 1354 scientific papers.

2.3 Inclusion and Exclusion Criteria

Once the preliminary list of papers has been obtained and the duplicates have been removed, i.e., altogether 3955 unique articles were identified, and the next step was to create an all-inclusive search with preset criteria. Studies were considered eligible if they satisfied a set of inclusion criteria (Fig. 2):

1. Academic paper is printed in English

Table 2 SEARCH B: List of refereed journals queried in transportation and urban planning

Rank	ID	Journal
		TOP 10 Transportation Journals (2020)
1	SMG^{T1}	*Tourism Management*
2	SMG^{T2}	*Transportation Research, Series B: Methodological*
3	SMG^{T3}	*Transportation Research, Part A: Policy and Practice*
4	SMG^{T4}	*Transportation Research Part C: Emerging Technologies*
5	SMG^{T5}	*Transportation Research, Part E: Logistics and Transportation Review*
6	SMG^{T6}	*Journal of Transport Geography*
7	SMG^{T7}	*Transportation Research, Part D: Transport and Environment*
8	SMG^{T8}	*Transport Policy*
9	SMG^{T9}	*Transportation Research Part F: Traffic Psychology and Behavior*
10	SMG^{T10}	*Journal of Air Transport Management*
		Top 10 Urban Planning Journals (2020)
1	SMG^{U1}	*Urban Studies*
2	SMG^{U2}	*International Journal of Urban and Regional Research*
3	SMG^{U3}	*Journal of Urban Economics*
4	SMG^{U4}	*Journal of the American Planning Association*
5	SMG^{U5}	*Computers, Environment and Urban Systems*
6	SMG^{U6}	*Journal of Urban Health*
7	SMG^{U7}	*Cities*
8	SMG^{U8}	*Environment and Planning B: Urban Analytics and City Science*
9	SMG^{U9}	*Regional Science and Urban Economics*
10	SMG^{U10}	*Habitat International*

Source: SCImago Journal Rank™, 2020

Inclusive Criteria:

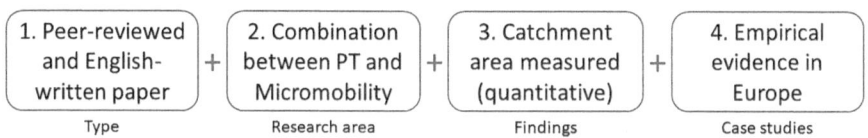

1. Peer-reviewed and English-written paper	+	2. Combination between PT and Micromobility	+	3. Catchment area measured (quantitative)	+	4. Empirical evidence in Europe
Type		Research area		Findings		Case studies

Fig. 2 Conditions for inclusion in the research protocol. (Source: Author elaboration)

2. The manuscript aims to investigate intermodality, and more specifically the combination of one or more types of micromobility with public transport
3. The paper production includes distance measures in relation to the influence area of the studied station(s)
4. The research field is in Europe, regardless of the scale of analysis.

The selection is made by reading the abstracts and contents of each document, excluding those that do not meet one of the four established criteria. For each scientific paper, preliminary relevance was determined by reading the title and abstract. The systematic selection was accompanied by the reason why the article displayed was listed or rejected.

More specifically, 1777 of the 3955 articles recorded were discarded because they did not address urban or transport-related issues. A total of 1599 of them did not address intermodality, while 334 articles addressing intermodality were rejected because they were not concerned with micromobility. Two articles could not be included in the final sample due to the unavailability of the full content. As a result, 243 articles were identified as discussing micromobility in combination with public transport (Fig. 3).

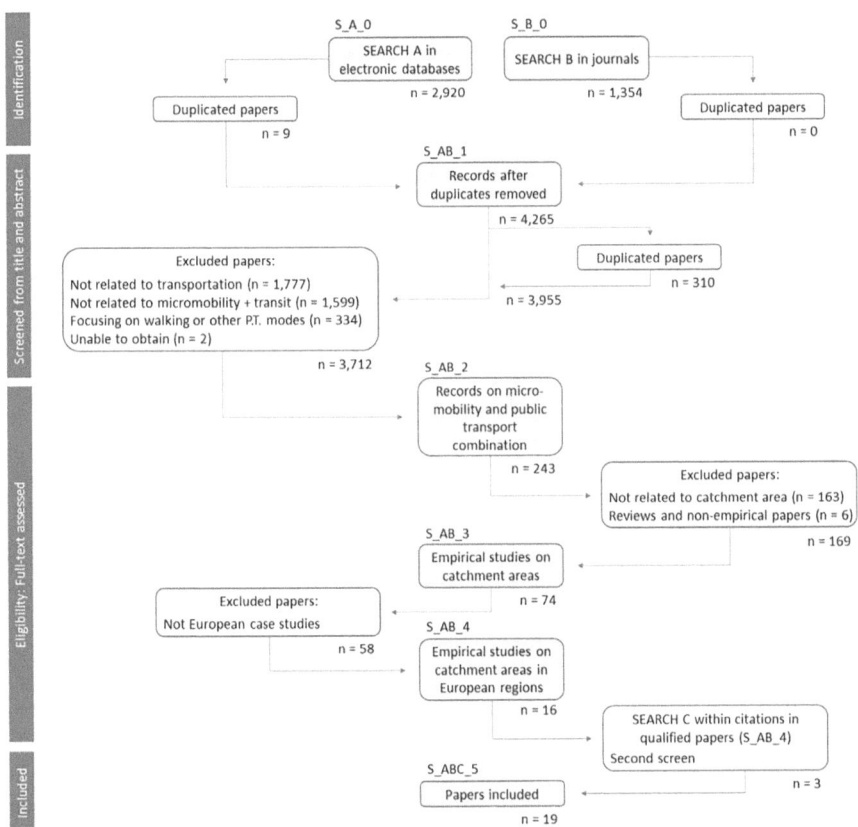

Fig. 3 Flow diagram representing the selection and review process. (Source: Author elaboration)

2.4   Verification and Snowballing Stage

Following the initial screening, the record's contents were examined to control and evaluate the selection process (Neilson et al., 2019, p. 36). Some 169 of them were removed, as they did not rely on distance measures. Finally, 58 of the 74 empirical articles related to transit catchment area by micromobility were not included, as they covered other regions. The selection process via the reading of abstracts revealed 16 scientific articles, as shown in Fig. 3.

At the same time, the references cited by papers and Google Scholar™ search were checked by the same reviewer to verify whether any other articles address the subject (Jain et al., 2020, p. 2545). To make sure to have not missed works, the snowballing phase was further supplemented by making use of the Connected Papers™ science mapping software tool. Overall, 3 scientific articles fulfilling the 4 criteria above were added, resulting in a final body size of 19 articles.

2.5   Aspects Considered

This systematic literature review will pay attention to the dimensions characterizing TOD in conjunction with the intermodal approach. Similar to Bertolini et al. (2012, p. 31), this work aims to review the factors driving revisited TOD areas in Europe as follows:

• Type of Integration
• Case Study Areas
• Methods
• D-variables: density, diversity, design, distance to transit, destination accessibility, and demand management

3   Results and Discussion

3.1   Research Publications on Micromobility and Transit-Oriented Development

By searching on Clarivate Analytics' WoS platform, it can be noted that the number of scientific publications has grown significantly since 2010 for "Transit-Oriented Development" and more recently since 2019 for "Micromobility" (Fig. 4).

In addition to the chart, a bibliometric search over the same period with the terms "Micromobility," "Bicycle," and "Bike" shows a similar trend starting in 2010. Consequently, there seems to be a significant and growing interest in these two subjects related to mobility and urban planning, as Jain et al. (2020, p. 2545) also observed on TOD based on the Google Trends platform. Figure 4 reveals the

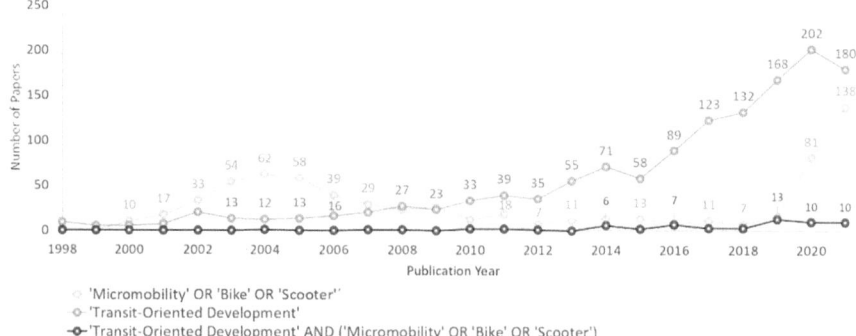

Fig. 4 Year-wise publication of bibliometric papers related to micromobility and/or TOD (March 16, 2022). (Source: Web of Science™ database)

gradual but still modest emergence of scientific studies across these subjects, with up to 7–13 papers published each year since 2019. The features of this recent pattern drive this chapter.

When looking at the geographical distribution of papers by author location, it clearly appears that one-third of the works on "Micromobility" since 1998 originate from Europe and North America each and one-quarter from Asia (Fig. 5a). France (9%), England (8%), Italy (7%), and Germany (6%) are the main contributors in Europe (36%); the United States (26%) and Canada (7%) in North America (33%); and South Korea (10%) and China (8%) in Asia (28%). This relatively wide distribution of a range of countries becomes more limited when focusing on "transit-oriented development." Research is highly specialized in North America (39%), particularly in the United States (34%) and Asia (38%), with China (24%) in the lead (Fig. 5b). This is further strengthened when crossing the two concepts, with most of the work coming from the United States (55%) and China (32%). Nevertheless, it should be noted that works from European countries account for 15% of this emerging literature. It should be mentioned that developed countries are at the top of the rankings, although the query is biased by English-only keywords (Fig. 5c).

3.2 Current State of International Studies on Cycling and Transit Coordination

As explained in Fig. 3, 74 articles were identified in the international literature on the subject of micromobility and PT integration, with specific consideration and measurement of the size of station areas. Taking a closer examination of this document collection, it can be observed that most articles come from Asia (38/74) and particularly from China (28/74), followed by the United States (18/74) and the Netherlands (10/74). These preliminary results are in line with the trend analyzed by browsing both keywords covered by this chapter: the three continents that

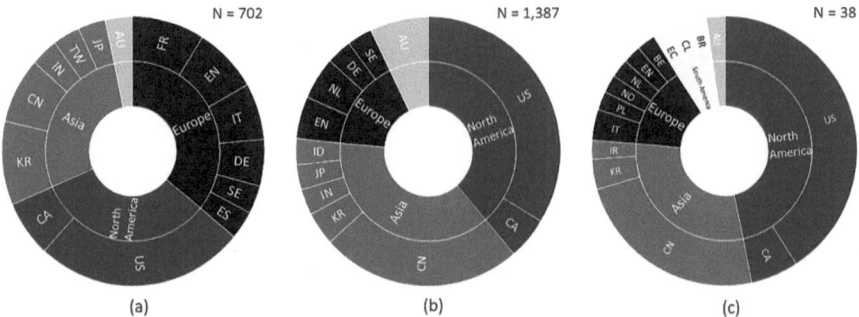

Fig. 5 Evolution of scientific production on micromobility and/or TOD (March 16, 2022). (**a**) "Micromobility," (**b**) "Transit-oriented development," (**c**) "Micromobility" and "transit-oriented development." (Source: Author calculation using Web of Science™ database)

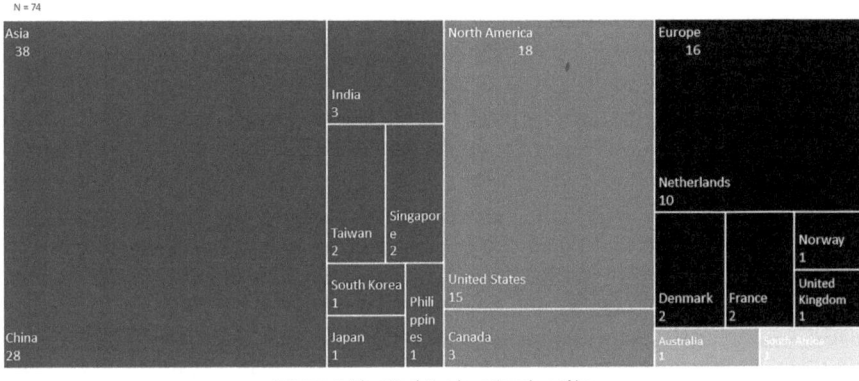

Fig. 6 Geographical distribution of international references addressing combined transit and micromobility. (Source: Author elaboration)

contribute most strongly to the knowledge regarding station area by micromobility can be found in Fig. 6.

By distinguishing the different types of micromobility investigated by this international database, private and nonelectric bicycles occupy a prominent place (47/74), which can be notably attributed to this mode's seniority and global availability. Accordingly, the first bar shows all the regions identified in the body (Fig. 7). Interestingly, shared micromobility embodies 27 of the 74 available items, being split between Asia-led dockless bike systems (13/74) and public bike-sharing systems (11/74). The electric scooter springs with two papers dealing exclusively with free-floating services, while one article focuses on pedicabs.[1]

---

[1] Pedicabs are a small pedal-operated vehicle serving as a taxi.

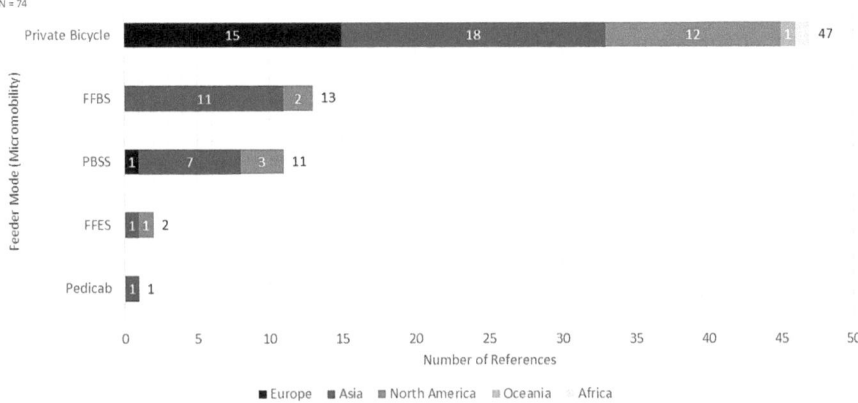

Fig. 7 Studied micromobility types across international references on intermodality. Note: FFBS: free-floating bike service; PBSS: public bike-share system; FFES: free-floating e-scooter service. (Source: Author elaboration)

Regarding the 16 European papers in the sample, 14 address a personal microvehicle: the bicycle. As a result, a literature gap may be detected with respect to the integration of shared micromobility schemes with public transport in relation to the scale of relevance around stations. Globally, no papers relating to the access distance of personal mobility devices (PMDs), such as standing scooters, to public transport are reported in the academic sphere. It should be noted that this corpus of documents does not include articles subsequently integrated through the snowball method. Thus, the geographical distribution of the literature indicated a modest emergence on the European side, with a hegemony of the Netherlands, which will have to be analyzed in depth in the next section.

3.3   Description of European Studies

3.3.1   Type of Integration Recorded

With further attention to the 16 European articles coupled with the 3 papers listed afterward, it is noteworthy that 17 out of 19 are oriented toward a private mode and particularly toward conventional cycling (16/19), as shown above. More specifically, this includes the combination of the personal bicycle and the train, which is highlighted in 13 works. Urban public transport is involved in 8 out of 19 papers, bus (5) and metro (4) in particular (Table 3). Only two publications deal with public bike-sharing systems (Adnan et al., 2019; Böcker et al., 2020), whereas a single one discusses electric bikes (Midenet et al., 2018) and another about electric standing scooters (Moinse et al., 2022). All four articles were published less than 4 years ago due to the recent spread of these microvehicles and the more difficult access to data.

Table 3 Academic articles included in the systematic literature review

ID	Authors (publication year)	System	Type of Integration
SLR[1]	Adnan et al. (2019)	Shared	PBSS+Rail
SLR[2]	Böcker et al. (2020)	Shared	PBSS+Metro/Rail
SLR[3]	Brand et al. (2017)	Private	Bike+Bus/BRT
SLR[4]	Debrezion et al. (2009)	Private	Bike+Rail
SLR[5]	Djurhuus et al. (2016)	Private	Bike+PT
SLR[6]	Geurs et al. (2016)	Private	Bike+Rail
SLR[7]	Hasiak (2019)	Private	Bike+Rail
SLR[8]	Heinen and Bohte (2014)	Private	Bike+PT
SLR[9]	Kager et al. (2016)	Private	Bike+Rail
SLR[10]	Keijer and Rietveld (2000)	Private	Bike+Rail
SLR[11]	Martens (2004)	Private	Bike+PT
SLR[12]	Midenet et al. (2018)	Private	(E-)Bike+Rail
SLR[13]	Moinse et al. (2022)	Private	(E-)Scooter+Rail
SLR[14]	Nielsen and Skov-Petersen (2018)	Private	Bike+Rail/Bus
SLR[15]	Nigro et al. (2019)	Private	Bike+Rail
SLR[16]	Rietveld (2000)	Private	Bike+Rail
SLR[17]	Rijsman et al. (2019)	Private	Bike+Tram
SLR[18]	Sherwin et al. (2011)	Private	Bike+Rail
SLR[19]	Ton et al. (2020)	Private	Bike+Tram

Note: Only microvehicles are considered in the Type of Integration category. "PBSS" stands for Public Bike-Sharing System, "BRT" for Bus Rapid Transit, and "PT" comprises Railway, Metro, Tramway, Bus, or/and Ferry

3.3.2 Case Studies and Publication Periods

Within the literature corpus, there is not only a variety of geographical scales and types of urban forms among the case studies but also a clear clustering in Northern Europe.

Five publications refer to metropolitan scales such as Oslo (Böcker et al., 2020), Amsterdam (Brand et al., 2017), the Capital Region of Denmark (Djurhuus et al., 2016), The Hague (Ton et al., 2020), and Delft, Zwolle, Midden-Delfland, and Pijnacker-Nootdorp (Heinen & Bohte, 2014). Four articles are set in urban areas and regions such as Belgian cities (Adnan et al., 2019), Randstad South (Geurs et al., 2016), medium-sized train stations in Provence-Alpes-Côte d'Azur (Moinse et al., 2022), and Campania Region (Nigro et al., 2019). For their part, three works focus on medium-sized areas as in French local cities (Hasiak, 2019), Amboise (Midenet et al., 2018), and Bristol (Sherwin et al., 2011). Table 4 further displays six scientific studies over 20 covering a national scale as described in the Netherlands (Debrezion et al., 2009; Kager et al., 2016; Keijer & Rietveld, 2000; Rietveld, 2000), with Germany and the United Kingdom (Martens, 2004), and Denmark (Nielsen & Skov-Petersen, 2018). The geographical distribution of the studied areas among the selected items clearly shows a spatial concentration around Western and Northern Europe in Fig. 8.

Table 4 Case studies of selected papers

Authors (Publication year)	Country	Geographical area
Adnan et al. (2019)	Belgium	30 small and medium cities
Böcker et al. (2020)	Norway	Oslo
Brand et al. (2017)	The Netherlands	Amsterdam
Debrezion et al. (2009)	The Netherlands	–
Djurhuus et al. (2016)	Denmark	Copenhagen
Geurs et al. (2016)	The Netherlands	Randstad South (Rotterdam-The Hague)
Hasiak (2019)	France	Local stations
Heinen and Bohte (2014)	The Netherlands	Delft, Zwolle, and surroundings
Kager et al. (2016)	The Netherlands	–
Keijer and Rietveld (2000)	The Netherlands	–
Martens (2004)	Germany, the Netherlands, and the United Kingdom	–
Midenet et al. (2018)	France	Amboise
Moinse et al. (2022)	France	Provence-Alpes-Côte d'Azur region
Nielsen and Skov-Petersen (2018)	Denmark	–
Nigro et al. (2019)	Italy	Campania region
Rietveld (2000)	The Netherlands	–
Rijsman et al. (2019)	The Netherlands	The Hague
Sherwin et al. (2011)	The United Kingdom	Bristol
Ton et al. (2020)	The Netherlands	The Hague

Note: Only microvehicles are considered in the Type of Integration category. "PBSS" stands for Public Bike-Sharing System, "BRT" for Bus Rapid Transit, and "PT" comprises Railway, Metro, Tramway, Bus, or/and Ferry

Beyond the geographical level, Table 4 draws a chronological scale, with a median set at the year 2018. While some publications are dated between 2000 and 2014 (Debrezion et al., 2009; Heinen & Bohte, 2014; Keijer & Rietveld, 2000; Martens, 2004; Rietveld, 2000; Sherwin et al., 2011), 13 of the papers were published from 2016 onward and at least three quarters of them after 2019, supporting the hypothesis of a very recent literature.

3.3.3 Research Methods

This SLR analyzed specific methods used among the 19 scientific works to study micromobility and public transport integration (Table 5). As identified by Oeschger et al. (2020, p. 8), the most common and the easiest method is the conduct and/or analysis of surveys to gain insight into practices and users' characteristics and

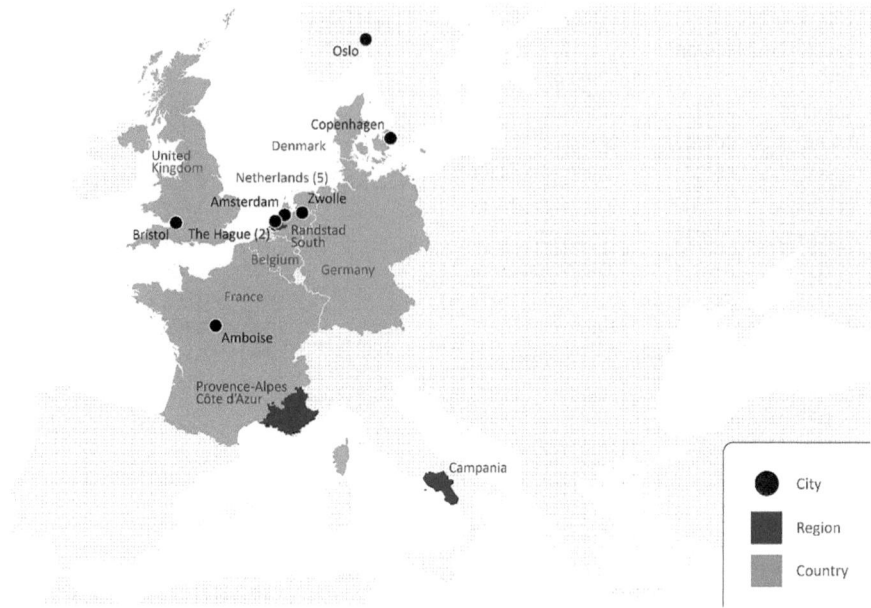

Fig. 8 Geographical distribution of areas investigated by the reviewed papers. (Source: Author elaboration)

Table 5 Methods used

Authors (publication year)	Methodology
Adnan et al. (2019)	Survey-based Hybrid Choice Model
Böcker et al. (2020)	Big Data with GIS-based Analysis
Brand et al. (2017)	Transit Model of VENOM
Debrezion et al. (2009)	Railway Station Quality Index (RSQI)
Djurhuus et al. (2016)	Multimodal Network Model and GIS Analysis
Geurs et al. (2016)	Multimodal Network Model and GIS Analysis
Hasiak (2019)	Survey's Analysis
Heinen and Bohte (2014)	Survey-based Analysis of Variance (ANOVA)
Kager et al. (2016)	System Perspective and Conceptual Analysis
Keijer and Rietveld (2000)	Survey-based Modal Choice
Martens (2004)	Surveys' Analysis
Midenet et al. (2018)	Access Mode Model
Moinse et al. (2022)	Survey's GIS-based Analysis
Nielsen and Skov-Petersen (2018)	Survey's GIS-based Analysis
Nigro et al. (2019)	Node-Place Model and GIS-based Analysis Piet
Rietveld (2000)	Survey's Analysis
Rijsman et al. (2019)	Survey-based Discrete Choice Models
Sherwin et al. (2011)	Monitoring of flow and Survey's Analysis
Ton et al. (2020)	Survey-based Discrete Choice and Access Mode Model

Note: GIS for a geographic information system

preferences. Various authors have appropriated national or regional survey results with an initial broader focus on mobility issues. The challenge lies in the identification of the relevant sample of research subjects. Keijer and Rietveld (2000, p. 231) and Rietveld (2000) undertook a statistical analysis based on the Dutch National Travel Survey from 1994. In the same way, Hasiak (2019, p. 12) analyzed 25 household travel surveys (HTS) carried out on different French urban areas and regions between 2009 and 2016 by using econometric methods, while Moinse et al. (2022, p. 10) analyzed the French national public rail network manager SNCF Réseau's survey data. Nielsen and Skov-Petersen (2018, p. 38) realized a regression analysis of the urban structure and cycling trips based on the Danish National Travel Survey. Martens (2004, p. 282) captures a multitude of multiscale surveys and papers from Germany, the Netherlands, and the United Kingdom but faces challenges due to the collection data methods focusing on the primary transport mode, which neglects access and egress modes.

For their part, Sherwin et al. (2011, p. 192) conducted a mixed methodological approach in Bristol by undertaking counts in addition to surveys with 135 bike-rail uses in October 2007. Correspondingly, Heinen and Bohte (2014, p. 113) collected and examined data from a 2008 Internet survey in Delft, Zwolle, Midden-Delfland, and Pijnacker-Nootdorp by sending 3500 email invitations to employees and 22,000 others to addresses held by local authorities. A multitude of academic papers led to a series of stated preference (SP) surveys, i.e., how respondents might behave in new situations. To study shared bikes in 30 small- and medium-sized Belgian cities, Adnan et al. (2019, p. 10) used a web-based stated preference questionnaire to provide nine hypothetical scenarios for the last mile of a rail trip. Rijsman et al. (2019, p. 1) and Ton et al. (2020, p. 827) completed a regression model by collecting data with an on-board transit revealed preference approach among tramway travelers in The Hague. In the same way, Geurs et al. (2016, p. 7) investigated the effects of bike–train integration policies by conducting an SP survey, subsequently implemented in a multimodal network model, extended from the Dutch National Transport Model (NVM).

Several scientific studies have applied accessibility indicators with multimodal models to analyze TOD and micromobility integration. Midenet et al. (2018, p. 11) analyzed the station's catchment area using time-indexed modal potential through an access mode spatial model based on a central zone close to the Amboise stations and an external zone devoted to motorized modes. Djurhuus et al. (2016, p. 6) proposed a multimodal network model from the Danish Geodata Agency combined with a spatial analysis in Copenhagen. From the 2011 Italian National Census, Nigro et al. (2019, p. 116) determined the extent of catchment areas by using the interconnectivity ratio and GIS analysis through the application of network distance analysis. A railway quality index based on the Dutch National Railway Company was estimated by Debrezion et al. (2009, p. 17) with a GIS analysis, while a VENOM transit model from the regional model of Stadsregio Amsterdam was used by Brand et al. (2017, p. 3) to assess the characteristics of access and egress modes with bus services.

Böcker et al. (2020, p. 395) used big data by analyzing the 2016–2017 records of 4.4 million trips by the Oslo public bike service, screening trips that start or end within 200 m of a PT station, and conducting multivariate modeling techniques.

3.4 Review of Distances Measured in Europe

The catchment area of a transit station designates the geographic area where residents are likely to access and egress the station to use public transport services (Hochmair, 2015, p. 15). Through the 19 analyzed papers, micromobility distances measured in the first or last trip chain leg tend to converge toward a cycling service area, in particular for railway stations, between 1 km minimum (3.14 km^2) and 3 (28.27 km^2) to 4 km (50.27 km^2) maximum (see Fig. 9). The overall range of these microvehicles has the capacity to widen TOD limits from 9 to 16 times the 1-km pedestrian pocket. In the Netherlands, cycling connects approximately 15 times as many people to main intercity stations and 4 times more people to local stations compared to walking (Kager et al., 2016, p. 213). Moreover, walking distance is found to be weighted 2.1 times more negatively than cycling distance to access tramways and bus stops due to the additional physical effort required for a lower speed (Ton et al., 2020, p. 832).

Distance intervals estimating the cycling subarea are computed by nine reviewed articles. Djurhuus et al. (2016, pp. 13–16) chose to study all PT stops in Copenhagen within a 1-km walking distance and within a 3-km cycling distance from home to cover cycling trips to stops. The same observations in Denmark are given by Nielsen and Skov-Petersen (2018, p. 42); the 3-km distance band reflects a convenient

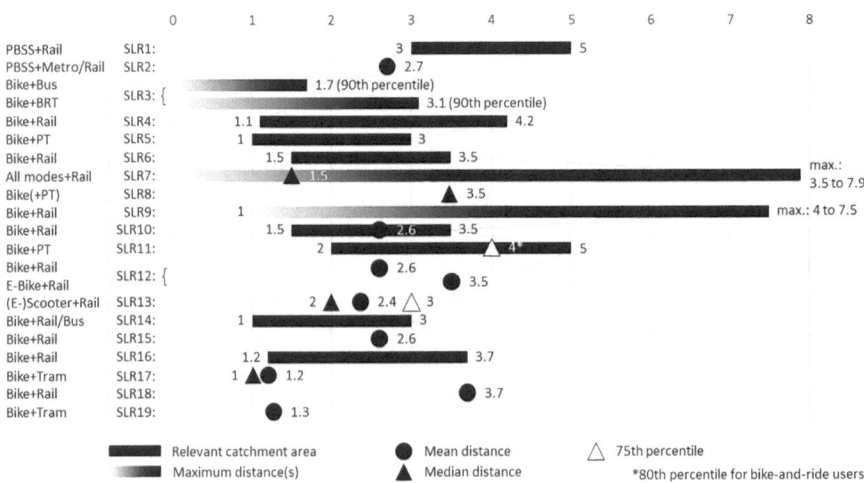

Fig. 9 Distances and influence areas in relation to public transport stations from the 19 reviewed articles. (Source: Author elaboration)

cycling range around a train station, whereas it is less competitive than walking and PT within 1 km. Debrezion et al. (2009, p. 23) note that the bicycle is the most likely access mode choice for distances between 1.1 and 4.2 km but that this maximum distance will decrease to 3.6 km if the frequency of urban public transport doubles in the Netherlands. Similarly, Keijer and Rietveld (2000, p. 234) determine that most people take the bike when the train station is situated between 1.5 and 3.5 km in the Netherlands, passengers opting for urban PT above this length. Furthermore, the rail patronage propensity remains relatively stable as long as the catchment area does not exceed 3.5 km for residential areas (Geurs et al., 2016, p. 9; Keijer & Rietveld, 2000, p. 233; Rietveld, 2000, p. 73). Rietveld (2000, p. 73) identifies an asymmetry for rail between access and egress trips by distinguishing home-end segments where the bike is dominant between 1.2 and 3.7 km and activity-end segments where bike is not attractive when compared to walking (up to 2.2 km) and urban transit. On average, cycling becomes more attractive than walking for distances of 1.31 km or more to The Hague tramway stations (Ton et al., 2020, p. 832). In Germany, the Netherlands, and the United Kingdom, Martens (2004, pp. 282–289) reports 2- to 5-km access trips by bike to PT stations, with longer distances (from 4 to 5 km) for faster collective modes and a 2-km threshold seen for the metro. These outcomes are in line with Adnan et al.'s (2019, p. 9) works, who note that the public-bike share is better for distances between 3 and 5 km in Belgium.

Four papers look at the maximum size of the catchment area delineation by estimating the 85th percentile key factor (Li et al., 2022, p. 14). In Germany, the Netherlands, and the United Kingdom, 20% of bike-and-bus users, mainly students, cycle more than 4 km (Martens, 2004, pp. 282–289), while 10% of cyclists travel more than 1.70 km by Comfornet[2] bus and 3.1 km by R-Net[3] BRT in Amsterdam (Brand et al., 2017, p. 5). In some medium-sized rail stations in a French region, 25% of scooter-and-rail users cycle more than 3 km (Moinse et al., 2022, p. 21).

In general, the access distance to train stations covered by bike is 2.6 km long for a 12-minute isochrone area in the Campania region (Nigro et al., 2019, p. 120). Cycle trips to and from Bristol railway stations have an average length of 3.7 km (Sherwin et al., 2011, p. 192). Similarly, passengers combining trains and standing scooters have an access range of 2.4 km, while half of these trips are 2 km long (Moinse et al., 2022, p. 21). On average, passengers accessing or egressing stations in Amboise by conventional bicycle would travel 2.5 km, while e-bike users would make trips of 3.5 km in a 2025 scenario (Midenet et al., 2018, p. 29). The mean cycling feeder distance in The Hague is fixed at 1.2 km (a 1-km median) for the tramway network (Rijsman et al., 2019, p. 4).

---

[2] "Comfornet [correspond to] conventional bus lines, which are feeder lines to other, hierarchically higher, modes of transportation" (Brand et al., 2017, p. 752).

[3] "R-Net [correspond to] high quality, high speed bus services in the region around Amsterdam, connecting different cities, towns and villages in Amsterdam region, and connecting these areas with the city of Amsterdam" (Brand et al., 2017, p. 752).

3.5   Review of TOD Aspects Studied in Europe

To examine the integration of the commonly recognized TOD-defining attributes, the six main criteria of the urban model were investigated in the analysis of the 19 selected articles. The purpose is to identify and assess the recurrence and relevance of some TOD dimensions and, conversely, specific literature gaps. Table 6 summarizes the incorporation of these variables based on two single options: "+" indicates the presence of the "D" criterion, while "−" suggests an explicit absence of this criterion as interpreted by the author. The following analysis attempts to outline the other five Ds, the Distance to Transit D being discussed earlier.

3.5.1   Density

Density is studied in seven articles within the database, despite being a key component of TOD station development. This can be explained in particular by the common area of study relating to cycling modes supposed to meet the first- and last-kilometer issue, whereas increasing the density of a larger perimeter seems more complex.

Table 6   Consideration of the conventional TOD 6Ds regarding micromobility and PT mix

Authors	Density	Diversity	Design	Destination	Demand	Distance
Adnan et al. (2019)	−	−	+	−	+	+
Böcker et al. (2020)	+	+	−	−	−	+
Brand et al. (2017)	−	−	+	−	+	+
Debrezion et al. (2009)	−	−	+	−	+	+
Djurhuus et al. (2016)	−	−	−	+	−	+
Geurs et al. (2016)	−	−	+	+	+	+
Hasiak (2019)	+	−	+	−	+	+
Heinen and Bohte (2014)	−	−	−	+	−	+
Kager et al. (2016)	+	−	−	+	−	+
Keijer and Rietveld (2000)	−	+	+	−	−	+
Martens (2004)	+	−	+	−	+	+
Midenet et al. (2018)	−	−	+	−	+	+
Moinse et al. (2022)	+	−	+	+	−	+
Nielsen and Skov-Petersen (2018)	−	−	+	+	−	+
Nigro et al. (2019)	+	−	+	−	−	+
Rietveld (2000)	−	+	−	+	+	+
Rijsman et al. (2019)	−	−	+	−	+	+
Sherwin et al. (2011)	+	−	+	−	−	+
Ton et al. (2020)	−	−	+	−	−	+
Total of "+"	7	3	14	7	9	19

Note: All articles in this table score the "Distance to Transit" heading since the empirical measurement of the catchment area for transit stations was one of the inclusion criteria for this systematic review

Martens (2004, p. 290) argue that bike-and-ride practices are significantly influenced by the location of transit stations in Germany, the Netherlands, and the United Kingdom by describing higher proportions in suburban neighborhoods. TOD in low-density areas needs to consider multiple feeder transports other than walking (Nigro et al., 2019, p. 119). Some authors address low-density locations in which improving the quality of urban PT services is challenging and for which micromobility is a suitable option. Sherwin et al. (2011, p. 191) focus on the Bristol Parkway rail station, located in a context of modern medium-density residential development and low-density office and retail development, where many access and egress trips exceed the walking range. Given the 1.5-km median access distance to rail stations, Hasiak (2019, p. 27) calls for reconsidering the primacy of car use to the benefit of walking and cycling in some French spread-out areas. Similarly, Böcker et al. (2020, p. 397) note that public bike-sharing near rail and metro correlates highly with lower job and population densities around the unconnected location in Oslo, suggesting that intermodal users move to low-density areas.

Other authors consider this intermodal approach to have effects on the densification of TOD areas and reciprocally. The synergy provided by this trip chain increases the urban densities of trip origin and destination locations, giving rise to a positive and reciprocal relationship between densities and proximity (Kager et al., 2016, p. 217). Moinse et al. (2022, p. 17) locate e-scooter access trips to French train stations in relatively dense places, whereas bike-and-ride trips are characterized from less dense areas, suggesting that the use of e-scooters in dense neighborhoods may be explained by car parking and traffic constraints. From radar diagrams, Nigro et al. (2019, p. 119) point out the possibility of urban intensification, as long as it is accompanied simultaneously by improvement of accessibility by either train or feeder transport. It should be noted that these articles on urban density all deal with railway stations, revealing a lack of knowledge on the role of density in the combination of urban PT and micromobility.

3.5.2 Diversity

Only three articles were identified as contributing to the urban mixed-use factor. To encourage the use of intermodal trips, it is essential to coordinate urban policies with respect to land use and transportation policies (Keijer & Rietveld, 2000, p. 234). Keijer and Rietveld (2000, p. 234) find a well-documented asymmetry in feeder mode use between access and egress, which reveals that cycling is mainly used during the first segment of the trip chain due to the location of the home in the Netherlands. They recommend priority for the construction of travel-intensive activities near railway stations, such as offices, education, cultural, and shopping facilities. Rietveld (2000, p. 75) supports this planning strategy, stating that residential areas can extend up to 3.5 km around a railway station to match the access range of the bike. This is consistent with Böcker et al. (2020, p. 397) results regarding the Oslo bike-share system, who report that users' routes serve higher building use diversity areas, particularly at the destination.

3.5.3 Design

Unsurprisingly, the third D associated with design emerges from the analysis, with 14 papers on this aspect linked to planning and infrastructures promoting cycling and ride. Three types of design-related development stand out: continuous, pleasant, and safe cycle routes; quality micromobility parking; and at the same time, restrictions on car use.

Nielsen and Skov-Petersen (2018, p. 41) note a robust effect of bike paths within an estimated 1 km of the residence, with the built environment close to home being the most important for accessing train stations. In their "ambitious" implemented scenario, Hasiak (2019, p. 26) set up a modal shift from car to walking within 1 km and to bike from 1 to 3 km. In this situation, with a condition based on quality cycling infrastructure including a network of cycle paths and parking facilities, 33% of drivers and 38% of passengers became cyclists (Hasiak, 2019, p. 26). Likewise, Geurs et al. (2016, p. 11) developed a logit model in which perceived connectivity and cycling route improvements on access time have similar sized effects as the train frequency increase scenario, especially in small railway stations from Randstad South. For low-density and natural areas where densification around the PT is complex, as in the Campania Region, the quality of feeder networks is crucial (Nigro et al., 2019, p. 110). Moinse et al. (2022, p. 22) make recommendations for designing a cycling-friendly environment as an opportunity to strengthen TOD to benefit micromobility, such as bicycles and standing scooters. However, findings suggest a minimized role of cycle paths in intermodal use for exurban areas: segregated bike lane availability is not significant in small-medium Belgian cities, due in part to the lower levels of traffic volume (Adnan et al., 2019, p. 8).

Bicycle parking facilities deserve more attention than they usually receive (Keijer & Rietveld, 2000, p. 234), and the availability of bicycle stands has a positive effect on the choice of Dutch departure railway stations accessed by bicycle (Debrezion et al., 2009, p. 26). In The Hague, Rijsman et al. (2019, p. 5) identify insufficient and unsafe parking places as one of the main barriers to accessing a tramway stop by bike. They recommend providing lockers and cameras at stops to potentially increase bike-and-ride users from 21.7% to 37.5%. Similar conclusions converge on high-quality bus systems, which would benefit from an increase in the share of users accessing a stop by bicycle with parking facilities (Brand et al., 2017, p. 7). This is also the case for railway stations, except that it is primarily large train stations that profit more from bicycle parking improvements (Geurs et al., 2016, p. 13). The impact of the installation of bicycle parking on the TOD perimeter is measured by Ton et al. (2020, p. 833), who show that these facilities increase the catchment area by 234 m and by 334 m for multimodal hubs (tramway and bus). Keijer and Rietveld (2000, p. 234) underline the role of bicycle parking facilities near rail stations for home-end segments in coordination with short walking distances at the activity end. The security of parked cycles at stations is an important aspect of TOD design, to the extent that the experience of theft and perceived poor security can encourage users to favor bicycles on board the train (Sherwin et al., 2011, p. 196). While it has been seen that the provision of bicycle parking spaces attracts the expected users, it is also true that park-and-ride (P+Rs) attracts drivers (Debrezion et al., 2009, p. 26), which raises issues about the integration of cars in TOD areas.

It seems necessary to rationalize the car parking supply around PT sites by evaluating the demand to enhance the TOD area, as mentioned by Hasiak (2019, p. 26). By promoting bike-and-ride at the expense of P+R, the station area has no negative externalities related to servicing costs, floor space, and environmental impact. Furthermore, Moinse et al. (2022, p. 20) illustrate that the e-scooter for access is competitive with car use only in the presence of car constraints such as parking regulations. With multiple scenarios, in a French local case study, Midenet et al. (2018, p. 30) highlight that the most significant initiative to stimulate a modal transfer from 53% of P+R users to bike or e-bike with PT lies in constraining the car parking size. The urban design approach should be considered as one of the solutions to promote lower carbon emission access to stations, as the modal shift from car to rail is significantly influenced by the connection quality to the station (Moinse et al., 2022, p. 21).

3.5.4   Destination Accessibility

Destination accessibility is studied in seven academic works within the corpus. The transport quality aspects for trains tend to increase with distance traveled, whereas for micromobility, its qualities tend to decrease as distance increases (Kager et al., 2016, p. 212). According to Nielsen and Skov-Petersen (2018, p. 42), bikeability can be divided into a local scale, a more urban scale with a range up to 4 km, and a regional scale that is up to 40 km in Denmark, while Heinen and Bohte (2014, p. 114) obtain an average of 36.5 km distance home-work by bicycle and PT.

The complementarity of micromobility with PT provides wider accessibility to destinations and consequently to resources, especially to employment areas. By accessing all PT stops within 1 km walking and 3 km cycling distances in the Copenhagen city area, Djurhuus et al. (2016, p. 13) demonstrate that a larger accessibility area is drawn as opposed to only accessing the nearest stop. Reflecting the destination benefits of micromobility and transit mix, the bicycle–train integration policy scenarios developed by Geurs et al. (2016, p. 11) provide greater job accessibility than the increased frequency scenario, with significant accessibility improvements toward small and medium-sized stations in Randstad South.

The use of these microvehicle feeders in conjunction with PT turns out to be effective compared to access or regional trips by private car. By studying the combination of the e-scooter with the regional train, Moinse et al. (2022, p. 17) note that the trip chain is time competitive with an unconstrained car scenario up to the threshold of 33 km. In their second scenario, characterized by ideal travel time and penalties, Kager et al. (2016, p. 216) reveal that cycling is an average 10-minute advantage over feeder transit services during a 10–80 km train trip.

3.5.5   Demand Management

The TOD's last D in relation to demand management is included in nine articles dealing with this aspect in various forms. The quality of the PT infrastructure and service is an important factor in attracting passengers, including micromobility riders.

Martens (2004, p. 292) states that bike-and-ride users are more inclined to favor faster and higher quality types of PT, such as train and intercity buses, unlike tram or local buses. However, while these elements may be necessary to build an attractive PT network, they appear to have different positive or modest effects in bringing in intermodal cyclists. First, reduced travel time on PT has a positive impact on the choice of departure station (Debrezion et al., 2009, p. 28). Although BRT stop distances do not influence the bicycle catchment area, their spatial spread affects the speed of the service, making the collective mode more efficient (Brand et al., 2017, p. 6). *In contrast*, transit frequency also has a moderate effect on the choice of departure stations by passengers accessing micromobility. The Geurs et al.'s (2016, p. 11) sixth scenario based on increased frequencies of local trains shows that rail–micromobility integration produces limited benefits. Moreover, doubling the frequency of local PT leads to a decrease in distance for bikes from 4.2 to 3.6 km (Debrezion et al., 2009, p. 28).

Managing the supply of parking spaces to influence demand around stations has a crucial role in the micromobility and PT mix. One of the most important attributes of bicycle access to the train station is the location of bicycle parking. In the Netherlands, providing free guarded bicycle parking and cycling spaces within 2 minutes from the train platform generates beneficial effects around the site, and users are even highly willing to pay for improvement of these facilities (Geurs et al., 2016, p. 9). From a systematic perspective, these effects interact with car use and parking space management. Debrezion et al. (2009, p. 24) evidence that car ownership of 0.60 per person would involve car domination over PT from 10 km. Midenet et al. (2018, p. 30) advocate the implementation of pricing policies on P+R depending on the reverse distance from the client's residence to the train station to encourage modal shift and optimization of car park size. Considering the potential modal shift from car to walking within 1 km and from cycling between 1 and 3 km, Hasiak (2019, p. 27) demonstrates that a savings of up to 40% in parking places emerges since the share of passengers reaching local rail stations by car would fall from 53% to 29%.

Demand management can also address specific challenges of micromobility–transit combination, such as the unavailability of bikes in access and egress. For instance, Rijsman et al. (2019, p. 3) recommend providing bicycle sharing schemes at tram stops in The Hague. This opportunity is arising and is being reinforced by the arrival of free-floating bicycle and scooter services and progressively by smart parking.[4] Brand et al. (2017, p. 7) focus more on the egress side, where riders are dependent on walking range, and call for providing bike-sharing and bike-renting opportunities. These statements follow the advice from Rietveld (2000, p. 75), which proposes services at the activity-end, such as bike renting, cycling facilities on board the train, or safe parking for second bikes. On-board folding micromobility solutions such as private e-scooters also represent an alternative for both access and egress sides, notably for exurban territories where the implementation of bike schemes seems complex (Moinse et al., 2022, p. 21). The articulation of the train with micromobility services is based on successful management and communication Mobility-as-a-Service (MaaS) applications (Adnan et al., 2019, p. 8). These guidelines can also be coupled with prevention and education programs for pedestrians and micromobility users (Hasiak, 2019, p. 27).

4   Revisiting the TOD Concept

4.1   A Hybrid and Smart TOD Adaptable to Spatial Contexts

Transit-oriented development is a promising model for building sustainable, smart urbanization, and mobility in the future (Cervero, 1998, p. 3). The author identified and identified four types of transit-oriented metropolises (Cervero, 1998, p. 7):

1. "Adaptive cities" that have invested in rail systems to guide urban growth
2. "Adaptive Transit" that has accepted spread-out with low-density areas and has adapted transit services to serve these regions
3. "Strong-core cities" that have integrated transit and development within a confined and central urban context
4. "Hybrid" adaptive cities and adaptive transit that are balanced between dense urbanization along transit corridors and suitable transit to serve suburbs

Recently, Cervero (2020, p. 131) sought to update the conceptualization given to the Transit Metropolises (TOD), which was written two decades ago and should be renewed as Hybrid Transit Metropolises from a twenty-first-century perspective. The dichotomy of adaptive cities and adaptive transit now gives way to a modern transit metropolis vision marrying well-designed TOD and flexible, door-to-door mobility options (Cervero, 2020, p. 144). Lee et al. (2016, p. 983) introduce the concept of bicycle-based transit-oriented development (B-TOD) as an alternative to walking-based TOD and shed light on the identification of the station impact area where bicycle is the primary access mode and on more or less relaxed density criteria (Fig. 10). This book chapter follows the revisited Transit Metropolis and B-TOD

---

[4]In response to deregulated micromobility operations, smart parking is considered as an urban management solution consisting of dedicated parking bays. Parking bays are a way to support and promote intermodal trips by creating shared micromobility hubs close to stations, when smart parking's density is sufficient.

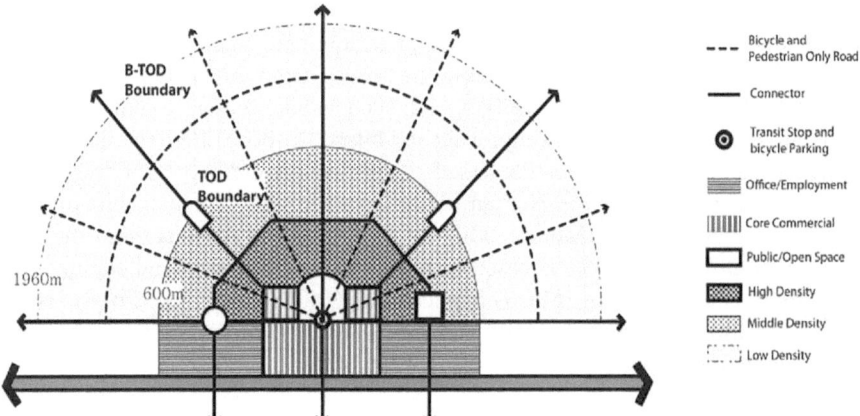

Fig. 10 Distances and influence areas in relation to public transport stations from the 19 reviewed articles. (Source: Author elaboration)

direction, highlighting 19 scientific papers that examine the role of feeder micro-modes in strengthening hybrid TODs, bringing sustainable mobility and urban forms to strategic auto-centric areas within enlarged transit catchment areas (Nigro et al., 2019, p. 38). Stransky (2019, p. 38) develops a simplified TOD framework for peri-urban locations around Paris, reinterpreting the 3Ds as "Walkability," "Variety," and "Treatment." These context-specific TOD criteria illustrate the adaptability of the model, according to the author, by favoring the ability of the area to be easily walkable, the variety of supply, and the quality of the feeder routes for walking and cycling at different scales.

4.2 15-Minute TOD-Friendly Areas

Bike-and-ride, and more broadly micromobility combined with transit, offers a number of environmental, social, and economic benefits over the use of private cars (Martens, 2004, p. 282), including reduction of the various forms of pollution, relative energy sobriety, better accessibility, or territorial and land use valorization. The emergence of an extended version of TOD seems to be in line with the recommendations of (Jamme et al., 2019, p. 421), which are based on the urge of renewed TOD with the original goal of developing inclusive and sustainable communities. Figure 11 highlights the advantages of the combination of PMDs and PT over other travel modes, allowing micromobility-and-ride to compete with auto, especially on congested roads. Kager et al. (2016, p. 212) emphasize that this hybrid system ensures both a relatively high speed (efficiency allowed by the mobility service) and the possibility of making a door-to-door journey (flexibility of micromobility). Recognizing the comparative advantages of this intermodal system, Bertolini (2017,

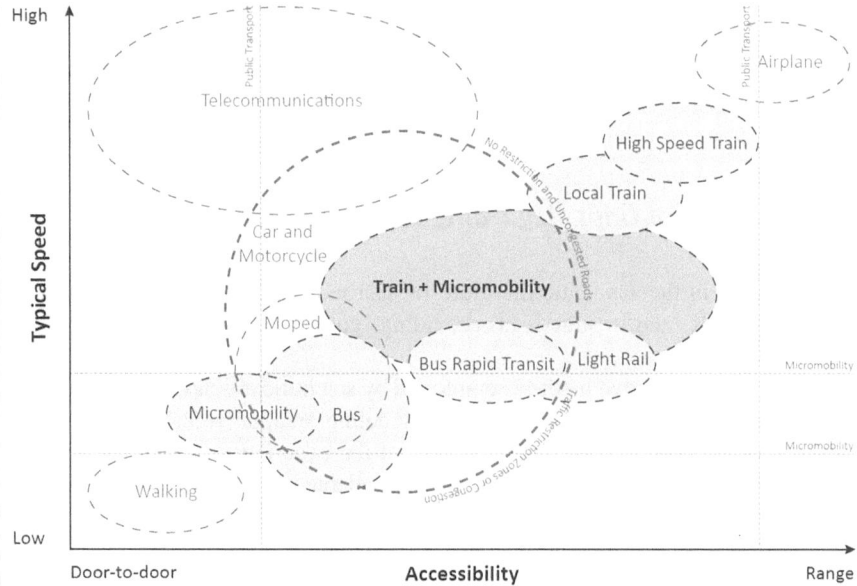

Fig. 11 Characterizing the bicycle train mode according to speed and level of accessibility (indicative, adapted from Meyer and Miller (2013)). (Source: Kager et al. (2016, p. 212))

p. 120) recommends placing interchange facilities as close as possible to the low-density and low-functional-mix areas, including bike- and park-and-ride areas.

From this analysis, the results suggest that two levels are integrated into the territorial system when considering urban development around PT and especially railway station areas. The walking catchment area of up to 1 or 1.5 km is covered by a 15-minute neighborhood, and the acceptable cycling range estimated to be 3 or 4 km is also within the scope of a 15-minute city, which is consistent with Brès (2014, p. 271) findings. Midenet et al. (2018, p. 29) further considers an access time by bicycle, with a cycling-friendly environment, set to 20 minutes. Nevertheless, this study follows the need for a site-specific conceptualization of TOD (Qvistrom & Bengtsson, 2015, p. 2531; Stransky, 2019, p. 38). Extending the railway station areas by promoting micromobility could reflect the emerging concept of the "15-Minute City" (Duany & Steuteville, 2021), where the combination of walking, cycling, scootering, and PT (Sadik-Kahn, 2021) would ensure that most commonly accessed services and activities can be reached within a 15-minute walk or cycling ride (Moreno & Hjelm, 2021). For this, more multimodal planning that can invest as much in active and micromodes, as would be spent on road and parking facilities for cars, is needed to create a least-cost planning 15-minute neighborhood (Litman, 2021, p. 34). From this broader point of view, the urban project oriented to structured PT, in particular the railway station, must simultaneously take advantage of a wide pedestrian zone appropriate to its real potential, as well as the micromobility and urban PT isochrones guaranteeing intermodality development (L'Hostis et al., 2009, p. 69).

Integrating auto-oriented Secondary Areas to benefit emerging micromobility options is a way of reinforcing the TOD model by considering European urban patterns and new mobility solutions complementing the transit network while bringing a more inclusive approach by appealing to populations that are more distant.

4.3   Knowledge Gaps Regarding Extended TODs

Existing gaps in the academic literature for European case studies were identified from this SLR and resulted in providing guidance for future research on extended TOD.

While the quantitative methods employed by scientific articles are diverse, varying from analyses of mobility surveys to modeling, qualitative methods are sorely lacking and prevent a better understanding of the experience and appropriation of users from secondary areas to favor policies adapted to an efficient modal shift. Moreover, the methods linked to the digital collection of big data are not very widespread in Europe (1/19), giving rise to a second gap.

As recognized by Oeschger et al. (2020, p. 17) through an international SLR on micromobility and PT, an evident gap is the lack of research focusing on new and electric micromodes in the context of integrated transit. Although research in North America and Asia related to new mobility services is gradually emerging, as demonstrated in Sect. 3, emerging shared and private PMDs integrated with transit are thus marginalized in the research. This can be explained by the novelty of these modes, such as electric bikes, scooters, skateboards, or folding bikes, as well as the difficulty of gathering data in specific contexts. Data are still scarce despite the establishment of open access platforms promoted by local authorities at different scales.

Regarding PT networks, gaps have also been identified with research mainly oriented toward rail, whereas some cases only exist for urban PT. Martens (2004, p. 292) notes that levels of bike-and-ride are much less clear for slower types of PT, shares in feeding trips to bus, tramway, or metro stops revealing to be poor partly because of a lack of policy attention for their integration.

This gap unfolds not only within micromodes or PT but also across case studies that are underrepresented in large parts of Europe, especially in Eastern and Southern Europe. Although research has focused on multiple types of urban patterns within metropolises, little attention has been given to comparing the characteristics that differ between neighborhoods with high and low levels of services (LOS), as well as access and egress legs.

At the same time, not all TOD criteria were treated equally, with some variables receiving little implicit or explicit consideration, including destination accessibility, urban diversity, and density. As identified by Knowles et al. (2020, p. 7), the reviewed subject devotes little space to the extended area effects on the founding assumptions of TOD, namely, density, diversity, and design, even though the design was regarded as one of the criteria privileged in the studies.

Last, more research is needed on the impacts of micromobility and PT integration on station areas through economic, social, and environmental dimensions, i.e., the potential for modal shift, regional accessibility and social inclusion provided by the combination of economic dynamics stimulated in territories as well as the energy balance in relation to modal shifts. As Oeschger et al. (2020, p. 17) also pointed out, the economic aspect related to extended TOD is the most important missing analysis dimension in Europe.

5   Conclusions

Through the framework of this literature body with respect to the 6Ds, it was possible to review such factors in these extended perimeters. First, it was possible to underline the critical place of "Distance to Transit" with the emergence of micromobility devices and services to support and accommodate the TOD application toward zones located 3 or even 4 km away. This SLR highlighted the importance of "design" and, in particular, the quality of the cycling network and parking in the feeder area, which has the capacity to extend the acceptable distance for passengers, those factors being expanders or contractors of the TOD walking radius. This can only be effective as long as the place of the car is questioned in this strategic secondary area, transportation "demand management" taking the form of spatial and ownership of car restrictions, the implementation of reverse distance-based pricing for P+R, and the provision of mobility services near station areas. "Diversity" is relevant to the extent that it promotes intensive activity clustering around the station and housing in the walking and cycling surrounding area, up to 3–4 km. This chapter underlined the unclear role of compacity in these secondary areas, with micromobility being a response to low-density territories. While some studies highlighted the positive correlation between micromobility use as a feeder mode and medium urban density, other variables may influence these observations, such as the presence of cycling infrastructure. Finally, "destination accessibility" proved the effectiveness of micromobility in providing a more inclusive connection to employment locations, guaranteeing local, urban, and regional accessibility.

This work aimed to better understand the growing interest in the extended TOD perimeter, whereby the association of new micromobility forms and public transport is expected to become more relevant in the future. It has been seen, by reviewing 19 scientific papers, that the catchment area of European transit stations, in particular train stations, is easily extended to 3 km with the assistance of microvehicles, especially personal bicycles, as a strengthening of the 1.6-km secondary area. TOD can be redefined for the twenty-first century through game-changing mobility paradigms, such as smart and sustainable mobility.

References

Adnan, M., Altaf, S., Bellemans, T., Yasar, A.-H., & Shakshuki, E. M. (2019). Last-mile travel and bicycle sharing system in small/medium sized cities: User's preferences investigation using hybrid choice model. *Journal of Ambient Intelligence and Humanized Computing, 10*(12), 4721–4731. https://doi.org/10.1007/s12652-018-0849-5

Banai, R. (1998). Transit-oriented development suitability analysis by the analytic hierarchy process and a geographic information system: A prototype procedure. *Journal of Public Transportation, 2*(1), 43–65. https://doi.org/10.5038/2375-0901.2.1.3

Bertolini, L. (2017). *Planning the mobile metropolis: Transport for people, places and the planet* (1st ed.). Red Globe Press.

Bertolini, L., Curtis, C., & Renne, J. (2012). Station area projects in Europe and beyond: Towards transit oriented development? *Built Environment, 38*(1), 31–50. https://doi.org/10.2148/benv.38.1.31

Böcker, L., Anderson, E., Uteng, T. P., & Throndsen, T. (2020). Bike sharing use in conjunction to public transport: Exploring spatiotemporal, age and gender dimensions in Oslo, Norway. *Transportation Research Part A: Policy and Practice, 138*, 389–401. https://doi.org/10.1016/j.tra.2020.06.009

Bozzi, A. D., & Aguilera, A. (2021). Shared E-scooters: A review of uses, health and environmental impacts, and policy implications of a new micro-mobility service. *Sustainability, 13*(16), 8676. https://doi.org/10.3390/su13168676

Brand, J., Hoogendoorn, S., van Oort, N., & Schalkwijk, B. (2017). Modelling multimodal transit networks integration of bus networks with walking and cycling. In *2017 5th IEEE international conference on models and technologies for intelligent transportation systems (MT-ITS)* (pp. 750–755). IEEE. https://doi.org/10.1109/MTITS.2017.8005612

Brès, A. (2014). Train stations in areas of low density and scattered urbanisation: Towards a specific form of rail oriented development. *Town Planning Review, 85*(2), 261–272. https://doi.org/10.3828/tpr.2014.16

Calthorpe, P. (1993). *The next American metropolis: Ecology, community, and the American dream.* Princeton Architectural Press.

Canepa, B. (2007). Bursting the bubble. Determining the transit-oriented development's walkable limits. *Transportation Research Record: Journal of the Transportation Research Board, 1992*, 28–34. https://doi.org/10.3141/1992-04

Cervero, R. (1998). *The transit metropolis: A global inquiry* (4th ed.). Island Press.

Cervero, R. (2001). Walk-and-ride: Factors influencing pedestrian access to transit. *Journal of Public Transportation, 3*(4), 23. https://doi.org/10.5038/2375-0901.3.4.1

Cervero, R. (2020). Chapter 7 – The transit metropolis: A 21st century perspective. In E. Deakin (Ed.), *Transportation, land use, and environmental planning* (pp. 131–149). Elsevier. https://doi.org/10.1016/B978-0-12-815167-9.00007-4

Cervero, R., & Kockelman, K. (1997). Travel demand and the 3Ds: Density, diversity, and design. *Transportation Research Part D: Transport and Environment, 2*(3), 199–219. https://doi.org/10.1016/S1361-9209(97)00009-6

Cervero, R., Murphy, S., Ferrell, C., Tsai, Y.-H., Arrington, G. B., Boroski, J., Smith-Heimer, J., Golem, R., Peninger, P., Nakajima, E., Chui, E., Dunphy, R., Myers, M., McKay, S., & Witenstein, N. (2004). *TCRP report 102: Transit-oriented development in the United States: Experiences, challenges, and prospects.* Transportation Research Board of the National Academies.

Cervero, R., Caldwell, B., & Cuellar, J. (2013). Bike-and-ride: Build it and they will come. *Journal of Public Transportation, 16*(4), 83–105.

Cottrell, W. D. (2007). Transforming a bus station into a transit-oriented development: Improving pedestrian, bicycling, and transit connections. *Transportation Research Record, 2006*(1), 114–121. https://doi.org/10.3141/2006-13

Debrezion, G., Pels, E., & Rietveld, P. (2009). Modelling the joint access mode and railway station choice. *Transportation Research Part E: Logistics and Transportation Review, 45,* 270–283. https://doi.org/10.1016/j.tre.2008.07.001

Dittmar, H., & Ohland, G. (2012). *The new transit town: Best practices in transit-oriented development.* Island Press.

Djurhuus, S., Sten Hansen, H., Aadahl, M., & Glümer, C. (2016). Building a multimodal network and determining individual accessibility by public transportation. *Environment and Planning B: Planning and Design, 43*(1), 210–227. https://doi.org/10.1177/0265813515602594

Duany, A., & Steuteville, R. (2021). Defining the 15-minute city | CNU. *Public Square. A CNJU Journal.* https://www.cnu.org/publicsquare/2021/02/08/defining-15-minute-city?fbclid=IwAR3pupE82_fzuMgLWzaeuRVYJKJo22xdVl9BjbRcF_8 9j1W8w2Z-vsDy6Ik

Ewing, R., & Cervero, R. (2010). Travel and the built environment. *Journal of the American Planning Association, 76*(3), 265–294. https://doi.org/10.1080/01944361003766766

Geurs, K. T., La Paix, L., & Van Weperen, S. (2016). A multi-modal network approach to model public transport accessibility impacts of bicycle-train integration policies. *European Transport Research Review, 8*(4), 25. https://doi.org/10.1007/s12544-016-0212-x

Guerra, E., Cervero, R., & Tischler, D. (2012). Half-mile circle. Does it best represent transit station catchments? *Transportation Research Record: Journal of the Transportation Research Board, 2276*(1), 101–109. https://doi.org/10.3141/2276-12

Hasiak, S. (2019). Access mobility to local railway stations: Current travel practices and forecast. *CyberGeo: European Journal of Geography.* https://doi.org/10.4000/cybergeo.33488

Heinen, E., & Bohte, W. (2014). Multimodal commuting to work by public transport and bicycle: Attitudes toward mode choice. *Transportation Research Record, 2468*(1), 111–122. https://doi.org/10.3141/2468-13

Héran, F. (2015). *Le retour de la bicyclette. Une histoire des déplacements urbains en Europe, de 1817 à 2050.* La Découverte.

Hochmair, H. H. (2015). Assessment of bicycle service areas around transit stations. *International Journal of Sustainable Transportation, 9*(1), 15–29. https://doi.org/10.1080/15568318.2012.719998

Ibraeva, A., de Almeida Correia, G. H., Silva, C., & Antunes, A. P. (2020). Transit-oriented development: A review of research achievements and challenges. *Transportation Research Part A: Policy and Practice, 132,* 110–130. https://doi.org/10.1016/j.tra.2019.10.018

Jain, D., Singh, E., & Ashtt, R. (2020). A systematic literature on application of transit oriented development. *International Journal of Engineering and Advanced Technology, 9*(3), 2542–2552. https://doi.org/10.35940/ijeat.C5415.029320

Jamme, H.-T., Rodriguez, J., Bahl, D., & Banerjee, T. (2019). A twenty-five-year biography of the TOD concept: From design to policy, planning, and implementation. *Journal of Planning Education and Research, 39*(4), 409–428. https://doi.org/10.1177/0739456X19882073

Jonkeren, O., & Kager, R. (2021). Bicycle parking at train stations in the Netherlands: Travellers' behaviour and policy options. *Research in Transportation Business & Management, 40,* 100581. https://doi.org/10.1016/j.rtbm.2020.100581

Kager, R., Bertolini, L., & Te Brömmelstroet, M. (2016). Characterisation of and reflections on the synergy of bicycles and public transport. *Transportation Research Part A: Policy and Practice, 85,* 208–219. https://doi.org/10.1016/j.tra.2016.01.015

Keijer, M. J. N., & Rietveld, P. (2000). How do people get to the railway station? The dutch experience. *Transportation Planning and Technology, 23*(3), 215–235. https://doi.org/10.1080/03081060008717650

Ker, I., & Ginn, S. (2003). Myths and realities in walkable catchments: The case of walking and transit. *Road & Transport Research, 12*(2), 69–80.

Knowles, R. D., Ferbrache, F., & Nikitas, A. (2020). Transport's historical, contemporary and future role in shaping urban development: Re-evaluating transit oriented development. *Cities, 99,* 102607. https://doi.org/10.1016/j.cities.2020.102607

Kostrzewska, M., & Macikowski, B. (2017). Towards hybrid urban mobility: Kick scooter as a means of individual transport in the city. *IOP Conference Series: Materials Science and Engineering, 245*, 052073. https://doi.org/10.1088/1757-899X/245/5/052073

L'Hostis, A. (2016). *Les périmètres du Transit Oriented Development: Caractérisation de la relation entre ville et transport collectifs* (pp. 1–9). HAL.

L'Hostis, A., Alexandre, E., Appert, M., Araud-Ruyant, C., Basty, M., Biau, G., Bozzani-Franc, S., Boutantin, G., Constantin, C., Coralli, M., Durousset, M.-J., Fradier, C., Gabion, C., Leysens, T., Mermoud, F., Olny, X., Perrin, E., Robert, J., Simand, N., et al. (2009). *Concevoir la ville à partir des gares, Rapport final du Projet Bahn.Ville 2 sur un urbanisme orienté vers le rail.* HAL. https://hal.archives-ouvertes.fr/hal-00459191

Lee, J., Choi, K., & Leem, Y. (2016). Bicycle-based transit-oriented development as an alternative to overcome the criticisms of the conventional transit-oriented development. *International Journal of Sustainable Transportation, 10*(10), 975–984. https://doi.org/10.1080/1556831 8.2014.923547

Li, X., Liu, Z., & Ma, X. (2022). Measuring access and egress distance and catchment area of multiple feeding modes for metro transferring using survey data. *Sustainability, 14*(5), 2841. https://doi.org/10.3390/su14052841

Litman, T. (2021). *New mobilities: Smart planning for emerging transportation technologies.* Island Press.

Martens, K. (2004). The bicycle as a feedering mode: Experiences from three European countries. *Transportation Research Part D: Transport and Environment, 9*(4), 281–294. https://doi.org/10.1016/j.trd.2004.02.005

Meyer, E., & Miller, E. (2013). Chapter 3: Urban travel and transportation system characteristics: A system perspective. In *Urban transportation planning: A decision-oriented approach.* McGraw-Hill.

Midenet, S., Côme, E., & Papon, F. (2018). Modal shift potential of improvements in cycle access to exurban train stations. *Case Studies on Transport Policy, 6*(4), 743–752. https://doi.org/10.1016/j.cstp.2018.09.004

Moinse, D., Goudeau, M., L'Hostis, A., & Leysens, T. (2022). *An analysis of intermodal use of electric and human-powered scooters with train in the Provence-Alpes-Côte d'Azur region, in France: Towards extended train station areas?* HAL. https://halshs.archives-ouvertes.fr/halshs-03523112

Moreno, C., & Hjelm, F. (2021, Juin 29). *Introducing the 15-minute city* [Séminaire].

Nasri, A., & Zhang, L. (2014). The analysis of transit-oriented development (TOD) in Washington, D.C. and Baltimore metropolitan areas. *Transport Policy, 32*, 172–179. https://doi.org/10.1016/j.tranpol.2013.12.009

Neilson, A., Indratmo, Daniel, B., & Tjandra, S. (2019). Systematic review of the literature on big data in the transportation domain: Concepts and applications. *Big Data Research, 17*, 35–44. https://doi.org/10.1016/j.bdr.2019.03.001

Nielsen, T. A. S., & Skov-Petersen, H. (2018). Bikeability – Urban structures supporting cycling. Effects of local, urban and regional scale urban form factors on cycling from home and workplace locations in Denmark. *Journal of Transport Geography, 69*, 36–44. https://doi.org/10.1016/j.jtrangeo.2018.04.015

Nigro, A., Bertolini, L., & Moccia, F. D. (2019). Land use and public transport integration in small cities and towns: Assessment methodology and application. *Journal of Transport Geography, 74*, 110–124. https://doi.org/10.1016/j.jtrangeo.2018.11.004

Oeschger, G., Carroll, P., & Caulfield, B. (2020). Micromobility and public transport integration: The current state of knowledge. *Transportation Research Part D: Transport and Environment, 89*, 102628. https://doi.org/10.1016/j.trd.2020.102628

Ogra, A., & Ndebele, R. (2014). The role of 6Ds: Density, diversity, design, destination, distance, and demand management in Transit Oriented Development (TOD). In *Proceedings of the NICHE-2014 Neo-international conference on habitable environments*, Jalandhar, India, 31 October–2 November 2014, pp. 539–546.

Padeiro, M., Louro, A., & da Costa, N. M. (2019). Transit-oriented development and gentrification: A systematic review. *Transport Reviews, 39*(6), 733–754. https://doi.org/10.1080/0144164 7.2019.1649316

Papa, E., & Bertolini, L. (2015). Accessibility and transit-oriented development in European metropolitan areas. *Journal of Transport Geography, 47*, 70–83. https://doi.org/10.1016/j.jtrangeo.2015.07.003

Park, K., Farb, A., & Chen, S. (2021). First-/last-mile experience matters: The influence of the built environment on satisfaction and loyalty among public transit riders. *Transport Policy, 112*, 32–42. https://doi.org/10.1016/j.tranpol.2021.08.003

Pojani, D., & Stead, D. (2015). Transit-Oriented Design in the Netherlands. *Journal of Planning Education and Research, 35*(2), 131–144. https://doi.org/10.1177/0739456X15573263

Pritchard, R. (2018). Revealed preference methods for studying bicycle route choice—A systematic review. *International Journal of Environmental Research and Public Health, 15*(3), 470. https://doi.org/10.3390/ijerph15030470

Qvistrom, M., & Bengtsson, J. (2015). What kind of transit-oriented development? Using planning history to differentiate a model for sustainable development. *European Planning Studies, 23*(12), 2516–2534. https://doi.org/10.1080/09654313.2015.1016900

Rietveld, P. (2000). The accessibility of railway stations: The role of the bicycle in the Netherlands. *Transportation Research Part D: Transport and Environment, 5*(1), 71–75. https://doi.org/10.1016/S1361-9209(99)00019-X

Rijsman, L., van Oort, N., Ton, D., Hoogendoorn, S., Molin, E., & Teijl, T. (2019). Walking and bicycle catchment areas of tram stops: Factors and insights. In *2019 6th international conference on models and technologies for intelligent transportation systems (MT-ITS)* (pp. 1–5). IEEE. https://doi.org/10.1109/MTITS.2019.8883361

Sadik-Kahn, J. (2021, février 18). From 15-minute cities to clutter control: Top trends from micromobility world 2021. *Joyride.* https://joyride.city/15-minute-cities-micromobility-trends/

Schneider, J. B. (1992). A PRT deployment strategy to support regional land use and rail transit objectives. *Transportation Quarterly, 46*, 135–153.

Schneider, J. (2012, février 21). *Describing and illustrating the extended transit-oriented development concept.* Faculty Washington. http://faculty.washington.edu/jbs/itrans/e-tod.htm

Sebban, A.-C. (2003). *La complémentarité entre vélo et transport public* [These de doctorat, Aix-Marseille 3]. http://www.theses.fr/2003AIX32060

Şengül, B., & Mostofi, H. (2021). Impacts of E-micromobility on the sustainability of urban transportation—A systematic review. *Applied Sciences, 11*(13), 5851. https://doi.org/10.3390/app11135851

Sherwin, H., Parkhurst, G., Robbins, D., & Walker, I. (2011). Practices and motivations of travellers making rail–cycle trips. *Proceedings of the Institution of Civil Engineers – Transport, 164*(3), 189–197. https://doi.org/10.1680/tran.2011.164.3.189

Stransky, V. (2019). Périurbain et transit-oriented development: Un couple invraisemblable ? *Flux, 115*(1), 33–57. https://doi.org/10.3917/flux1.115.0033

Thomas, J., & Harden, A. (2008). Methods for the thematic synthesis of qualitative research in systematic reviews. *BMC Medical Research Methodology, 8*(1), 45. https://doi.org/10.1186/1471-2288-8-45

Ton, D., Shelat, S., Nijënstein, S., Rijsman, L., van Oort, N., & Hoogendoorn, S. (2020). Understanding the role of cycling to urban transit stations through a simultaneous access mode and station choice model. *Transportation Research Record, 2674*(8), 823–835. https://doi.org/10.1177/0361198120925076

Transportation Research Board of the National Academies. (2015). *Literature searches and literature reviews for transportation research projects. How to search, where to search, and how to put it all together: Current practices* (E-C194; Transportation Research Circular, pp. 1–84). Conduct of Research Committee, Library and Information Science for Transportation Committee, Transportation Research Board. https://www.trb.org/Publications/Blurbs/172271.aspx

Yang, R., Yan, H., Xiong, W., & Liu, T. (2013). The study of pedestrian accessibility to rail transit stations based on KLP model. *Procedia – Social and Behavioral Sciences, 96*, 714–722. https://www.researchgate.net/publication/273538379_The_Study_of_Pedestrian_Accessibility_to_Rail_Transit_Stations_Based_on_KLP_Model

Open Access This chapter is licensed under the terms of the Creative Commons Attribution 4.0 International License (http://creativecommons.org/licenses/by/4.0/), which permits use, sharing, adaptation, distribution and reproduction in any medium or format, as long as you give appropriate credit to the original author(s) and the source, provide a link to the Creative Commons license and indicate if changes were made.

The images or other third party material in this chapter are included in the chapter's Creative Commons license, unless indicated otherwise in a credit line to the material. If material is not included in the chapter's Creative Commons license and your intended use is not permitted by statutory regulation or exceeds the permitted use, you will need to obtain permission directly from the copyright holder.

Disposing of Daily Life Resources by Active Modes

Analysis Based on Ergonomics of Access Applied to the Eurometropole de Strasbourg

Maxime Hachette, Eliane Propeck-Zimmermann, and Alain L'Hostis

Abstract Today, many cities are promoting sustainable mobility. Their policies have already reduced the car's place, developed pedestrian and bicycle facilities, or renewed public transport. This raises the question of the effects of these policies on the conditions of access to everyday resources. Are the facilities for sustainable mobility configured in such a way as to enable the population's needs to be met? Globally or selectively? Do active modes (walking and cycling) offer a credible alternative to the car in order to effectively provide the resources necessary for daily life throughout the urban agglomeration?

To answer these questions, this chapter presents a geographical analysis approach based on the concept of spatial ergonomics. The application to 12 test areas, using a geographic information system, has revealed cleavage situations within the Eurométropole de Strasbourg, to study finely differentiated situations and to put them in perspective with socio-demographic profiles to analyze socio-spatial disparities.

The various levels of information shed light on leeway available to inhabitants, wherever they are located, to change their mode of travel. The method makes it possible to produce territorial diagnoses and to help local authorities to promote effective sustainable development policies.

Keywords Spatial ergonomics · Access ergonomics · Sustainable mobility · Active modes · Daily resources · Socio-spatial disparities

M. Hachette (✉) · A. L'Hostis
LVMT, Ecole des Ponts, Univ Gustave Eiffel, Marne-la-Vallée, France
e-mail: maxime.hachette@univ-eiffel.fr; alain.lhostis@univ-eiffel.fr

E. Propeck-Zimmermann
Laboratoire Image Ville Environnement (LIVE), UMR 7362 CNRS, University of Strasbourg, Strasbourg, France
e-mail: eliane.propeck@live-cnrs.unistra.fr

© The Author(s) 2024   205
F. Belaïd, A. Arora (eds.), *Smart Cities*, Studies in Energy, Resource and Environmental Economics, https://doi.org/10.1007/978-3-031-35664-3_13

1 Introduction

In urban planning, the slogan "always faster, always further" is now outdated (Papon, 2003; Piombini, 2006; Saint-Gérand et al., 2019). As far as public policy is concerned, the car is no longer the symbol of progress and modernity that it once was. Indeed, in addition to being one of the largest CO_2 emitters on an international scale, the automobile has various negative externalities, especially in cities: noise pollution, congestion, massive use of space, accidents, etc. The time has come for peaceful and virtuous mobility on an ecological level.

Within the framework of sustainable development, particularly in urban areas, one of the objectives is to slow down the automobile system to make way for more sustainable mobility in accordance with the aims of the ecological and energy transition. Such a policy is now announced at all levels: local, national, and international. Limiting the damaging impact of cars appears to be a main guideline in all policies.

This priority has been reflected not only in a set of regulatory and normative measures (vehicle emissions, speed, etc.) but above all in urban planning in favor of sustainable modes of transport. Public transport, led by the tramway, and active modes, such as walking and cycling, are the main beneficiaries of these developments. Currently, many cities have developed facilities for pedestrians and cyclists (pedestrian zones, meeting areas, street furniture, play areas, cycle lanes, bicycle racks, etc.), they also shared vehicles (scooters, bicycles, or electric or nonelectric cars), carpooling, intermodality, multimodality, etc. Many measures in favor of sustainable mobility are combined with restrictive measures for private cars (regulated or paid parking, traffic restrictions, urban tolls, etc.). These sustainable mobility policies have already led to or accompanied a reduction in the use of cars in most major cities.

Trips are necessary for each person for different reasons (professional, leisure, and other activities). From this perspective, we may be curious if the various forms of mobility, especially active mobility, can meet their needs. Our interest in this research has focused on access to current resources (shops, education, health, public services, and leisure). Although policies recommend sustainable mobility over the whole territory, they seem to focus on strategic areas of cities (urban centers, eco-neighborhoods, upper tertiary centers, etc.) at the risk of increasing socio-spatial inequalities.

Under what conditions can inhabitants access daily resources through active modes?

Are the measures in favor of sustainable mobility configured in such a way as to meet the needs of the population? Globally or more narrowly focused?

Do sustainable modes, and in particular active modes, offer a reliable alternative to cars as a means of disposing effectively of the resources of everyday life? For all and wherever they are?

To tackle these questions, this research is based on the operational concept of spatial ergonomics in the sense of a conceptual approach that can be modeled in spatial analysis methodology. Spatial ergonomics is defined as "the ability of a territory to provide its population with the socioeconomic resources they need at the lowest cost/effort/risk" (Saint-Gérand, 2002). The basic hypothesis is that "the suitability of a space for the life of its population depends largely on the ease it offers to the inhabitants to appropriate the resources they need, according to their specificities and where they are located" (Saint-Gérand et al., 2021).

Within the framework of this research, we developed the first exploration of spatial ergonomics through mobility and the appropriation of everyday resources by populations using active modes. We have called it "access ergonomics." It adopts an approach that focuses particularly on the fulfillment of needs and on conditions of access to resources. Spatial ergonomics and access ergonomics undeniably play a role in the smart city (as presented by Cerema), related to the first component of collective intelligence but not so much to the technological part (Brussels Smart City, 2022; Cerema, 2020; CNIL, 2022).

The objective of this research is to develop an approach to analyze and evaluate the "ergonomics of access" to everyday life resources at each point in space, in a reasonable time, according to different modes of travel (on foot, by bicycle, and by car), and to analyze spatial disparities. On the one hand, the approach takes into consideration the overall functioning of the territory through the availability and distribution of the potential resources and, on the other hand, a panel of criteria characterizing their access conditions (service conditions, safety, comfort, monetary cost).

First, the chapter focuses on the theoretical framework of the concept of "spatial ergonomics" to distinguish it from related concepts such as accessibility. To ensure the reproducibility of the results, the method used for the calculation of each indicator will be detailed. Finally, the application focuses on 13 test areas within the Eurométropole de Strasbourg, which has a very proactive policy in favor of active mobility. It is based on a geographic information system (GIS) and an associated database. The calculation of a synthetic score, which is then declined by mode of travel, by time step, and according to different criteria, aims to characterize differentiated situations within the urban space. In fine, putting the indicators into perspective with the socio-urban environment allows revealing socio-spatial disparities.

2 Theoretical Approach

2.1 The Concept of Ergonomics in Geography as a Result of Conceptual Transfers

Ergonomics was originally and is currently practiced in workshops, factories, and companies. It aims to adjust the workspace, equipment, and process to the physical and behavioral capabilities of the workers to improve efficiency. The International Ergonomics Association defines ergonomics as follows:

Ergonomics (or human factors) is the scientific discipline concerned with the understanding of interactions among humans and other elements of a system, and the profession that applies theory, principles, data, and methods to design in order to optimize human well-being and overall system performance. [...]. [Ergonomics] is a multidisciplinary, user-centric integrating science. The issues [ergonomics] addresses are typically systemic in nature; thus [ergonomics] uses a holistic, systems approach to apply theory, principles, and data from many relevant disciplines to the design and evaluation of tasks, jobs, products, environments, and systems. [Ergonomics] takes into account physical, cognitive, sociotechnical, organizational, environmental and other relevant factors as well as the complex interactions between the human and other humans, the environment, tools, products, equipment, and technology. (International Ergonomics Association, 2022)

Ergonomics aims to optimize the well-being of the person and the overall performance of a system. It therefore adopts a systemic approach to analyze the interactions between humans and other system components (configuration of workshops, materials, handling process, rhythm) to obtain the best individual work efficiency for lower overall cost (energy, time, money, effort, stress, exposure to danger) (Fleury, 2009). The analysis is thus based on economic and cost/performance type models.

A systemic approach must also take into account the complexity of the regulations that take place at two levels:

at the level of the individual who regulates his activity, according to his external environment and his internal state (tiredness, for example),
at the level of the company, which reviews the configuration of the workshops, the material equipment, the handling processes, and the rhythms for greater efficiency at the lowest cost.

Gradually, ergonomics began to be applied to different processes that can be assimilated into work. Conceptual transfers have led to their application to different fields, particularly in geography.

As in geography, ergonomics thus attaches great importance to planning, i.e., to spatial configuration (understood as a reasoned arrangement aiming at a general quality of connectivity). By considering the city as a man-machine system (De Montmollin, 1967), it seems to be able to fit into the field of study of ergonomics, and its methods could contribute to a reflection on its planning in connection with the activities and characteristics of the users of the territory.

Several researchers have been investigating the application of ergonomics to cities. Different concepts have thus appeared, such as "spatial ergonomics" (Saint-Gérand, 2002), "urban ergonomics" (Antoni, 2014), "ergonomics of daily mobility" (Lanteri & Ignazi, 2005), "ergonomics at the service of public space" (Bouché, 2014), and "ergonomics of the city" (Lejeune, 2004).

J-P. Antoni presents "urban ergonomics" as "the design of a given space in compatibility with the various characteristics of activities or users in order to achieve greater comfort or efficiency" (Antoni, 2014). Ergonomics reviews the way in which individuals "move" to perform a job or task. It is more specifically interested in accessibility through the study of distances, urban landscapes, and risks in the city. The aim of this type of ergonomics is to maximize proximity to improve comfort and reduce effort. The author proposes a theoretical reflection on three-dimensional mobility in urban space, referring in particular to the work of (Reymond, 1998) on the "tridiastatic city" and the research of (Frankhauser, 1994) on fractal urban planning.

From a more practical point of view, the "ergonomics of the city" has been raised in the debates of the Centre d'Études sur les Réseaux, les Transports, l'Urbanisme et les Constructions Publiques (CERTU) to evaluate existing urban facilities or to pre-evaluate an urban project at the design stage (Lejeune, 2004). The CERTU bases its work on the postulate that the city, and in particular public space, is like a machine that city inhabitants used to achieve their goals. The objectives of city ergonomics include, among other things, improving accessibility, practices, and uses of urban space. This type of ergonomics emphasizes the human being as the main factor in the space and urban design with which he interacts.

In the same approach, other authors or professionals, such as G. Bouché, a consultant in ergonomics and architectural project management, focus on the psychological dimension of urban space, life scenarios, the study of people's real expectations, and the quality of urban furniture (Bouché, 2014).

Therefore, there are different notions of ergonomics in geography. These are generally the result of conceptual transfers from the notion of ergonomics in its original meaning. The common objective between the different notions of ergonomics in geography is to take into consideration the human scale, the human becoming therefore an element among the major elements of urban conception. The main goals are then to reduce the costs and efforts of citizens in the accomplishment of their daily tasks. Ergonomics is multiscalar and can study different processes for the realization of a task in the city while focusing on the interactions of citizens with their environment (physical, psychological, etc.).

Many authors have made connections between ergonomics and geography, but T. Saint-Gérand was the first author to establish the concept of "spatial ergonomics" in the most holistic and systemic way. This research mobilizes this overall concept.

2.2 Spatial Ergonomics as a Founding Concept and Operating Model

Ergonomics of access to the resources of daily life by the population, developed in this research project, is based on the broader concept of spatial ergonomics introduced by T. Saint-Gérand in 2002. For this author, "spatial ergonomics is the expression of the adequacy of space to its occupants, which translates into the ease with which the territory offers its occupants access to the resources they need at the lowest cost/effort/risk" (Saint-Gérand, 2002). "Cost" is considered in a broad sense and covers all the constraints associated with the mobilization of resources: distance, time, money, security, and comfort.

Spatial ergonomics uses reasoning based on classical ergonomics applied to geographical space.

A semantic translation and a change in conceptual scale were proposed by this author. The work action is assimilated to a daily life action, the worker to the population, the workshop to the daily living area, the materials to the resources, the tools to the equipment, the process to the socio-spatial behavior, and the efficiency to the adequacy to the urban needs (Fig. 1) (Hached, 2019; Hached & Propeck-Zimmermann, 2020; Saint-Gérand, 2002).

The main idea of spatial ergonomics, as mentioned above, is the adequacy of the territory/space to the life of its population. The hypothesis stated is that this adequacy depends to a large extent on the ease with which the territory offers all the territorial users to obtain the resources they need and to carry out all their activities. It must take into account spatial constraints, social constraints, the availability of resources, and their access at the lowest cost/effort. All these elements dynamically interact and form a complex system (Saint-Gérand et al., 2021).

The analysis of the ergonomics of a territory therefore refers to a systemic approach, which takes into account the overall functioning of the territory with the interactions and regulations that take place, at different scales, between different elements of the urban system. The population, the socioeconomic resources, and the space of the community form a triad of objects that are to be integrated into a data structure aimed at modeling geographical phenomena from an ergonomic perspective (Fig. 2):

Ergonomics	Spatial ergonomics
Workshop	Geographic space
Worker	Population
Working materials	Socio-economic resources
Working tools	Equipment / infrastructure
Working action	Daily life action

Fig. 1 From ergonomics to spatial ergonomics, according to (Saint-Gérand, 2002)

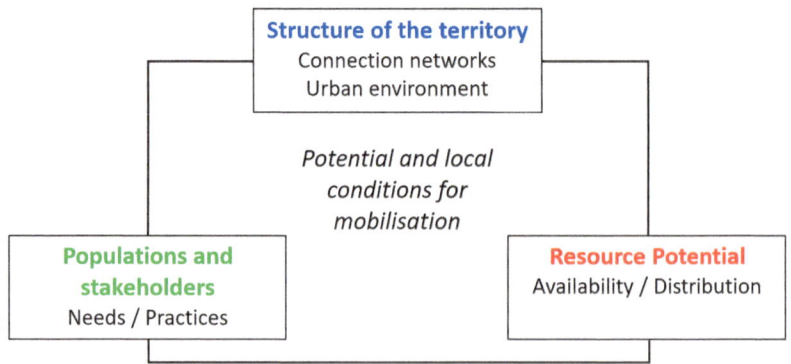

Fig. 2 General model of access ergonomics. (Source: authors)

- Populations (individuals, households, or other actors) have mobility needs and regulate their activities and practices, depending on their own characteristics, the availability and distribution of resources, and their environment, which is more or less suitable for travel by different modes.
- The availability and distribution of socioeconomic resources generate trips, allow the optimization of activity programs, or depending on the economic model (in particular with the help of digital technology), can be conveyed in part to consumers.
- The structure of the territory (the space of the community) is configured by the spatial distribution of populations, resources, functional mix, connection networks, urban environment, etc., which influences the conditions of access to local resources. It also determines the alternatives of resources and access that the territory can offer to users according to their socioeconomic profiles and the constraints of the moment. All the criteria linked to the territory's layout interfere with the demands/needs of the populations (all the territorial actors) and the potential of the territorial resources to form the potential and the local conditions of mobilization.

  Every territory has a level of ergonomics due to the way it is structured at a given time. An ergonomic territory can then be understood as a territory designed and developed to provide the society that inhabits it (individuals, households, companies, or other territorial actors) with the resources it needs at the lowest cost/effort/risk of mobilization. (Saint-Gérand et al., 2021)

Spatial ergonomics has connections with other concepts such as accessibility, capability, motility, walkability, etc. Their similarities and differences are explained in the following section.

2.3   Spatial Ergonomics, Accessibility, Capability: Close Links But Different Objectives

The investigation of spatial ergonomics raises the question of its links with notions such as accessibility, capability, motility, walkability, etc. While the data and methods used may have similarities, their purposes differ.

2.3.1   Spatial Ergonomics and Accessibility

Accessibility is a widely used concept, especially in the fields of transport, urban planning, and geography. It seems simple until one tries to define or measure it. To exist formally, in geography, "accessibility only requires a topographic space" (Dumolard, 1999), and it is a measure of spacing that determines the distances between different entities in space. Accessibility can be established as "the possibility, the capacity of a place or anything else to be accessible to an individual; that is, that one is able to reach, use, understand…" (Richer & Palmier, 2011) or as "the greater or lesser ease with which this place [can] be reached from one or more other places, by one or more individuals likely to travel using all or part of the existing means of transport" (Chapelon, 2004). It is therefore a broad concept that implies the accessibility of something or somewhere.

While ergonomics stems from the fields of improving working conditions and work efficacy, accessibility comes from the intersection of geography and spatial economics. While ergonomics focuses on the adequacy of peoples and territories, accessibility more broadly encompasses the spatial potential for developing existing or new activities from or at the destination of given places.

The investigation of accessibility makes it possible to evaluate the capacity of the urban environment, with its various components and infrastructure, to "reach a place in order to carry out an activity" (Richer & Palmier, 2011) or to fulfill a need.

We have referred to accessibility as the characteristic of a place that an individual can reach. However, for a place to be considered effectively accessible, several conditions must be met. These conditions can then be studied to evaluate the accessibility of places, especially in an urban space.

2.3.1.1   Connection Between Two Points

First, there must be a point of departure and a point of arrival, the latter corresponding to the destination or resource that the individual wishes to reach.

For the individual to get from their departure point to their destination, a link must exist that ensures "a spatial crossing between two points that respond to the person's reason for traveling" (Cerema, 2015). The path from the point of departure to the point of arrival may, however, be faced with impassable obstacles: rivers without bridges, no crossing of a railway line, or a major road. Nonmotorized soft modes are the most sensitive to the effects of barriers. Cyclists and "pedestrians as a whole appear to be the first victims of the effects of the barrier. As they are not very mobile, they are forced to make deviations or cross sloping passages" (Héran, 2011).

2.3.1.2   Means of Transport Adapted to the User

To get to their destination, individuals must also be able to move around by means that are adapted both to themselves and to the environment in which they are located: walking, cycling, public transport, cars, etc. This condition then depends on various parameters, such as the existence and quality of transport infrastructure (frequency of transit, safety, operating hours, etc.) and their match with the user's capacities in terms of time, distance, cost, quality, facilities for people with impaired mobility, physical capacities (age, disability, etc.) and financial means. Thus, "Accessibility can be measured by evaluating the area individuals can potentially reach within their time and mobility limits, or their (PPA) Potential Path Area" (Weber & Kwan,

2003). The "quality of service of the transport offer and the understanding of the complete travel chain of user" (Baptiste, 2003) are an essential part of the accessibility evaluation. This evaluation "also reflects the difficulty of the journey, the difficulty of connecting, which is most often measured by spatial and temporal constraints" (Chapelon, 2004).

2.3.1.3 Taking into Account the User Constraints

Finally, the accessibility of a location depends on the user's ability to withstand the constraints of the journeys offered by urban networks (physical capacity, effort needed, distance, duration, etc.) and his or her ability "to reach the goods, services and activities desired by an individual" (Cerema, 2015).

Considering all these elements, we can therefore notice that accessibility can vary in a given location. Indeed, it depends, first of all, on the individual, on his or her requirements, on his or her own constraints (in terms of physical and intellectual capacities and his or her schedule) which may not correspond or not be compatible with the targeted resource (e.g., its opening hours), thus resulting in a loss of accessibility.

> In addition, a temporal component should be integrated in the interpretation of accessibility, since it is influenced by the opening times that govern access to goods and services at different time periods of the day, by the amount of time that individuals allocate to these activities and by the quality of the transport system according to the different periods of the day (peak period, off-peak period, evening…). (Cerema, 2015)

Furthermore, accessibility also depends on variations in the connecting infrastructure system (closed tunnel or bridge, lack of lighting at night for a pedestrian, etc.) and the constraints caused by the means of transport used (public transport timetables, lack of parking facilities, etc.).

> There is particularization because instead of being an eternally true measure (the Euclidean distance between a and b will always be the same), accessibility results from the conjunction of elements that can be modified in time and space. It implies a travel operator (characterized by speed and energy consumption); a travel infrastructure (sophisticated infrastructure or simple paths); knowledge of the place to be reached and the path to do so. Each mode of travel has its own properties (an operator, a graph, a speed, a cost). Its accessibility can therefore be modified by changing the operator or the network or the time of travel. (Dumolard, 1999)

Accessibility is a huge field. There is a large body of literature providing a comprehensive approach to the history, definitions, measures of accessibility, and practical applications. However, four categories of accessibility measures have been identified (Geurs & van Wee, 2004; Salze et al., 2011):

Infrastructure-based measures: This type of measure is mainly used for the planning of transport networks. Its aim is to assess the efficiency of transport networks through simulations or observations. For this purpose, studies often take into account indicators such as "degree of congestion" and "average speed on the network."

Location-based measures: These measures are often used in the fields of urban planning and geography. They take into account the availability and spatial distribution of amenities in a given area, such as the number of bakeries accessible within 20 minutes. Other more complicated parameters can also be taken into account, such as the characteristics of the activities provided by the amenities or the consideration of competition between different resources.

Person-based measures: Often used in space-time geography, this is concerned with the assessment of space-time accessibility at the individual level, such as "the activities that an individual can participate in at a given time." This type of measure is based on the work of Hägerstrand (1970). They assess the limits of an individual's freedom of action according to, for example, his or her location, the duration of the activity to be performed, the travel-time budget, and the speed allowed by the existing transport system.

Utility-based measures: often used in economic studies, this measure assesses the ability of an individual or group of individuals to carry out a maximum number of activities in a given program. It also analyzes the (economic) benefits that individuals gain by accessing the activities distributed in the territory.

Depending on its goal, each of the four categories of accessibility measures focuses on well-defined components of accessibility but ignores others. K. Geurs and B. Van Wee propose a comparative table between the different categories of accessibility measures according to the indicators they take into account (Table 1):

Table 1 Perspectives on accessibility and components

Measures related to:	Transport element	Land-use element	Temporal element	Individual element
Infrastructure	Speed; time spent in traffic congestion		Peak period duration per day	Stratification by travel, such as home-to-work, commercial, etc.
Location	Time and costs of commuting to and from activities	Quantity and distribution of supply and demand of facilities	Journey duration as well as costs could fluctuate, from time of day, weekday, or season	Population fragmentation (e.g., according to household income, educational level)
Person	Journey duration to and from the locations of the activities	Quantity and distribution of facilities provided	Time-related limitations for activities and time to perform activities	The issue of accessibility is addressed at the individual level
Utility	Costs of commuting to and from the activity locations	Quantity and distribution of provided facilities	The duration and cost of journeys may vary, in particular from one hour of the day to another, from one day to another in the same week, or from one season to another	Benefits are considered at the individual level or at the level of a uniform population cluster

Adapted from Geurs and van Wee (2004)

To summarize, the concept of spatial ergonomics and the concept of accessibility have links and similarities. Accessibility in its broadest sense considers the greater or lesser ease for the inhabitants of a territory to carry out activities (Conesa & L'Hostis, 2010; Huriot & Perreur, 1994). However, despite probable similarities in the methods and data used, the objectives of ergonomics and accessibility differ. On the one hand, ergonomics seeks to understand the overall functioning of the territory and its capacity to meet needs through different elements, in particular the arrangement of its resource potential and its ability to respond to disturbances, whereas accessibility focuses on a very specific aspect of the functioning of the territory, that of travel and transport. While elements such as individual time use (Fosset et al., 2016) or accident risks (Cui & Levinson, 2018) can be taken into account in accessibility studies, the full systemic implications at the scale of the territory are generally not considered.

2.3.2   Spatial Ergonomics and Capability

Concepts such as capability, motility, walkability, and cyclability also have connections with ergonomics. Capability (in geography) is seen in this context as the ability (in the broadest sense: physical, psychological, cognitive, cultural, etc.) of an individual or a group of individuals to access a place that matches their needs. Indeed, an individual may dispose of resources (cinema, theatre, etc.), but he does not take advantage of them because these resources do not meet his needs, or they do not take into account his capacities (e.g., for a wheelchair user, a shop that is not adapted to welcome people with impaired mobility). Vincent Kaufmann considers that "Each individual has potential for mobility, the premises of movement, which he or she may or may not transform into movement according to desires and circumstances. This potential may not be strongly linked to mobility...." He then introduces the concept of motility, which he defines as "the capacity of a person or a group to be mobile, spatially and virtually" (Kaufmann, 2007). Walkability and cyclability are the pedestrian/cycling potential of an area. They reflect the capacity of a place to facilitate access on foot or by bicycle. The concept of walkability emerged in the early 2000s when American researchers began to focus on what they called the walkability of cities, their pedestrian potential. [...]. In Northern Europe, where cycling is widespread, research has been carried out on cyclability (Misery, 2013). The pedestrian (or cycling) potential of a place is determined by five main elements: housing density, diversity of activities, good location of activities, urban design, and location of public transport (VIVRE EN VILLE, 2016). It can be assessed at different scales: the parcel, the street or neighborhood unit, and the district.

Like accessibility, the concepts of capability, motility, walkability, or cyclability are concerned with a particular aspect of the functioning of the territory or its inhabitants. Ergonomics seeks to cover the different factors to characterize the ability of the territory to facilitate the real appropriation of the range of resources by the user in demand.

2.4 Territorial Modeling of the Ergonomics of Access to Daily Life Resources

Spatial ergonomics, as described above, integrates the potential of the territory in terms of availability of resources and conditions of appropriation by the population. It aims to build a global logic of description of the territory (physical characteristics of the territory, distribution of resources, practicability, alternatives, etc.) to establish multicriterion measures of the space in terms of provision of resources. In the framework of this research, the question raised is that of the conditions of access to the resources of daily life in active modes. Do active modes (walking, cycling) offer a credible alternative to cars for accessing the resources necessary for daily life?

To respond to this issue, the accessibility approach and related concepts (capability, walkability, etc.) have their limits. We propose a more global approach based on ergonomics.

The ergonomics of access to resources is the first step in the implementation of the global concept of spatial ergonomics. It focuses on the distribution of resources in the city and the conditions for accessing them in a reasonable time at the lowest cost/effort/risk. It underlines the importance of a quality urban space for access to resources and in this sense reminds us of the HQE (High Environmental Quality)[1] approach in architecture, which aims to improve the comfort and health of the users of a building while limiting its impact on the environment.

The ergonomics of access to resources cannot exist without the condition of accessibility being validated. In this study, accessibility is understood as the potential for travel, i.e., the greater or lesser ease with which a place can be reached from one or more other places, using all or part of the existing means of transport (Bavoux et al., 2005; Huriot & Perreur, 1994). However, ergonomics goes beyond accessibility criteria. Its evaluation is not limited to the capacity to reach a type of resource but includes a set of conditions of access to the resources that the population needs, according to different modes of travel and according to the offer of local resources, such as alternatives offered by the territory (Eliane Propeck-Zimmermann et al., 2018a, b). An ergonomically accessible resource must correspond to the needs of the citizen, be located at the shortest distance, offer an efficient connection, at the lowest cost, in comfortable conditions, and a quality environment.

The ergonomics of access to resources focuses as much on the material and physical conditions of the mobility system (infrastructure, resources, distances, etc.) as on the more immaterial variables of the urban space (landscape, safety, environment, "pleasant" quality of an urban space, etc.), accidents and monetary costs.

Within the framework of sustainable mobility policies, spatial ergonomics can contribute to the production of diagnostic tools to improve knowledge of the functional morphologies of the urban environment, of socio-spatial inequalities, and, in

---

[1] This approach includes 14 targets relating to eco-construction, eco-management, comfort, and health. The targets include, for example, hygrothermal comfort, acoustic comfort, visual comfort, olfactory comfort, air quality, water quality, and quality of spaces.

fine, to help define operations and developments adapted to a local level. The objective of this research work is to develop a method for evaluating the ergonomics of access to the resources of daily life by the population at each point in the territory and to analyze the socio-spatial disparities potentially induced by sustainable mobility policies.

3 Development and Implementation of an Evaluation Approach of the Ergonomics of Access to Resources

The development of a method for analyzing and evaluating the ergonomics of access to everyday resources is an exploratory and experimental approach that opens up a way of understanding the complexity that prevails in the field in the practice of everyday life. However, the study conducted initially required making fundamental choices to simplify the approach, choices that will obviously have to be taken into consideration when evaluating these first results.

3.1 Methodological Approach and Hypotheses

Geographic location is clearly a key factor in conditioning access to resources for any population. Ergonomics therefore considers the conditions of access to resources at the lowest cost/effort that a territory provides to its occupants in the location where they live (Eliane Propeck-Zimmermann et al., 2018a, b).

Access ergonomic level for everyday resources can be assessed on the basis of two sets of complementary criteria (Fig. 3):

1. The spatial distribution of resources and the conditions of access to proximate resources. In the common sense that proximity evokes neighborhood, contiguity, and short distance (Huriot & Perreur, 1998). Access ergonomics is conditioned by the spatial distribution of resources and by the means of accessing them and therefore by the structure of the territory.
2. Resources and access alternatives within an acceptable range. This means a potential choice of alternative resources and access modes considered as "close substitutes, i.e., which are capable of satisfying the same need" (Huriot & Perreur, 1998). The ergonomic of access is conditioned here by the notion of ductility/plasticity of space, or "plastic space" (Wood, 1978).

The resources of everyday life are targeted by this study by considering that they respond to a universal need. Ergonomics of access is then analyzed to evaluate the access of citizens to shops, schools, leisure activities, health services, and public services.

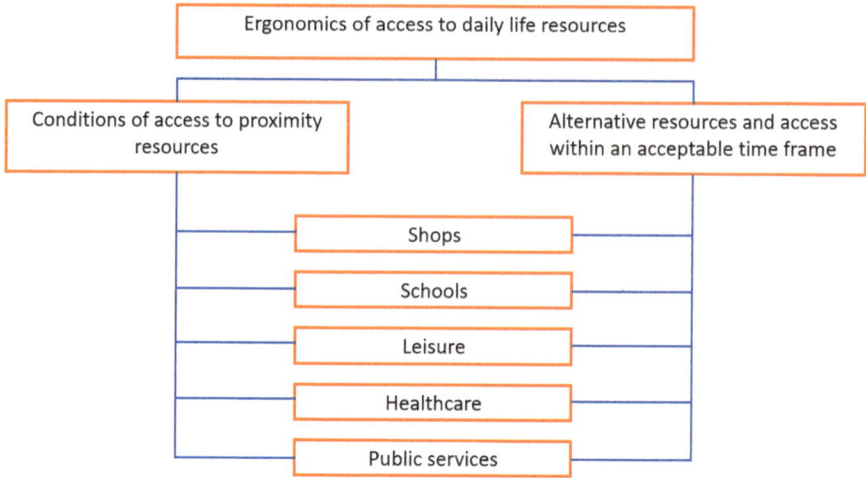

Fig. 3 Synthetic indicator of access ergonomics (Hached, 2019, 2020)

The social characteristics of the populations must be integrated into the model, and they also play an important role in assessing the needs (or demands) and the capacity of individuals to access and appropriate these resources according to their situation (average age, lifestyle, income, etc.). The characteristics of the population are taken into account in the second step.

The access ergonomics level assesses the more or less capacity to minimize the costs (in the broad sense) of appropriating the resources that the population needs in daily life.

Although access to resources may concern different categories of actors (individuals, workers, companies, managers, etc.), we were only interested in the access of inhabitants to the resources they need on a daily basis.

Currently, thanks to the development of the Internet, there are offers to deliver resources (services, goods, etc.) to the inhabitant or to a collection point. This type of offer, which is increasing but still little explored, although it can meet certain daily needs, has not been taken into account. Furthermore, only a standard range of equipment and services have been considered. For example, clothing and aesthetic shops have not been taken into account.

Different populations, from different social or professional categories, may have different needs or demands for resources. In addition, the level of demand (daily, weekly, etc.) may vary significantly from one category to another. The selection of daily life resources in this study ignores these differences, as it is supposed to match most inhabitants of the territory. However, a person's age, health condition, or disabilities have an impact on their ability to travel in daily life. This is particularly true for journeys made by soft modes: the speeds of trips made by these modes can vary, particularly according to age, while accessibility can be compromised in the case of disability and the absence of adapted facilities. We therefore assume—temporarily—in this research that the individuals considered are adults, in good health and without disabilities. The specificities of the older age groups, which will become increasingly numerous, will be introduced later.

We assume that the resources considered are available and therefore take into account travel during the daytime and on a working day.

We also assume that urban inhabitants make a trip from a starting point (home) to reach one resource at a time. Activity programs that combine several resources at once, although often part of the daily routine of households, have not been taken into account for the moment. They would require the development of a displacement model beyond the time constraints of this research. However, an indicator of resource dispersion was taken into account to partially remedy this limitation, reflecting the fact that the most grouped resources improve the ergonomics of access and that the most dispersed resources deteriorate it.

We also assume that the trips of the inhabitants of the city are made in a logical, thoughtful, and least effort manner (Lynch, 1960). The routes studied are then the shortest paths to the nearest resources (Zipf, 1949). These routes are studied to find the resources from the points of departure, assuming that the return path is identical to that of the original path.

The modes of travel studied are active modes, such as walking and cycling. They have benefited from specific facilities in the cities. Other light individual modes exist, such as scooters, rollerblades, and electric bikes. Scooters and electrically assisted bicycles can have an impact on access ergonomics, particularly on access times. Despite the fact that their use is growing rapidly today, this work does not take them into account. The method developed here could nevertheless be easily adapted to these other modes of travel by adapting, among other things, to the travel speeds.

3.2   Synthetic Indicator of the Ergonomics of Access to Resources and Its Variation in Different Levels of Information

Considering the predefined framework of the study, the approach consists of calculating a synthetic indicator at each point of the territory, according to different modes (walking and cycling compared to the car), based on criteria applied to the two sets above (proximity and alternatives). The synthetic indicator puts into perspective the sustainable mobility facilities and the types of socio-spatial environments and allows to evaluate the impact of the sustainable mobility policy on the potentialities of access to resources and characterizing the socio-spatial disparities.

3.2.1 Implementation of Two Joined-Up Approaches

To study the ergonomics of access, two different but complementary approaches were developed in parallel:

1. The first approach is global and aims to establish an overall diagnosis and a vision at the scale of the Eurometropole de Strasbourg of the ergonomics of access to current resources. It was developed within the framework of the National Research Agency (ANR) project on Emerging Risks of Sustainable Mobility (RED) (Saint-Gérand et al., 2021).
2. The second method developed in this chapter is more detailed. It is route-based and takes into account more criteria but on a limited number of test areas (Hached, 2019, 2020; Hached et al., 2018; Hached & Propeck-Zimmermann, 2020).

In this chapter, we will focus only on the second method.

3.2.1.1 Local Detailed Approach

This approach consists of refining the analysis of the ergonomics of access at a more local scale. It analyzes the distribution and conditions of access to resources and puts them into perspective with socio-urban environments to analyze the social disparities within the territory.

The territory is divided into 200 m × 200 m grids, which represent the most detailed INSEE grid for French socio-spatial data. From a starting point (centroids of an inhabited mesh), the overall approach involves four stages:

The first step consists of calculating the shortest path to the nearest resources in the chosen panel of resources. The calculation is performed for each transport mode (walking, cycling, cars) and for different time steps (5, 10, and 20 minutes). The closest resources for each type, with a defined time step, form a proximity zone. The "5, 10, and 20 minutes" travel times were defined on the basis of the relevance of the travel modes, the physical capacity of citizens, the travel-time budget, and the number of trips in the study area from the 2009 household travel survey.

The second step is the computation of the area of alternatives representing the whole space accessible by the networks, from the starting point, in a given time (5-minute, 10-minute, and 20-minute isochrones) and where additional resources are likely to be found, beyond the closest ones (Fig. 4).

Different criteria are then calculated, which refer, for a given time step, to

- The amount and diversity of resources in the catchment area
- The conditions of access to local resources on the paths (distribution of resources, safety, comfort, and monetary cost)
- The amount and diversity of alternative resources

Access alternatives to the resources (different paths, different modes of transport, especially public transport, or multimodality) can be integrated at a later stage.

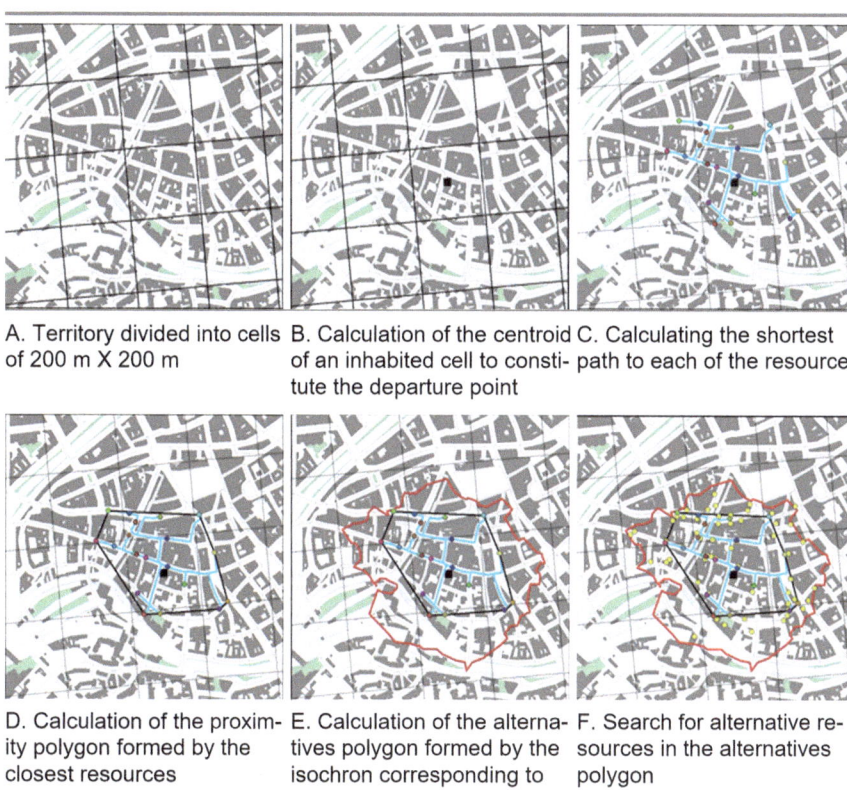

A. Territory divided into cells of 200 m X 200 m

B. Calculation of the centroid of an inhabited cell to constitute the departure point

C. Calculating the shortest path to each of the resources

D. Calculation of the proximity polygon formed by the closest resources

E. Calculation of the alternatives polygon formed by the isochron corresponding to the accessible area for the chosen time interval

F. Search for alternative resources in the alternatives polygon

Fig. 4 Calculation of the proximity zone and the zone of alternatives within a 5-minute walking distance (the resources studied are common to both approaches, by isochrone and by path). (Source: Author, Hached, 2020). (**a**) Territory divided into cells of 200 m × 200 m. (**b**) Calculation of the centroid of an inhabited cell to constitute the departure point. (**c**) Calculating the shortest path to each of the resources. (**d**) Calculation of the proximity polygon formed by the closest resources. (**e**) Calculation of the alternatives polygon formed by the isochrone corresponding to the accessible area for the chosen time interval. (**f**) Search for alternative resources in the alternatives polygon

In the last phase, a normalized score between 0 and 100 is attributed to each criterion calculated previously, with 100 representing, for a given criterion, the highest result of all the meshes of the study territory. For example, the score for the "number of resources in the proximity zone" will be equal to 100 for the grid cell from which the highest number of resources in the study area can be accessed and 0 for the mesh from which no resources can be accessed. The average of the scores of all the criteria (as presented below) provides the synthetic indicator of usability at each point of the territory.

The approach required reflection on the indicators and the development of a large localized database including the distribution of resources, infrastructure, roads, urban environment, etc., and its exploitation using the functionalities of a GIS.

3.3 Implementation of the Approach Within a GIS

Implementing the approach to assessing the ergonomics of access to everyday life resources by itinerary explained in the previous sections requires several steps:

- Selection of daily life resources.
- Calculation of itineraries based on the principle of the shortest paths.
- The selection and calculation of numerous indicators contributing to the ergonomics of access.
- The elaboration of a synthetic indicator translated into scores.
- To do this, an extensive localized database had to be created.

3.3.1 Selecting Everyday Life Resources

Work-related trips represent 26% of all trips in the Eurometropole de Strasbourg, according to the 2009 Household Travel Survey (ADEUS, 2010). Despite the significant proportion that this represents for daily travel, access to employment is not taken into account at this stage but should be integrated into future research. Daily life resources are the main concern of this study. They are defined as the resources, other than work, that people may need in their daily lives. They are intended to correspond to a universal need, without distinction as to social or professional categories. The chosen resources are then classified according to a typology that aims to group them by category and/or class.

Different typologies exist (Boudouda, 2019). The INSEE typology distinguishes, for all municipalities or irises,[2] ten main categories, classified according to their frequency of establishment (local, intermediate, higher-level equipment):

- Commercial
- Services to individuals
- Social action service
- Medical and paramedical functions
- Health services
- Primary education
- Second-level education
- Higher education, training, and education services

---

[2] Iris is a geospatial subdivision used for socioeconomic data in France.

- Transport, tourism
- Sports, leisure, and culture

The Institut d'aménagement et d'urbanisme Île-de-France lists five main categories (education, care and health, sport, market services, leisure), each of which contains several types of resources that we will not detail here (Mangeney et al., 2014).

Finally, the choice and classification of resources were the object of in-depth reflection and numerous debates, notably within the framework of the ANR RED project. The selection of everyday resources from the 2015 SIREN data was based on several parameters. First, we want to consider the resources that city dwellers frequently use in their daily lives on a daily, weekly, or monthly basis. In addition, there are other resources that are perhaps less frequently used, but their presence near the household is a desirable asset. Examples include public services (municipality, job center, etc.) and health services (hospital activity, dentistry, etc.). Consequently, resources that are occasionally required or are not necessary for everyday life have not been taken into account (clothing, cosmetics, etc.). Future developments could include attributing weight to resources according to their probability of a visit. It would have been interesting to include other resources related to sociability, such as children's playgrounds, green spaces, associations or cash dispensers, banks, relay points, etc. However, various practical difficulties were encountered: nongeographical referencing, classification confused with other resources, difficult data verification, etc., which led to their exclusion. The access ergonomics to everyday resources is then investigated to evaluate the accessibility of city inhabitants to the following five categories of resources: shops, schools, public services, health services, and leisure. Classes (subcategories from A to O) are defined for each category, taking into account more precisely the nature of the activities, the surface area of the activity, the number of employees, and the levels of demand by the population (Tannier, 2014; Tannier et al., 2014). These classes are useful for calculating the diversity indicator.

The list of everyday resources selected for this study (Table 2) can be expanded according to needs and data availability.

3.3.2 Creation of a Geographical Information System (GIS)

Analyzing the ergonomics of access to resources requires investigating the movement of a population with its own demographic, professional and income characteristics, mobility habits, etc., in a chosen territory that itself has intrinsic properties at both the administrative (boundaries, public policies, etc.) and geographical (relief, land use, etc.) levels. This territory is distinguished by its urban organization, whether it is built up or not. Travel to access resources is carried out using specific means of transport and networks adapted to the needs of the population and the characteristics of the area. The daily travel of the population in the territory creates traffic and flows while generating various accidents. The state and local authorities

Table 2 Daily life resources (Hached, 2019, 2020; Hached & Propeck-Zimmermann, 2020)

Categories	Class	Daily life resources
Commercial	A	Bakery and patisserie
		Patisserie
		General food trade
		Retail sale of fruit and vegetables in specialized shops
		Retail sale of meat and meat products in specialized shops
	B	Mini-market
		Supermarket
	C	Hypermarket
	D	Retail sale of tobacco products in specialized shops
		Retail sale of newspapers and stationery in specialized shops
	E	Retail sale of pharmaceutical products in specialized shops
	F	Marché
Schools	G	Preprimary education
		Primary education
	H	General secondary education
Healthcare	I	Hospital activities
		Surgical activities
	J	Activities of family doctors
		Activities of nurses and midwives
	K	Diagnostic and radiotherapy activities
		Dental practice
		Medical analysis laboratories
Public services	L	Municipal office
		Police station—Gendarmerie
		Medico-social centers
		Pôle Emploi (Job Centre)
Leisure	M	Cinema
		Museum
		Performance hall
	N	Library—Media library
	O	Activities of sports clubs
		Sports infrastructure

act on the whole system of territorial mobility through public policies applied to the territory, thus causing social, economic, and other changes.

The synthetic structure of this data model is composed of five families of data (hyperclasses): resources, territorial actors, territories (specific features of the territories and land use), infrastructure and facilities, and finally access costs. The elaborated database then includes the different elements necessary for the evaluation of the ergonomics of access in the area under study, taking into account the analysis method. The data used for this study concern the year 2015 and come mainly from INSEE or the Eurometropole de Strasbourg. Collecting and organizing this information into a structured database required several months:

- The data related to the population are provided by INSEE
- The data related to the territory and the network (boundaries, roads, etc.) were obtained from Eurometropole de Strasbourg.
- Some resources were georeferenced manually (markets), but most of them were extracted from the SIRENE file of INSEE.
- The accident data come from the BAAC file administered by ONISR and supplied by the SIRAC of Eurometropole de Strasbourg.

3.3.3 Itinerary Calculation

The application of the method of assessing the ergonomics of access to resources required the use of ESRI's "ArcGis 10.6" GIS software, and more specifically the "Network Analyst" module, which specializes in the calculation of itineraries. The starting points and resources were linked to the nearest roads within a 300 m radius. The following criteria were used to calculate the itineraries:

- The trips made are considered door-to-door, from the point of departure to the resource, as if the chosen mode of travel is immediately available. However, parking is included in the criteria for assessing the ergonomics of access. For car drivers, the journey to the nearest parking space on foot was taken into account.
- For cars and bicycles, it was necessary to respect the direction of circulation.
- Speeds were chosen by mode. Indeed, the speed adopted for walking is 4 km/h, for cycling 15 km/h, and for driving 30 km/h in the extended city center and regulatory speed elsewhere.
- The possibility of turning at junctions (intersections) was taken into account.
- The hierarchy of lanes has been ignored.
- The impedance chosen for the calculation of the shortest path is time. The calculated itineraries are the shortest in terms of time-distance (Hached, 2019).

3.3.4 Indicators for Evaluating the Ergonomics of Access to Resources

Many criteria are used to evaluate the ergonomics of access to resources. The diagram in Fig. 5 summarizes the indicators taken into account and explains how they are prioritized in order to obtain a synthetic indicator of the ergonomics of access. In practice, the indicators were chosen for their divisive nature in the study area (e.g., slope was not taken into account in the Eurometropole de Strasbourg, as our itineraries are all located in a flat zone). They are also adapted to the mode studied (e.g., parking is not considered for pedestrians).

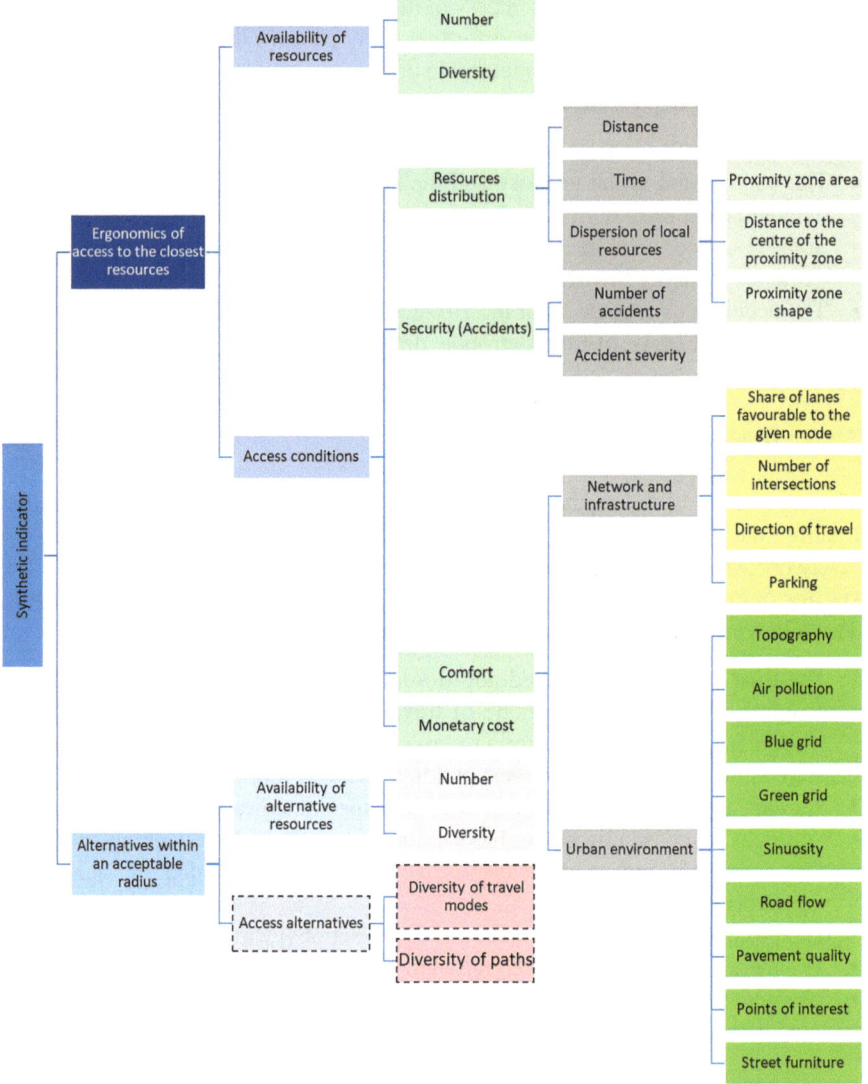

Fig. 5 Indicators of ergonomics of access to resources (Hached, 2019, 2020; Hached & Propeck-Zimmermann, 2020; Saint-Gérand et al., 2021)

The synthetic indicator of ergonomics of access depends on the combination of two indicators:

1. The first is related to the ergonomics of access to the nearest resources.
2. The second is related to the alternatives within a given radius (5, 10, or 20 minutes).

These are the result of the combination of the lower levels (Hached, 2019):

1. **The indicator of the ergonomics of access to the nearest resources**. It depends on two criteria. The availability of resources in the proximity area and the conditions of access to these resources.

   1.1. *The availability of resources in the proximity area* is defined by:

      1.1.1. *The number of resources available* from the panel of 32 everyday resources

      1.1.2. *The diversity of these resources*. The latter takes into account the number of accessible classes in relation to the total number of classes (15 classes of resources from A to O).

   1.2. *The conditions of access to the nearest resources* through 4 variables: distribution of resources, security, comfort, and monetary cost:

      1.2.1. *The resource distribution indicator* aims to investigate the distribution of resources in relation to the selected study points. It takes into account:

         1.2.1.1. The total distance to reach all accessible resources in a given time interval (5, 10, and 20 minutes). The further the resources are located, the more effort it takes to access them.

         1.2.1.2. The total time to access all resources as quickly as possible. The access time is proportional to the distance on foot and by bicycle but can vary considerably with the speed for cars.

         1.2.1.3. The dispersion of resources within the proximity polygon. If the resources are clustered together, it is easier to establish a program of activities.

            1.2.1.3.1. *The dispersion of resources* then investigates the area of the proximity zone. The wider the area, the further away the proximity resources are from the starting point and the greater the distances they have to travel to reach them.

            1.2.1.3.2. *The distance to the center of gravity* of the proximity zone informs about the location of the greatest concentration of resources in relation to the starting point. Thus, if the center of gravity of the proximity resources is located at a short distance from the starting point, this implies mostly short-distance daily trips.

            1.2.1.3.3. *The shape of the proximity zone* reflects the homogeneity of the distribution of resources around the study point. A circular shape means that resources are distributed in a homogeneous way around the starting point, indicating that the inhabitants have a higher probability of finding resources, regardless of the direction of their journey. A more elliptical shape shows that resources are concentrated in a main direction, which must be taken to access the majority of resources.

1.2.2. ***The safety indicator*** takes into account accidents involving the mode under study on access routes to local resources. Accidents are evaluated for each study point for each mode (walking, cycling, or driving) with regard to the following:

    1.2.2.1. Their number. The accidents taken into account are those involving at least one user of the investigated mode,

    1.2.2.2. The accident rate (number of accidents per kilometer)

    1.2.2.3. Their severity, which takes into account accidents resulting in serious injury or death.

    In addition to accidents, the safety indicator can have broader dimensions by taking into account other parameters such as incivilities (assaults, thefts, damage to bicycles, etc.), but these data were not available in our study area.

1.2.3. ***The comfort indicator*** evaluates, on the one hand, the comfort linked to networks and infrastructure and, on the other hand, the quality of the urban environment.

    1.2.3.1. Comfort linked to networks and infrastructure includes 6 variables defining the ease of traveling by a mode:

        1.2.3.1.1. *The proportion of facilities dedicated to each mode*, i.e., those designed specifically for a particular mode: bus lanes for buses, cycle facilities for bicycles (cycle tracks, cycle lanes, etc.), pedestrian streets for pedestrians and roads for cars (motorways and roads where the speed is generally higher than 70 km/h). These facilities attempt to reduce conflicts between different modes with different vulnerabilities and to improve the performance, including speed, of each mode by giving it more space in public areas. From a political point of view, this would ensure equitable sharing of the street between the different users.

        1.2.3.1.2. *The proportion of lanes favorable* to a given mode. These are the lanes that allow one mode to be favored over another. For soft modes, in addition to dedicated facilities, favorable lanes are those that have been the subject of policies to reduce the dominance of cars. Thus, all lanes with a reduced speed of 30 km/h or less (30 km/h zones, meeting zones, pedestrian zones) are favorable to soft modes (walking and cycling), and all lanes with a speed of more than 30 km/h are considered favorable to cars.

1.2.3.1.3. *The proportion of physically practicable lanes* that enable a flow of traffic that meets the current normative requirements in terms of width or number of lanes. For infrastructure (roadway, pavement, etc.) to be considered practicable, it must have a width greater than or equal to 1.4 m for pavements (Legifrance, 2007; Bruyere, 2014), 1.5 m for bicycle facilities (Bruyere, 2014; Fédération française de cyclotourisme, 2019) and 3.5 m for a car lane (Grandlyon, 2010).

1.2.3.1.4. *The number of intersections.* A high number of intersections, although contributing to the porosity of the urban space and offering users opportunities to change routes, is considered negative. Indeed, this study is based on the shortest paths (in terms of distance-time), and each intersection represents a slowing down and an additional effort of attention, in a way similar to the "space syntax" approach (Hillier & Hanson, 1984), as well as a higher risk of accident.

1.2.3.1.5. *The direction of traffic* was taken into account for both bikes and cars. For example, a two-way cycle facility is favorable to cyclists, as it provides extra width to facilitate overtaking. On the other hand, a one-way system is favorable to cars because of the absence of crossing with cars in the opposite direction.

1.2.3.1.6. *Parking* is a parameter for bikes and especially for cars. Users often try to park as close to resources as possible. The number of bicycle racks within 50 m of the resources was taken into account. For cars, the number of on-street parking spaces, the number of car parks within 50 m of the resource, and, inversely, the number of resources located in a tariff zone (where parking is charged) were taken into consideration.

1.2.3.2. <u>Comfort related to the urban environment</u>. The notion of adherence to the territory refers to the idea that interaction with the urban environment increases with the decrease in user speeds (Appleyard, 1980; Conesa, 2010). In this sense, soft modes would be the most sensitive to the quality of their immediate environment (weighting could be introduced in the analysis):

1.2.3.2.1. *Natural elements*: The green and blue grids bring nature into the city and partially hide some nuisances. For example, trees provide shade for walkers and cyclists (in summer, on sunny days). They reduce the perception of noise (psychoacoustics) by partially hiding the sources of noise pollution. In addition, trees planted between the pavement or cycle path and the roadway provide a feeling of safety by forming a barrier that protects soft modes from motorized modes. Watercourses, often lined with greenery, contribute to the animation of the urban space (swans, ducks, etc.). For example, water partially masks traffic noise. The green and blue framework is also important for motorists. They provide landmarks as well as a pleasant, nonmonotonous landscape. Air quality is an important element, even if it is more difficult to perceive, except for people with respiratory problems. Pedestrians and cyclists are more exposed to pollutants because their physical activity requires deeper breathing. Car drivers are less sensitive to air pollution because the car is typically a closed capsule with filtered air entering the cabin. The topography of the terrain can make mobility difficult. Cyclists and pedestrians are much more sensitive to slopes than motorists. The sharper and longer they are, the more physical effort they require.

1.2.3.2.2. *Sinuosity* is the ratio of the actual length of routes relative to the straight-line distance from the starting point to the resources. The more sinuous the routes, the longer they are and therefore the less efficient they are in providing connections between the starting point and the resources. Furthermore, the more sinuous the paths, the more difficult it is for the urban user to locate himself in space.

1.2.3.2.3. *Flow* is a data item that is often available for cars but rare for soft modes. It can be investigated in two ways. The flow of the same mode as the one studied, crossed with the number of traffic lanes, reflects the degree of congestion. The flow of modes other than the one studied reflects the inconvenience/risk experienced. For the purposes of this study, only car flow data were available for the entire study area. The higher the car flow, the more it was considered to be negative for car drivers (traffic jams, risk of accident, etc.) and for soft modes (noise, air pollution, risk of accident, etc.), consistent with Appleyard observations (Appleyard, 1980).

1.2.3.2.4. *The quality of pavements* has an impact on travel comfort but also on safety. For example, a good-quality surface reduces the risk of falling for pedestrians and slipping or loss of control for bicycles and allows better braking for cars.

1.2.3.2.5. *Landmarks* such as monuments, sculptures, and historical buildings contribute to the aesthetics of the urban space and form landmarks for moving around the city (Hillier & Hanson, 1984).

1.2.3.2.6. *Street furniture* has also been taken into consideration in terms of number and diversity. This equipment ensures the comfort of those who use soft mobility. Indeed, public seats allow people to take a break, to sit down, to put down their shopping, etc., fresh water fountains allow people to hydrate and refresh themselves in hot weather, while dustbins and public toilets (rare) provide appreciable comfort.

1.2.3.3. The monetary cost is a difficult indicator to assess. It is estimated according to the distance traveled, with a cost per kilometer per mode. It is strongly correlated with the distribution of resources. Furthermore, it is mainly of interest in the comparison between modes. This cost is difficult to evaluate. On the basis of scientific literature and local studies (Beauvais, 2012; Papon, 2002; SMTC, 2008; STIF, 2005) and excluding clothing costs for pedestrians and cyclists, an average price has been adopted for the different modes: on foot 0 euro/km, bicycles: 0.036 euro/km, and cars: 0.41 euro/km. This average price is subject to discussion. Indeed, walking is not free, but clothing costs (for pedestrians and cyclists) are difficult to assess. Moreover, whatever the mode, the figures may vary according to the studies carried out in different contexts. The main point here is to put the prices of the different modes into perspective for comparison.

2. **The indicator of alternatives within an acceptable radius.** It takes into account two parameters:

   2.1. *The availability* (number and diversity) of alternative resources, i.e., all resources accessible within a given time interval (zone defined by a network distance of 5, 10, or 20 minutes)
   2.2. *Alternative accesses* have not been developed in this research project, but they should take into account the diversity of travel modes (presence of public transport, multimodality, etc.) and the diversity of itineraries (choice of possible itineraries between the starting point and the studied resource in a given time interval).

3.3.5   Scores and Synthetic Indicator of Access Ergonomics

A score between 0 and 100 is attributed to each criterion calculated previously, where 100 represents, for a given criterion, the highest (i.e., favorable) result of all the meshes of the study area. For example, the score for the "number of resources in the proximity zone" will be equal to 100 for the mesh from which one has access to the highest number of resources in the territory studied and 0 for the mesh from which one has access to no resources. In contrast, for negative criteria such as accidents, a score of 100 is assigned to the grid cell with the fewest accidents and a score of 0 to the grid cell with the highest number of accidents. The average of the criteria scores (as presented below) provides the synthetic ergonomics indicator at every single location in the zone (Hached, 2019, 2020; Hached & Propeck-Zimmermann, 2020; Saint-Gérand et al., 2021).

The diagram in Fig. 6 summarizes all the criteria taken into account and explains how they are combined to produce a synthetic indicator of ergonomics of access. The latter represents the average of the scores of the indicator "Ergonomics of access to proximate resources" (EAPR) and the indicator "Alternatives within an acceptable radius" (AAR).

The scores for the overall indicators (EAPR and AAR) are obtained by the nested averages of the scores for the lower-level criteria. For example, the score for "Ergonomics of Access to Proximate Resources (EAPR)" is the average of the scores for "Resource Availability" (defined as the average of the scores for Number and Diversity) and the scores for "Access Conditions." The latter is defined by the average of the scores for four criteria: resource distribution, safety, comfort, and monetary cost.

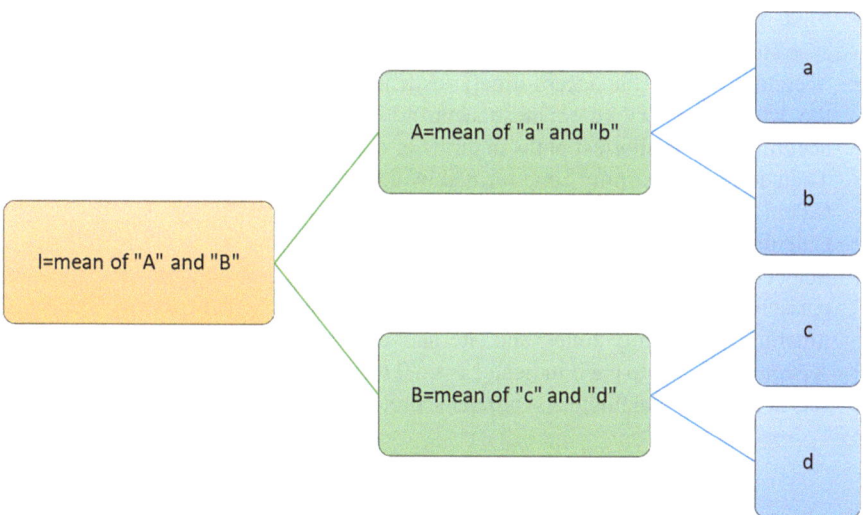

Fig. 6 Method of calculating nested indicators (Hached & Propeck-Zimmermann, 2020)

The "alternatives within acceptable radius" score is the average of the "availability of alternative resources" scores (the "access alternatives" are not taken into account in this study).

Access alternatives, requiring additional data and a specific and heavy processing chain, will be the object of further research. In fact, to not give too much weight to the availability of alternative resources, weighting was adopted: a factor of "2" for the indicator "Ergonomics of access to proximity resources" (EAPR) and a factor of "1" for the indicator "Alternatives within an acceptable radius" (AAR).

Finally, the synthetic indicator, calculated for each mesh on a scale of 0–100, allows the analysis of spatial disparities in terms of ergonomics of access to resources, all modes combined, and according to different modes or different time steps. In a second step, an in-depth study allows us to understand the combinations of criteria leading to a particular level and thus to provide information on the contribution of each criterion to the overall level. Putting the levels of ergonomics into perspective with a typology of socio-urban environments ultimately allows us to study the socio-spatial disparities in access to resources within the territory.

To summarize, spatial ergonomics, and more particularly access to resources in this context, is a geographical concept with an operational purpose. This means that the global and complex vision of territories, which is the basis of this concept, makes it suitable for experiments in spatial analysis that can be applied to several scales of territory and issues. In the subfield studied here, which focuses on the local conditions of appropriation of everyday resources using active modes, its assessment required the development of a methodology based on a GIS. This method consists of the following:

- First, define the resources that fit the needs (resources of daily life in this case study).
- Second, calculate the access itineraries from the starting points distributed over the whole territory (e.g., mesh centroids) to the closest selected resources, according to the shortest paths in terms of distance-time.
- Then define a proximity zone formed by the closest resources.
- Calculate the access itinerary to all alternative resources.
- Calculate the alternative paths of access.
- Select discriminating indicators adapted to the study area and allowing the comparison of the different zones. These indicators are applied to proximity resources, itineraries, proximity zones, and alternatives.
- Assigning scores to each indicator from 0 to 100, with 100 representing the best score, to finally calculate a synthetic score.

4 Application to the Eurometropole de Strasbourg

The implementation focused on the Eurometropole de Strasbourg. The territory has 2645 inhabited meshes of 200 m × 200 m. To test the developed method, 12 study areas were selected (Fig. 7). The choice of the test areas is crucial for making comparisons within the territory and revealing socio-spatial disparities. Two selection criteria were used: on the one hand, the geographical distribution throughout the Eurometropole, taking into consideration the urban morphology structuring the territory (center, planned center, first and second peri-urban ring), and on the other hand, the socio-environmental characteristics of the different neighborhoods resulting from a typology based on population data (age, household size, socio-professional categories, etc.) in their respective urban environments (land use, blue grid, green grid, etc.). The typology (carried out within the framework of ANR RED) distinguishes 8 classes represented on map 2. All the test areas were validated by urban planning experts at the Eurometropole de Strasbourg.

5 Results

The method of assessing the ergonomics of access (detailed and on a local scale) to the resources of daily life by active mobility was applied to the selection of selected study points distributed in the Eurometropole de Strasbourg, presented above. The assessment considered 5-, 10- or 20-minute walking and cycling trips and 5- and 10-minute car trips (Hached, 2019, 2020; Hached & Propeck-Zimmermann, 2020). For each location, a synthetic indicator of access ergonomics is obtained in the shape of a score between 0 and 100. This synthetic indicator can be analyzed and developed according to all the lower-level criteria that make it possible.

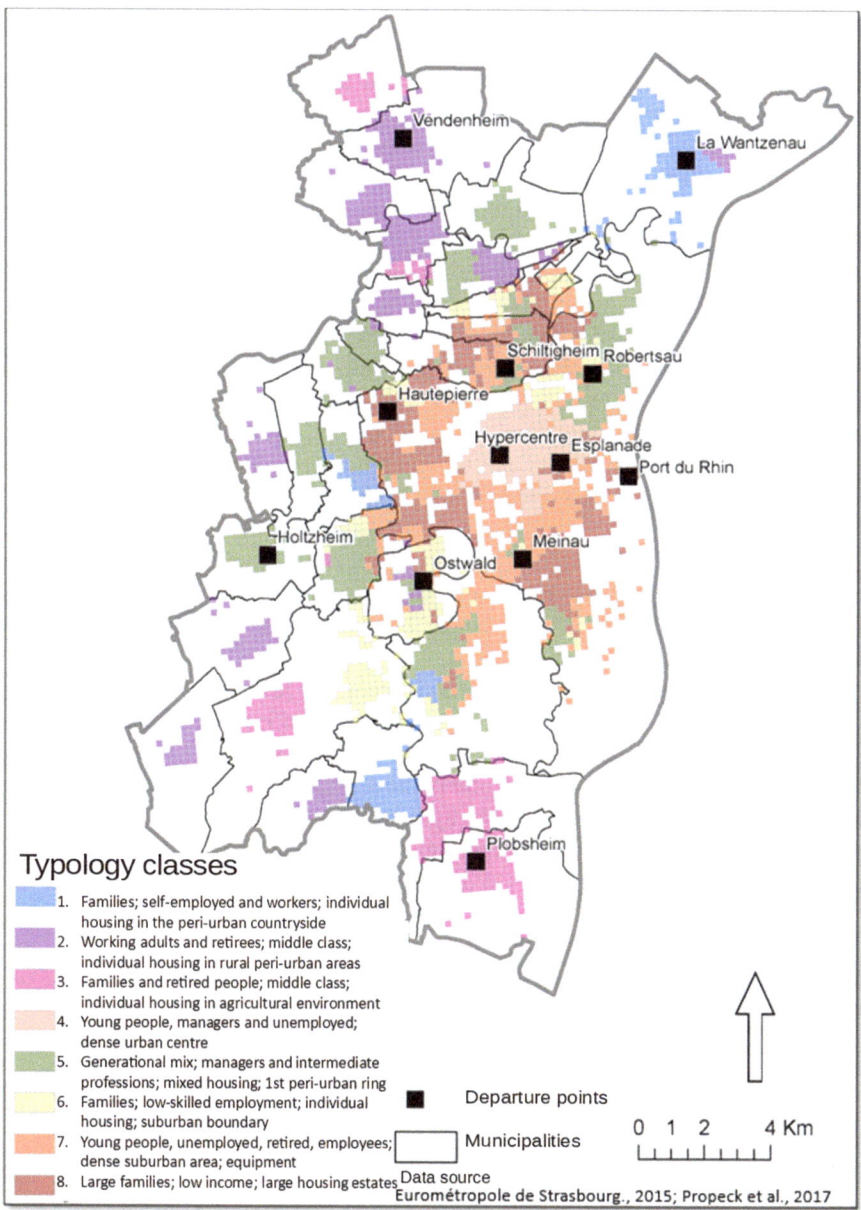

Fig. 7 Typology and investigation areas in the Eurometropole de Strasbourg (Hached, 2019; Propeck-Zimmermann et al., 2018a, b)

The results can be presented in two forms. The first, synthetic, allows a comparison of the level of access ergonomics between the different modes studied (walking, cycling, and cars) for each time step. The second, analytical, allows the analysis and comparison of the ergonomics profiles, i.e., the combinations of the different criteria leading to a given level of ergonomics (Hached, 2019, 2020; Hached & Propeck-Zimmermann, 2020).

The application to the 12 test areas shows disparities at the level of the Eurometropole de Strasbourg. Indeed, the synthetic indicator, for a time step of 10 minutes, varies from 41.0 to 86.2 for the bicycle and from 38.9 to 76.9 for the car (Table 3 and Fig. 8).

By bicycle, the highest ergonomics score (86.2) is located in the hypercenter of Strasbourg. This can be explained by long-standing developments in favor of soft modes (large pedestrian area, numerous cycle facilities), a high number and diversity of resources, and favorable access conditions with regard to the criteria selected. Conversely, travel by car has the lowest ergonomic score (38.9) due to facilities that discourage the use of this mode (study point located in the pedestrianized hypercenter).

Ergonomics by bicycle are lower in the inner ring and even lower in the outer ring. By car, the level of ergonomics is relatively homogeneous. Map 3 and Table 3 therefore show a center-periphery gradient, but this gradient is not perfect, and some places have very distinct characteristics.

The Robertsau, to the northeast of the city center, is less ergonomic to cycle than other points located at an equivalent distance. It is an affluent neighborhood with a generational mix where managers and middle occupations are overrepresented. The resources grouped together in the center of the district are numerous and diversified, but the potential resources accessible by bicycle within 10 minutes are generally lower due to a low-density urban environment that is relatively far from the other districts. The proportion of lanes dedicated to cycling is low, but conditions are favorable to the car.

Table 3 Scores of ergonomics of access to daily life resources in 10 minutes by bicycle and cars (Hached & Propeck-Zimmermann, 2020)

Point of departure	Ergonomics of access to resources by bicycle	Ergonomics of access to resources by car	Distance to city center in km
Hypercenter	86.2	38.9	0.0
Esplanade	73.6	76.3	1.8
Schiltigheim	69.4	76.9	2.6
Meinau	65.2	71.6	3.2
Hautepierre	63.9	69.6	3.6
Robertsau	61.4	66.8	3.7
Port du Rhin	41.0	44.6	3.9
Ostwald	54.1	63.4	4.4
Holtzheim	43.1	62.9	7.6
Vendenheim	49.9	57.2	10.0
La Wantzenau	60.6	64.1	10.6
Plobsheim	57.6	59.8	12.3

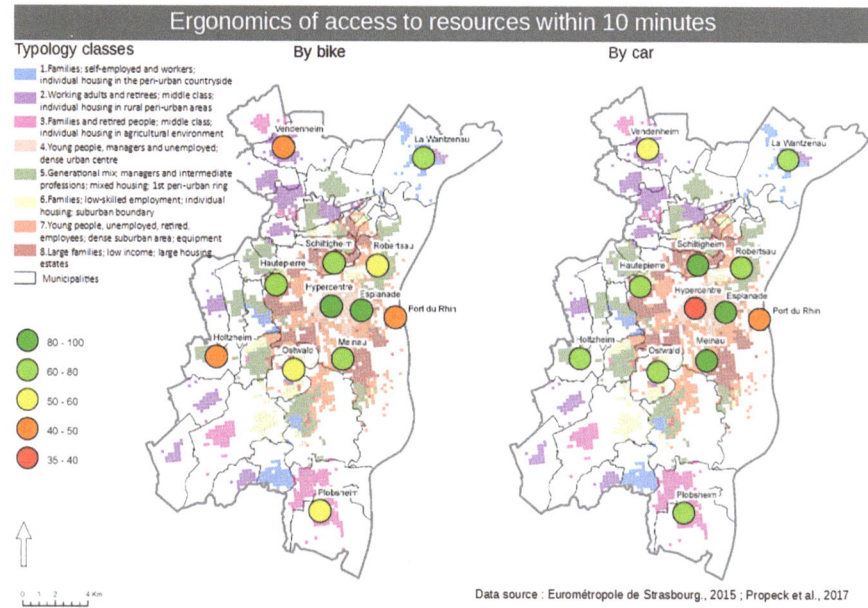

Fig. 8 Ergonomics of access to everyday resources within 10 minutes by bike and car (Hached, 2019, 2020)

The Port du Rhin (red dot in the east) also presents an atypical situation in the first suburban ring. It is a popular neighborhood with large families, low incomes, and large housing estates, with the lowest ergonomics by bicycle and one of the lowest scores by car. Resources are very limited, and alternatives are almost nonexistent due to the isolation of the area between an industrial zone and Germany. However, the result in this respect should be treated with caution due to the lack of data on the German side, which does, however, offer local resources (particularly by bicycle since the creation of a new bridge over the Rhine to the city of Kehl).

In the most densely populated peri-urban districts and suburban areas, the bicycle seems to be an alternative to the car for accessing everyday resources (numerous accessible resources with networks favorable to the use of the bicycle, bicycle racks, and urban furniture, especially in Esplanade, Schiltigheim, and Hautepierre). However, the ergonomic scores by car are still slightly higher.

The communes of the second ring are favorable to the use of bicycles thanks to the relative safety of travel and low exposure to road traffic.

In the second ring, the situation is more heterogeneous between municipalities of comparable size. In some cases, the overall score is relatively good, the resources are numerous and well distributed in the center of the town, and the urban environment is of good quality (La Wantzenau, Plobsheim). In others, the resources are more fragmented between retail outlets and large shopping centers, the network is more complex, and the number of serious accidents is higher (Vendenheim). Finally, in some communes characterized by a certain isolation from Strasbourg (such as

Holtzheim in the west), resources are limited and dispersed, although their ergonomics improve rapidly by car because of their integration into the network of surrounding communes.

The examples presented above show that although the synthetic indicator gives an overall level of usability that reveals disparities on the scale of the Eurometropolis, it is necessary to use combinations of criteria to explain these disparities. Moreover, the same score can correspond to very different ergonomic profiles.

5.1 From the Synthetic Indicator to the Exploration of Combinations of Criteria

Consider an example to illustrate spatially variable situations leading to similar ergonomic scores but with different profiles. These profiles can be related to the socio-urban characteristics of the neighborhoods.

Two dense suburban areas, La Meinau in blue and Schiltigheim in orange, are located approximately 3 km from the city center. They are both inhabited by managers and intermediate professions with a generational mix in mixed housing areas. Their scores, for 10 minutes by bicycle, are relatively close (65.2 and 69.4, respectively) but hide contrasting ergonomic profiles (Fig. 9). While the availability of

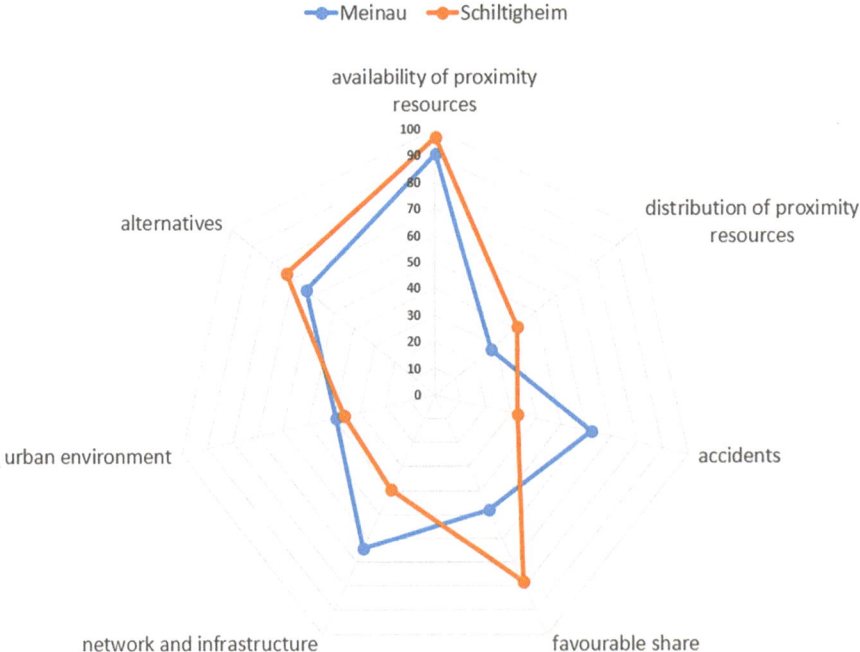

Fig. 9 Comparison of the ergonomics of 10-minute bicycle access at two study points with similar profiles but different scores (Hached, 2019; Hached & Propeck-Zimmermann, 2020)

resources appears to be equivalent, the two neighborhoods differ greatly in terms of networks and safety. Schiltigheim has more facilities for cycling but is included in a denser traffic network (in terms of road flow, intersections, share of two-way road network), and the number of accidents is higher. Conversely, in Meinau, the proportion of facilities in favor of cycling is lower, and there are fewer accidents in an environment marked by many one-way roads. This analysis needs to be completed, in particular by looking at the flow of bicycle traffic. It should also be borne in mind that even if particular care was taken in selecting the study points, the scores for the various criteria could vary significantly over short distances.

6 Conclusion

The research presented in this chapter is part of the contemporary issue of urban policies for sustainable mobility, particularly active modes. It questions the effectiveness of these policies in terms of access to the daily resources that people need. In this context, spatial ergonomics seems to be a relevant operational concept to reveal in a global and detailed way the situations favorable and unfavorable to these modes and to investigate the socio-spatial disparities induced by sustainable mobility policies.

On the one hand, a global approach to evaluate the level of ergonomics of access to current resources has been elaborated, taking into account the general functioning of the territory through the availability and distribution of resources and, on the other hand, a panel of criteria characterizing their access conditions.

Compared to accessibility approaches, our own contribution, based on spatial ergonomics, shares many criteria and methods related to the efficiency of the mobility system but places more emphasis on several criteria related to user welfare, in particular road safety and comfort. Resource accessibility criteria (number and diversity of resources accessible in a given time) reflect the effectiveness of a spatial configuration in optimizing travel and meeting the mobility needs of populations. Comfort can be included in accessibility approaches in the way that it contributes to influencing itinerary choices, but in ergonomics approaches, its justification is different: comfort will be considered part of the urban design, with the intention of favoring sustainable and active modes (share of lanes dedicated to soft modes, continuity of infrastructure, parking facilities, etc.). Road safety plays a role in the choice of the mode, but it also reveals problems of incoherence in the design of urban and peri-urban spaces, a dysfunction of a territory linked to the choices of development and urban organization. The criteria of alternatives/choice in terms of resources (number and diversity) and modes of access (in particular public transport to be integrated into the study) are intended to reveal the possibilities that the territory provides to respond to the constraints of the moment or to the differentiated needs of a population with different socioeconomic profiles (range of goods and services).

The application of the approach to 13 test areas, with the help of a GIS and an adapted database, has made it possible to detect disparities within the Eurometropole de Strasbourg to analyze the differentiated situations in detail, and to put them into perspective with the sociodemographic profiles. The different levels of information (maps that can be presented by mode of transport, by time step, and criteria by criteria) show how certain sectors allow easier adoption of active modes, point out locally the barriers to be removed to adapt the facilities and the urban environment to the mobility needs of the inhabitants, and provide information on the scope of residents to modify their mode of transport. The results clearly show, as expected, that the ergonomics are more favorable to walking and cycling than to the car in the hypercenter and that there are very different degrees of ergonomics for cycling in suburban areas. It is worth mentioning that the city's policy priority area has higher-than-average scores. More unexpected is the fact that active mode access can correspond to car access in peri-urban contexts, where urban developments have been closely associated with cars.

Implementation throughout Eurometropole has been initiated on the basis of certain key indicators (distribution of local resources, share of roads favorable to active modes, number of accidents, etc.). Many developments are still necessary to get closer to the real complexity on the ground. In particular, it seems crucial to include the issue of access to employment in the analysis. In the context of a sustainable mobility policy, it is also essential to take into account the public transport offer to evaluate the possibilities of intermodality with soft modes. Household travel surveys provide information on mobility practices, but they must be supplemented by field surveys to identify specific needs and practices and to assess the adequacy of urban planning to the needs and practices of people in their daily lives. These developments are envisaged within the framework of a research program of the A.N.R. URFé.

The ergonomics approach implemented therefore makes it possible to carry out territorial diagnoses to help local authorities develop effective sustainable mobility policies that meet the needs and expectations of users. However, beyond mobility, ergonomics aims to take into account more globally the issue of territorial ecology through a reflection on resources and alternative practices (e.g., short circuits, digital place in the process of supply of resources) at the heart of actual developments.

Acknowledgments This research was conducted within the framework of the A.N.R RED project. Our sincere thanks to all the participants in this project for their help, comments and remarks.

References

ADEUS. (2010). Enquête ménages déplacements. Résultats essentiels CUS (1988 - 1997 - 2009) [WWW Document]. Adeus. URL https://www.adeus.org/publications/resultats-essentiels-cus-1988-1997-2009/. Accessed 13 Apr 2022.

Antoni, J. -P. (2014). Modélisation et anticipations urbaines - éléments théoriques pour une approche géo-érgonomique.

Appleyard, D. (1980). Livable streets: Protected neighborhoods? *The Annals of the American Academy of Political and Social Science, 451,* 106–117.

Baptiste, H. (2003). Evaluer la qualité d'un service de transport collectif interurbain: L'exemple du réseau ferroviaire régional. In: Sixièmes Rencontres 2003. Presented at the Théo Quant, Besançon, p. 13.

Bavoux, J. -J., Beaucire, F., Chapelon, L., & Zembri, P. (2005). Géographie des transports, Armand Colin. ed, Colin U. Armand Colin, Paris.

Beauvais, J. -M. (2012). Estimation des dépenses unitaires selon les différents modes de transport en 2011.

Bouché, G. (2014). « Aller chercher les attentes des usagers » : entretien avec Gérard Bouché, consultant en ergonomie.

Boudouda, A. E. (2019). L'impact d'une centrale nucléaire sur l'offre en équipements et services à la population : analyse spatiale du territoire de Fessenheim.

Brussels Smart City. (2022). Definition | Brussels Smart City [WWW Document]. URL https:// smartcity.brussels/the-project-2-definition. Accessed 23 Mar 2022.

Bruyere, L. (2014). L'accessibilité à la voirie: L'essentiel de la réglementation et son actualité. Poitiers. https://docplayer.fr/63625921-L-accessibilite-a-la-voirie.html

Cerema. (2015). Mesurer l'accessibilité multimodale des territoires - État des lieux et analyse des pratiques.

Cerema. (2020). Définition: qu'est-ce qu'une smart city? [WWW Document]. Cerema. URL https://smart-city.cerema.fr/territoire-intelligent/definition-smart-city. Accessed 23 Mar 2022.

Chapelon, L. (2004). Accessibilité. HyperGeo.

CNIL. (2022). Smart city | CNIL [WWW Document]. URL https://www.cnil.fr/fr/definition/ smart-city. Accessed 23 Mar 2022.

Conesa, A. (2010). Modélisation des réseaux de transports collectifs métropolitains pour une structuration des territoires par les réseaux : applications aux régions Nord-Pas-de-Calais et Provence-Alpes-Côte d'Azur (These de doctorat). Lille 1.

Conesa, A., & L'Hostis, A. (2010). Définir l'accessibilité intermodale. In D. A. Banos & T.T (Eds.), Systèmes de Transport Urbain (Vol. IGAT, p. 24). Hermès.

Cui, M., & Levinson, D. (2018). Full cost accessibility. Journal of Transport and Land Use, 11, 661–679. https://doi.org/10.5198/jtlu.2018.1042

De Montmollin, M. (1967). Les systèmes hommes-machines. PUF.

Dumolard, P. (1999). Accessibilité et diffusion spatiale. Espace Géographique, 28, 205–214.

Fleury, D. (2009). L'ergonomie spatiale, réflexions sur une avancée conceptuelle. Transp. Urbains, 116, 3–8.

Fosset, P., Banos, A., Beck, E., Chardonnel, S., Lang, C., Marilleau, N., Piombini, A., Leysens, T., Conesa, A., André-Poyaud, I., & Thévenin, T. (2016). Exploring intra-urban accessibility and impacts of pollution policies with an agent-based simulation platform: GaMiroD. Systems, 4, 1–22. https://doi.org/10.3390/systems4010005

Frankhauser, P. (1994). La fractalité des structures urbaines. Economica.

Geurs, K. T., & van Wee, B. (2004). Accessibility evaluation of land-use and transport strategies: Review and research directions. Journal of Transport Geography, 12, 127–140.

Grandlyon. (2010). Cohérence des dimensions: Référentiel, conception et gestion des espaces publics. Granglyon Communauté urbaine. https://www.grandlyon.com/fileadmin/user_upload/ media/pdf/voirie/referentiel-espaces-publics/20091201_gl_referentiel_espaces_publics_ dimensions_coherence_dimensions.pdf

Hached, W. (2019). Ergonomie d'accès aux ressources de la vie quotidienne en mobilité douce : application à l'Eurométropole de Strasbourg [phdthesis]. Université de Strasbourg, Strasbourg.

Hached, W. (2020). Accès aux ressources du quotidien en modes doux à l'eurométropole de Strasbourg : une approche basée sur le concept d'ergonomie spatiale. Urbia Hors série, 7, 17–37.

Hached, W., & Propeck-Zimmermann, É. (2020). Mobilité douce et disparités socio-spatiales : évaluation de l'ergonomie d'accès aux ressources du quotidien. Territ. En Mouv. Rev. Géographie Aménage. Territ. Mov. J. Geogr. Plan.

Hached, W., Propeck-Zimmermann, E., & Piombini, A. (2018). L'ergonomie d'accès aux ressources de la vie quotidienne en mobilité douce à l'Eurométropole de Strasbourg. In *1ères Rencontres Francophones Transport Mobilité*. Lyon/Vaulx-en-Velin, France.

Hägerstrand, T. (1970). What about people in Regional Science? *Papers of the Regional Science Association 24*(1), 6–21. https://doi.org/10.1007/BF01936872

Héran, F. (2011). Comment Strasbourg est devenue la première ville cyclable de France.

Hillier, B., & Hanson, J. (1984). *The social logic of space*. Cambridge University Press.

Huriot, J.-M., & Perreur, J. (1994). L'accessibilité. In *Encyclopédie d'économie spatiale. Concepts - Comportements - Organisations*. Economica.

Huriot, J.-M., & Perreur, J. (1998). Proximités et distances en théorie économique spatiale. In *La Ville Ou La Proximité Organisée* (pp. 17–29). anthropos.

International Ergonomics Association. (2022). *What is ergonomics?*

Kaufmann, V. (2007). De la mobilité à la motilité. In *Enjeux de la sociologie urbaine* (p. 94). Presses Polytechniques et Universitaires Romandes [PPUR].

Lanteri, R., & Ignazi, G. (2005). *Accessibilité des espaces publics urbains, outil évaluation ergonomique*. Certu.

Legifrance. (2007). *Arrêté du 15 janvier 2007 portant application du décret n° 2006-1658 du 21 décembre 2006 relatif aux prescriptions techniques pour l'accessibilité de la voirie et des espaces publics*. https://www.legifrance.gouv.fr/loda/id/JORFTEXT000000646680

Lejeune, S. (2004). Et l'ergonomie de la ville, vous y avez pensé ? Pour une ville accessible à tous. CERTU, Lyon.

Lynch, K. (1960). *The image of the city*. The MIT Press.

Mangeney, C., Michel, Y., & Philippon, A. (2014). *Les polarités d'équipements et services en Île-de-France*. IAU île-de-France.

Misery, Y. (2013). *La « marchabilité », paramètre méconnu du milieu urbain [WWW Document]*. Lefigaro. URL https://sante.lefigaro.fr/actualite/2013/05/24/20580-marchabilite-parametre-meconnu-milieu-urbain. Accessed 13 Apr 2022.

Papon, F. (2002). La marche et le vélo : quels bilans économiques pour l'individu et la collectivité ? - 1ère partie : le temps et l'argent. Transp. Rev. Bimest., Transports : revue bimestrielle. - Paris : Les Ed. Techniques et Economiques, ISSN 0564–1373, ZDB-ID 8619591. - 2002, 412, pp. 84–94.

Papon, F. (2003). La ville à pied et à vélo. In D. D. Pumain & M.-F. Mattei (Eds.), *Données urbaines 4* (pp. 75–85). Anthropos.

Piombini, A. (2006). Modélisation des choix d'itinéraires pédestres en milieu urbain. Approche géographique et paysagère. Géographie. Université de Franche-Comté, Franche-Comté.

Propeck-Zimmermann, E., Liziard, S., Conesa, A., Villette, J., Kahn, R., & Saint-Gerand, T. (2018a). New mobilities and territorial complexity: Is the promotion of sustainable mobility risk-free for cities? The case of Strasbourg, France. In *Town and infrastructure planning for safety and urban quality*. CRC Press.

Propeck-Zimmermann, E., Saint-Gérand, T., Haniotou, H., Liziard, S., & Medjkane, M. (2018b). Ergonomie spatiale pour territoires résilienciels : approches et perspectives. VertigO - Rev. Électronique En Sci. Environ.

Reymond, H. (1998). Approches nouvelles de la coalescence. In *L'espace géographique des villes: pour une synergie multistrates* (pp. 21–48). Anthropos.

Richer, C., & Palmier, P. (2011). Mesurer l'accessibilité en transport collectif aux pôles d'excellence de Lille Métropole. : Proposition d'une méthode d'évaluation multi-critères pour l'aide à la décision.

Saint-Gérand, T. (2002). *S.I.G. : Structures conceptuelles pour l'analyse sapatiale, H.D.R.* Université de Caen.

Saint-Gérand, T., Propeck-Zimmermann, E., Hached, W., Liziard, S., Medjkane, M., Conesa, A., Piombini, A., Kahn, R., & Villette, J.-P. (2019). Ergonomie des espaces urbains : quel accès aux ressources du quotidien en mobilité durable ? In *Colloque RED, Les Risques Émergents de La Mobilité Durable*. Aix-en-Provence, France.

Saint-Gérand, T., Propeck-Zimmermann, É., Hached, W., Liziard, S., Medjkane, M., Conesa, A., Piombini, A., & KAHN, R. (2021). Mobilité durable et mobilisation des ressources : Une approche par l'ergonomie. In *Les Faux-Semblants de La Mobilité Durable* (p. 288). Éditions de la Sorbonne.

Salze, P., Banos, A., Oppert, J.-M., Charreire, H., Casey, R., Simon, C., Chaix, B., Badariotti, D., & Weber, C. (2011). Estimating spatial accessibility to facilities on the regional scale: An extended commuting-based interaction potential model. *International Journal of Health Geographics, 10*, 2.

SMTC. (2008). Plan de Déplacements Urbains de l'agglomération clermontoise. Syndicat Mixte des Transports en Commun de l'agglomération clermontoise, Clermont-Ferrand.

STIF. (2005). Compte déplacements de voyageurs en Ile-De-France pour l'année 2003.

Tannier, C. (2014). Evaluation de l'attractivité des commerces et services pour la simulation des mobilités quotidiennes sur l'agglomération de Besançon [2010–2030] avec la plateforme MobiSim. In: Projet ODIT [Observatoire Des Dynamiques Industrielles et Territoriales] 2012–2015, Chantier « Construction de l'espace Urbain et Périurbain ». Maison des Sciences de l'Homme et de l'Environnement.

Tannier, C., Houlot, H., & Epstein, D. (2014). Attractivité [masse] des activités de commerces et services - Version 2.

VIVRE EN VILLE. (2016). Potentiel piétonnier [WWW Document]. Collectiv. Viables. URL https://collectivitesviables.org/articles/potentiel-pietonnier.aspx#references-content. Accessed 13 Apr 2022.

Weber, J., & Kwan, M.-P. (2003). Evaluating the effects of geographic contexts on individual accessibility: A multilevel approach. *Urban Geography, 24*, 647–671.

Wood, D. (1978). Introducing the cartography of reality. In *Humanistic geography: Prospects and problems* (pp. 207–219). Maaroufa Press.

Zipf, G. K. (1949). *Human behavior and the principle of least effort: An introduction to human ecology*. Martino Fine Books.

Open Access This chapter is licensed under the terms of the Creative Commons Attribution 4.0 International License (http://creativecommons.org/licenses/by/4.0/), which permits use, sharing, adaptation, distribution and reproduction in any medium or format, as long as you give appropriate credit to the original author(s) and the source, provide a link to the Creative Commons license and indicate if changes were made.

The images or other third party material in this chapter are included in the chapter's Creative Commons license, unless indicated otherwise in a credit line to the material. If material is not included in the chapter's Creative Commons license and your intended use is not permitted by statutory regulation or exceeds the permitted use, you will need to obtain permission directly from the copyright holder.

Mobility Hubs, an Innovative Concept for Sustainable Urban Mobility?

State of the Art and Guidelines from European Experiences

Maxime Hachette and Alain L'Hostis

Abstract Mobility hubs bring together, connect and provide users with several modes of transport. Cities adopt them to help reach many objectives simultaneously, mainly the reduction of air pollution, congestion, and car ownership. Each mobility hub is unique, but many of them have similar characteristics that allow them to be classified. Various typologies exist. Although the mobility hub concept is flexible, the implementation of a mobility hub adapted to the needs and objectives can sometimes be complicated, as it requires many steps and may face difficulties at each step. Despite the simplicity of the mobility hub concept. They seem to be an interesting, complex, and challenging topic to investigate. As part of the Interreg Mobi-Mix project, we have taken a close look at mobility hubs. Based on bibliographic research and discussions with experts and cities, we established a state of the art that will help to better understand the concept. Without focusing on economic aspects, cities will benefit from different European experiences and a number of recommendations for a better implementation of mobility hubs.

Keywords Mobility hubs · Sustainable mobility · Shared mobility · CO_2 · Car reduction · Modal shift

1 Introduction

Mobility represents one of the main pillars of the smart city concept (Brussels Smart City, 2022). In recent decades, mobility policies and transportation services have been evolving rapidly, mainly in large and medium-sized cities. The aim of public authorities is to meet the targeted objectives of sustainable mobility. In this context, we can notice many changes in the urban environment, infrastructure, amenities, and services. In addition to the reinforcement of public transport, cities also

W. Hachette (✉) · A. L'Hostis
LVMT, Univ Gustave Eiffel, IFSTTAR, Ecole des Ponts, Marne-la-Vallée, France
e-mail: maxime.hachette@univeiffel.fr

© The Author(s) 2024   245
F. Belaïd, A. Arora (eds.), *Smart Cities*, Studies in Energy, Resource and
Environmental Economics, https://doi.org/10.1007/978-3-031-35664-3_14

encourage the use of active modes, mobility that moves away from vehicle trips, and ownership of shared vehicles. Public–private partnerships are also being organized to better engage the transition to more sustainable mobility. Within this framework, it seems that *"mobility hubs present an opportunity to integrate different sustainable transportation options to enhance connectivity. [...]. Mobility hubs have the potential to become a catalyst to prioritize low emission transportation options [...]"* (Aono, 2019).

Mobility hubs are perceived to be one of several solutions or a mix of solutions that cities and regions could consider for more sustainable mobility to overcome the "all-car model." These urban facilities are likely to offer significant advantages over already existing solutions, such as encouraging the use of public transport, multimodality, walking, cycling, and shared mobility. Indeed, locating various modes of mobility in the same place would increase the visibility of the modes provided. In addition, other advantages can be mentioned, such as helping to make transit easier, allowing the possibility of multimodality, giving a wider and more flexible choice, improving accessibility, and compensating for the lack of public transport in many areas (CoMoUK et al., 2019).

The mobility modes provided by the mobility hubs can be integrated into MaaS (Mobility as a Service) applications to contribute to a more efficient, modern, and digital transport system to facilitate users' transit, access to information, reservations, or payment, for example. Mobility Hub implementation is one of the results and a representation of mobility policies. This mobility policy itself is the consequence of wider ideological, social, economic, and environmental orientations. Mobility hubs could offer cities new concepts of urban planning and can be seen as a form of implementation of TOD (Transit Oriented Development). Mobility hubs therefore contribute to the connection between two dimensions of the smart city, the human-centered dimension (collective intelligence: placing people at the heart of the city, needs-centered approach, low-tech) and the technology-centered dimension (artificial intelligence: technological solutions, techno-centered approach, high-tech...) (Cerema, 2022).

Therefore, what are the mobility hubs? What are their main objectives? What are the different types of mobility hubs? What are the most relevant mobility hub projects, and what should we learn from them?

In this chapter, we are seeking to answer these questions. We will start with a summarized review of the literature. Then, without focusing on economic aspects, we will suggest lessons to be learned from European experiences, followed by recommendations for better mobility hubs. We will finish by briefly presenting two ongoing projects within the framework of the Interreg 2 seas project Mobi-Mix in Norfolk and Valenciennes and will introduce the method used for analyzing their impacts, particularly in terms of CO_2. The first results, which seem promising, will be reported succinctly.

Fig 1 "Mobility hub" research trend (Google, 2022)

2 Understanding the Mobility Hub Concept

Using the term "mobility hub" is estimated to have emerged in the 2000s but has become increasingly well known in the last two decades. The first searches for the term "mobility hub" appeared in the mid-2000s (Fig. 1). The frequency of searches for this term fluctuated, with increasingly less frequent breaks until 2011. Since then, interest in the concept has not stopped and seems to be increasing (Google, 2022).

2.1 Mobility Hub Definition

It is very likely that the term mobility hub in its beginnings was not based on a theoretical concept. To the best of our knowledge, no specific author claimed to be the founder. However, it would seem, although not with absolute certainty, that Michael Glotz-Richter from the city of Bremen is among the first to adopt the concept since the early 2000s (Gray, 2017; IMS, 2019).

To date, there is still a lack of scientific literature on this subject. Several definitions of the term "mobility hub" are used simultaneously. However, in the corpus of operational literature, we identified approximately a dozen definitions of the mobility hub. Therefore, various definitions have been and continue to emerge. Each proposed definition depends on the status of the author and his experience, plans, and goals. Mobility hubs can be defined based on different parameters, such as their use (private/professional/both), location, and the service they provide. One of the major goals is to contribute to more sustainable mobility by reducing the predominance of private cars (especially internal combustion engines) and helping to change mobility behaviors toward more sustainable practices. The latter favors soft/active modes, shared modes, public transport, etc.

We did our best to cover all the available literature, which, despite its rareness, continues to evolve rapidly. We have therefore tried, within the framework of an insight report during the Mobi-Mix project,[1] to retain only the definitions that we

---

[1] Third Mobi-Mix insight report: "Mobility Hubs, a lever for a more sustainable mobility?"

felt were the most relevant. Some of the definitions can be confused with other facilities that have existed for a long time (multimodal hub, shared vehicle station, etc.).

To bring more clarity to the subject, it would be necessary to open a debate between different stakeholders to establish a consensus-based definition that will be considered as a reference. At this stage, we believe that the keyword in mobility hubs is "shared mobility." Public transport is also important. In addition, a hub is a central point that brings together several elements. For greater inclusiveness and ease of implementation, we propose to consider simply that *mobility hubs are urban infrastructures that provide at least several (two or more) shared modes of transport in the same place*. The mobility hub can be connected to public transport and provide other modes or services, features, and additional considerations. These are positive features that will be considered quality factors.

The fact that a location with only one shared mode can be considered a mobility hub can be discussed. Indeed, it can be seen as a classic shared-mode station (bike, car scooter). In this particular case, it seemed to us that the term "mobility point" or "shared vehicle station" could be more appropriate. Otherwise, to be considered a mobility hub, it has to have at least public realm improvements that might help with the differentiation of some vehicle-sharing stations. Equally, if the public transport is very close and well connected to the shared modes provided, it is possible to consider this as a mobility hub. In this case, several mobility hubs already exist, even though they may have not been designed/labeled as such.

It is important to remember that to improve the articulation and efficiency of networks, planners traditionally and most often connect and articulate the modes of transport to each other. Bringing together different modes of transport in the same place, such as multimodal hubs, is already a classic and logical approach. The originality of mobility hubs can be seen above all in the field that they provide shared mobility and other accompanying services.

It should be noted that when we mention the same location, it should not be restrictive. It could be the same place well delimited and distinguished from the surroundings that offer, among others, several modes. It could also be a more fragmented structure where various modes are located in different spatially separated places (due to lack of space, for example). However, these locations should remain very close and easily identifiable with a continuous visual link between them. This is not necessarily the best configuration from the functional point of view.

2.2 Mobility Hubs Requirement

The definition of the term "mobility hub" we adopt assumes a minimalist approach. This makes the concept more inclusive and easier to implement in the local environment. We consider the fact that at least two shared mobility modes are located in the same place to be sufficient to constitute a mobility hub. However, there are several other parameters that could affect the quality of the mobility hub and its success.

In the literature, some components can be necessary to consider/label an area as a mobility hub. Metrolinx identified recurring essential features that are required for an area to be designated a mobility hub (Metrolinx, 2011; Aono, 2019). In the Metrolinx approach, which differs from our proposal, a mobility hub should surround a major transit station (airport, train stations, public transport), provide more sustainable transportation options than solo used private cars, and be located in areas with high residential and employment density. Therefore, an important feature of a mobility hub is that it is serviced by one or more higher-level public transport modes that constitute its core. This core is bounded by a catchment area that takes advantage of the services that the mobility hub provides (Metrolinx, 2011). Mobility Hubs should also provide services and offer the possibility of accessing nearby amenities within a 5-minute walk or by using the proposed travel modes. *"Therefore, vehicle sharing options are highlighted as a key component to incorporate into mobility hubs"* (Metrolinx, 2011). Mobility Hubs will also need to be located in areas with high residential and employment density, but precise values are rarely mentioned in the literature.

According to the literature, urban facilities that can be considered "mobility hubs" should, in addition to providing shared travel modes, also fulfill four additional criteria:

1. Providing sustainable transportation options
2. Providing services and offering the possibility to attend nearby amenities
3. Surrounding major transit station
4. Located in high-density areas

Although we agree with the first two criteria, we must stress that the last two criteria do not seem to us to be needed.

Being located near a major transit station is certainly a qualitative consideration. On the one hand, it guarantees more travel options, but, on the other hand, it is also a very limiting factor. By definition, major transit stations are rare. Therefore, linking mobility hubs to them is very restrictive, especially as they mostly offer short- and medium-distance efficient modes (bikes and scooters) and aim to compensate for the effects of the first and last mile. In addition, the definition of a mobility hub requiring a major transit station will be redundant with the definition of a more classical multimodal hub. Being located close to a major transit station or public transport station in general is, however, encouraging for multimodal trips.

Locating mobility hubs only in areas with a high population or employment density is probably more cost-effective than locating them in low-density areas. However, high-density areas have often benefited from successive mobility and service policies, particularly with regard to public transport. The shared modes provided by mobility hubs, in this case, erode more modal split from active modes and public transport than from the car. In addition, replacing journeys usually made on foot or by public transport with journeys made on electric scooters could be far from sustainable.

In addition, low-density areas are often far from city centers and are usually relatively neglected by mobility policies and relegated to the second level. The inhabitants of these areas, who regularly make long journeys, often have no choice but to use their private car. The solutions proposed to restrict this phenomenon are frequently limited to encouraging carpooling and setting up park-and-ride facilities at the entrance to cities. However, a well-meshed network of mobility hubs, with mobility hubs of different scales and with different services adapted to the location, could offer a real and reliable alternative to private cars.

2.3 Mobility Hub Objectives

When cities are planning mobility hubs, they usually establish a number of key objectives. The authors agree that the most important objective is *"the reduction of car ownership, car use and car use-related emissions"* (Aono, 2019; Claasen, 2020; Interreg NWE, 2019; SANDAG, 2019). Mobility hubs also provide alternative shared modes to private cars to help inhabitants to be mobile without needing a private car (Claasen, 2020; Miramontes et al., 2017). This then brings with it a number of benefits, such as fewer parking spaces needed on the street and a better and more efficient use of the available space (shareNL, 2018; Claasen, 2020) or enhancing equity and inclusivity, especially among seniors, people with disabilities and reduced income groups (SANDAG, 2019). Furthermore, it can lead to more connections between individuals within the same neighborhood on the basis of sharing (Claasen, 2020; ShareNL, 2018).

Aono has focused on seven main objectives:

1. Integration of sustainable transportation options
2. Improving user experience
3. Ensures safety and security
4. Develop a meaningful place-based identity through the introduction of significant and efficient placemaking strategies
5. The capacity to be flexible in introducing technological innovations and increasing resilience
6. Equity by ensuring that the accessibility and the availability of transport options within the various neighborhoods are being considered
7. Ability to forge meaningful partnerships (Aono, 2019)

The South–East Scotland Transport Partnership (SEStran) presented four main groups of objectives, each made up of various objectives. These groups are economy, accessibility, environment, and safety and health. The economic dimension aims not only to enhance connectivity through the inclusion of transport options and additional services but also to incorporate shared mobility to be complementary to the already established transport network (GO SEStran et al., 2020). In regard to the accessibility aspect, the aim is to promote inclusivity, especially for people with mobility impairments. The objective is also to improve accessibility for those with limited transport choices or no access to a car and to support people in their choices of transport by better integration and provision of information. With regard to the environmental objective, the aim is to primarily support low-carbon choices and reduce emissions. Then, it is about enhancing the use of shared mobility as an appropriate alternative to the use of private cars and making it easier to migrate to more sustainable and active modes of transport and therefore to lower the number of cars, as well as encouraging behavioral change. This is due to an easier and fluent modal shift to more sustainable modes of transport. Finally, the health and safety criteria remain important in two aspects. The first is to guarantee the safety of people who are using the hub. The second is to build a feeling of place and of community and to reassign space in the public realm in place-making and efficient use of land (GO SEStran et al., 2020).

We can therefore deduce that mobility hubs can be considered urban and political tools at the disposal of cities. Its aim is not only to consolidate their environmental policies in terms of mobility but also to achieve broader social, security, and economic objectives. This was confirmed by the expressed intentions stated by the partner cities of the Mobi-Mix Interreg project.

2.4   Mobility Hub Types

Although we can consider "mobility hubs" as places where at least two shared modes are provided, all mobility hubs are not equal. This broad definition extends the acceptance limits of mobility hubs. It offers mobility hubs many possibilities and combinations of roles, sizes, and quality. In this case, it seems legitimate to classify mobility hubs into different categories. Many cities and authors already classify them according to many parameters, such as size, energy used, and target users. *"These distinctions are essential in understanding mobility hubs as a multifaceted concept, where the local context shapes the hub typology. Additionally, these existing typologies can help inform how to classify hubs [...] in a way that suits the local transportation network"* (Metrolinx, 2008).

In this regard, based on their urban location and function, we can then distinguish between "regional mobility hubs," "community mobility hubs," and "neighborhood mobility hubs" (RTP, 2022):

- *Regional mobility hubs* serve multiple communities and regional activity centers. They have strong population and employment potential, resulting in substantial travel needs both coming from and going to these hubs. Among the public transport modes that can be considered are high-capacity public transport services (train and/or bus rapid transit), as well as both express and local bus systems. Regional hubs are distinguished according to their size, specification, availability, and type of public transport service, as well as their function.
- *Community mobility hubs* connect to major regional destinations and/or important functional entry points that provide interregional linkages, such as airports, emerging activity centers, universities and colleges, major parks and stadiums, and regional shopping *centers.*

- *Neighborhood mobility hubs* are placed alongside a high-capacity public transport line, which essentially ensures that residents of low-density, single-use areas not covered by the previous definitions will be able to access both high-capacity public transport services and local public transport services (RTP, 2022).

Another method to categorize mobility hubs is possible. The Regional Transportation Plan (RTP) for the Greater Toronto and Hamilton Area created by Metrolinx prefers to distinguish mobility hubs by their role in the transportation network. Metrolinx identified two types of mobility hubs. Anchor hubs and gateway hubs:

- The *anchor hubs* offer the potential to evolve the regional urban configuration and to be convergence nodes in the regional transport system. They incorporate the perimeter of the main public transport station and neighboring areas in the urban growth centers...
- The *Gateway hubs* are major nodes in the regional transportation network that are located at the junction of two or more currently operating or planned regional rapid transit lines and where significant passenger activity is expected (Metrolinx, 2011).

It is also possible to classify mobility hubs according to urban context and the transportation function the area serves. This method is used by Metrolinx. The goal is to make it easy to identify the *"specific needs and characteristics of the area"* (Aono, 2019; Metrolinx, 2011). Regarding the urban context, the mobility hubs are classified as follows: city center, urban transit nodes, emerging urban growth centers, historic town centers, suburban transit nodes, and unique destinations:

- *City Centre*: Such areas are densely populated regional centers with several destinations and therefore generate a large amount of employment and population. As a key destination, a multimodal environment with a high-quality walkable network is already in place. Due to the density of the surroundings, there is limited development land, and most of it will be on fill-in sites.
- *Urban Transit Nodes*: They refer to both major and local centers with moderately high to high density and a mixture of uses.
- *Emerging Urban Growth Centers*: In contrast to city centers and urban transport nodes, these areas have the possibility of development as land becomes more available. Unlike the two previous types of areas, they are typically more car-oriented.
- *Historic Town Centers*: Smaller town centers characterized by low to medium-density urban development. Such areas provide a combination of mixed development and a network of pedestrianized streets.
- *Suburban Transit Nodes*: These areas offer potential for further development because of the growing pressure for mixed-use facilities and the greater land availability. As with the emerging urban growth centers, these areas are usually auto-oriented.
- *Unique Destinations*: These areas, which are similar to the typology of destinations that are identified as Gateways and Anchor Points, both attract and engender

a large volume of activity and travel. For example, universities and airports are considered to be in this category (Aono, 2019).

Considering the transportation functions, the following sorting is used. Entry, Transfer, and Destination

- *Entry*: Those stations that have a considerable proportion of outgoing trips in the morning rush hour. Such areas are generally local public transport terminals, which have parking facilities for commuters' cars as well as bicycles.
- *Transfer*: They are transfer areas along the regional rapid transit system. Transfers can be made between rapid transit lines or other services delivered by several service operators.
- *Destination*: In contrast to the entry areas, the destination areas have a significant proportion of entering trips in the morning peak hours. Such areas are major destination zones with a high density of employment, recreational and institutional functions. Consequently, they are frequently covered by a high number of rapid transit lines (Aono, 2019).

LA Urban Design Studio (2016) employs these three typologies to categorize mobility hubs, which are neighborhood, central, and regional hubs. Such typologies represent the requirements for both the surrounding urban environment and the components of the hub (GO SEStran et al., 2020):

- *Neighborhood hubs* are smaller hubs that are based in low-density districts. These stations offer basic features along the street.
- *Central mobility hubs* are set within the urban context and provide more commodities, such as shared cars and bikes. These services are found all along the intersection and embedded in the district.
- *Regional mobility hubs* are the most significant hubs with regard to their scale and are within the context of densely populated urban areas. As a core area linked to other regional transit providers, these hubs have most of the features that are integrated into the station itself.

The Future Mobility Network is a knowledge and consultancy agency in the Netherlands, which is made up of a team of independent advisors and partners in actual and future mobility. They have been developing a number of different mobility hubs with a particular interest in electric mobility in Amsterdam, Nijmegen, Leuven, and Manchester. These so-called eHubs provide electric mobility and host infrastructure for local residents, commuters, and leisure travelers. Based on their model, there are four main eHubs that vary in size, location, and services provided. Their idea underlying these four categories is that the services of a hub should match the existing transport demand within that location. Therefore, the four eHubs are defined as minimalist, light, medium, and large (Aono, 2019).

- *Minimalistic*: This type of hub refers to a small-scale facility, where there is a minimum of one mode offered. The objective is to take advantage of the already existing infrastructure to have a minimal physical impact on the environment. The goal is then to use the existing infrastructure with minimum physical effect.

This hub includes components that can be easily set up or displaced and is appropriate for demonstrator projects. This type is not considered a mobility hub according to our proposed definition since it possesses a single mode of transport.

- *Light*: Similar to minimalistic eHubs, this category of hubs should be rather simple to implement or expand. However, they do involve at least two different options in regard to modes.
- *Medium*: Such hubs have a variety of different modes and are more permanent due to a mobility infrastructure with high physical impact. Furthermore, more space is required to host the various modes.
- *Large*: This refers to a large-scale hub that also includes multiple modes of transport but on a more extensive scale than medium-category hubs. These hubs tend to be oriented toward commuters and visitors.

Another hybrid topology combining size, location (urban environment), and services provided can also be used. This typology distinguishes large interchanges/city hubs, transport corridor/linking hubs, key destinations (business parks, hospitals, etc.), mini-hubs (or a network of mini-hubs), and market towns/village hubs (CoMoUK et al., 2019; GO SEStran et al., 2020):

- *Large interchanges/city hubs are characterized by the following*:

  (a) High transportation needs – a high volume of travelers to begin or to end a journey or to move from one mode to a different one.
  (b) Opportunity to decrease private car and taxi journeys by improving supply and raising awareness of sustainable modes of transport and by better connectivity of transport.
  (c) There may be limited availability of space, which may require a greater focus on prioritizing more sustainable and more efficient modes of mobility, as well as connections with both first- and last-mile modes of transport.
  (d) This category may also cover touristic destinations in urban areas.
  (e) Because of their large size, significant upgrades of the public area would probably only be achievable through a more ambitious project.

- *Transport corridor/linking hubs*

  (a) The main focus is on services that will connect inhabitants of the neighboring areas to the main transport infrastructure.
  (b) This is an opportunity to offer people more options for first- and last-mile travel.
  (c) Such a hub may also be implemented on Park and Ride (P&R) facilities and may also involve car parking.

- *Key destinations (business parks, hospitals, etc.)*

  (a) High user density.
  (b) It is necessary to ensure commuting links and back-to-basic options.
  (c) Based on areas that regularly and continuously attract a high number of visitors.

(d) *"Key destinations can include the following places: Universities and colleges; Hospitals; Tourist destinations; Business parks and key areas of employment; Industrial estates; Stadiums and event venues; Shopping centers; and Community centers."*

- *Mini-hubs (or a network of mini-hubs)*

    (a) Transport services are more limited, and demand is lower.
    (b) It is essential to guarantee that there are connections and solutions for returning to the base.
    (c) *"Mobility hubs may be developed to meet local needs, e.g., car clubs places to resolve parking issues, bike sharing or secured bike parking for flats without space to park bikes, or demand-responsive transport (DRT) to complete the limited bus network."*
    (d) The locations of the mini-hubs can be established in suburban environments or in new housing developments.
    (e) The network of mini-hubs has been successfully implemented in Bremen, Germany, but the concept has been expanding gradually, beginning with pilot projects for larger hubs.

- *Market towns/village hubs*

    (a) Assess local requirements such as restricted public transport with shared electric bike fleets.
    (b) Where space is available, these areas can be used to carry out a much wider spectrum of services, as long as there is a critical mass to guarantee viability.
    (c) Some small town/market town hubs may also serve as tourist centers (consider those services where visitors can register with ease, which may subsequently boost the viability of the service on a seasonal basis for rural inhabitants) (GO SEStran et al., 2020).

Through these different classification approaches, Table 1 shows that it seems that the most recurrent and influential element on which the choice of the type of mobility hub is based on size and the urban environment (location). Particular attention is rightly paid to the existing transport infrastructure and offer. In addition to these parameters, it is necessary to point out that the population density and activities (work, leisure, education, health…) have a significant impact on the classification of mobility hubs and defining their size and of the services they provide.

To achieve more sustainable and fairer mobility, and above all in line with the various social, economic, and environmental policies of the city or region, it would be more judicious to proceed with the development of a global but evolving action plan. It will serve as a thoughtful guide with a long-term vision for setting up a network of mobility hubs. It will take into account the different characteristics of the urban environment, the population, the current or planned transport infrastructure, and other parameters if necessary. A single mobility hub, whatever its size, will certainly only have a one-off impact, and it is unlikely that a tangible change in

Table 1 Different classifications of "mobility hub"

Author	Classification	Criteria
RTP (2022)	1. Regional 2. Community 3. Neighborhood mobility hubs	Size, location, density, levels of travel generated, existing public transport...
METROLINX (2011)	1. Anchor 2. Gateway mobility hubs	Role in urban structure, existing transit lines...
METROLINX (2011) Aono (2019)	1. City center 2. Urban transit nodes 3. Emerging urban growth centers 4. Historic town centers 5. Suburban transit nodes 6. Unique destinations	Location, density, urban context, land availability, volume of activity and travel...
Aono (2019)	1. Entry 2. Transfer 3. Destination	Trip direction (outgoing, transit, entering), facilities and features, public transport...
Aono (2019) e-hubs	1. Minimalistic 2. Light 3. Medium 4. Large	Size, features (number of modes), ease to implement, physical impact...
LA Urban Design Studio (2016) GO SEStran et al. (2020)	1. Neighborhood 2. Central 3. Regional	Size, location, density, urban context, features...
GO SEStran et al. (2020)	1. Large interchanges/City hubs 2. Transport corridor/linking hubs 3. Key destinations (business parks, hospitals, etc.) 4. Mini-hubs (or a network of mini-hubs) 5. Market towns/village hubs	Size, location, density, level of travel, urban context

travel habits and modal shares will result at the city or region scale. For example, *"in the UK, Nexus had implemented the local hub idea in a single free-standing location at Ryton, west Gateshead, in 2002, but this proved to have a number of problems with it in practice and was closed in the late 2000s. We believe that this was another factor in the UK failing to embrace the mobility hub concept in the 2000s/2010s"* (mobihub.com, 2022). However, a network of mobility hubs will cover more territory, inhabitants, and passengers. Each of the mobility hubs that make up the network must obviously be adapted to the local context and to the role it plays in the global network. This adaptation could be visible through the size of the mobility hub, the modes, links, and the services it offers. The establishment of such a network could begin with the installation of test mobility hubs, which will be used to better adjust future mobility hubs.

3   Insights to Be Gained

Following the analysis of different European mobility hub projects,[1] we have tried to draw essential recommendations to cities and to make them benefit from the relevant practices: Bergen[2] and Stavanger in Norway,[3] Amsterdam in the Netherlands,[4] Flanders[5] and Leuven in Belgium[6] and Bremen in Germany.[7] This selection of projects is the result of a literature review as well as participation in the e-HUBS academy event of 2021, which allowed us to learn more about each project and to discuss in more detail with different stakeholders. Of course, many other mobility hub projects outside Europe can be mentioned, such as in Hong Kong (Zielinski, 2007), Vienna (GO SEStran et al., 2020), Scotland (Intelligent Transport, 2021), Plymouth City Council (Plymouth, 2022), Manchester (Tague, 2021), Linz (GO SEStran et al., 2020), San Diego (SANDAG, 2022), Toronto (Aono, 2019), Denver (Aono, 2019), Chicago (Aono, 2019), and Vyttila.

Each territory is unique. The space and the environment in the broad sense (natural, urban, political, legal, social, economic) that it offers for each mobility hub is just as specific. Therefore, it seems necessary to recall that to achieve the assigned objectives, each of these facilities must be adapted to its own specific context. An adapted solution should be provided. In this sense, each mobility hub will then be unique (size, vehicles offered, number of vehicles, services). *"There is not a perfect*

---

[2] Børjesson (2022a, b), ESPON (2022), Ove Kvalbein (2021), Stavnes Hisdal (2021), The Explorer (2020a), SHARE North et al. (2019), and Karbaumer (2018).

[3] ESPON (2022), Henrik Haaland (2022), Stavanger Kommune (2020, 2021a, b, c), Dirks Eskeland (2021), University of Stavanger (2021), e-MOPOLI (2020), The Explorer (2020b), Thorsnæs (2020), and Kleiner (2020).

[4] City of Amsterdam (2020, 2022a, b, c, d, e), City Ratings (2022), ESPON (2022), FUB (2022), eHUBS (2022), N-W Europe (2022), Basta (2021), Gemeente Amsterdam (2021), I amsterdam (2021), Intertraffic (2021), Liao and de Almeida Correia (2021), Copenhagenize Index (2019), Coya (2019), and Gemeente Amsterdam (2019)

[5] Be. Brussels (2022), Flandre (2021, 2022a, b, c, d), Statistics Flanders (2021, 2022a, b), Mpact (2022), SHARE North (2020, 2022), VISITFLANDERS (2022a, b), Belga (2021), CoMoUK (2021b), De Muelenaere (2021), Intertraffic (2021), Roelant (2021), Saelens (2021), Times (2021), Bailey (2020b), eHUBS (2020a), GO SEStran et al. (2020), CoMoUK et al. (2019), and Flanders Environment Agency (2018).

[6] Citypopulation (2022), eHUBS (2020b, 2022), European Commission (2022), ESPON (2022), KU Leuven (2022), Leuven (2030), NuMIDAS (2022), VISITFLANDERS (2022a), Evenepoel (2020, 2021a, b), Schmalholz (2021), Asperges (2020), VisitLeuven (2020), Leuven MindGate (2019), and Ripa (2019).

[7] Bremen (2022a, b, c, d, e, f, g, h), ESPON (2022), Karbaumer (2018, 2020, 2021a, b, c, 2022), Universität Bremen (2022), Transit Forward (2022), Wegweiser Kommune (2022a, b), Austin (2021), Chamberland et al. (2021), CoMoUK (2021a), Intertraffic (2021), Movmi (2021), ARUP (2020), Bailey (2020a), Bremer et al. (2020), GO SEStran et al. (2020), Lanagarth (2020), Actionfigure (2019), Aono (2019), European Commission (2019), IMS (2019), Pais (2019), SHARE North (2018a, b), Gray (2017), Miramontes et al. (2017), Frei Hansestadt Bremen (2014), Fairfax County, Virginia (2013), ITDP (2012), Britaninica (2009), and The Big Move (2008).

solution for mobility hubs, and the approach to planning and implementation of each hub will need to be tailored" (GO SEStran et al., 2020).

A network of strictly identical mobility hubs may not be ideal. This solution, which is typically designed to fit the majority of users, could prove to be effective on a city-wide scale. However, it does not consider the disparities and specificities of each area, which nevertheless are very diverse in the city. It could therefore create or reinforce inequalities. With an increasing concern for equity and to overcome disparities, it would be more appropriate to provide a multiscale solution. This means the development of a network with hierarchical mobility hubs (in terms of importance in the network, size, services offered), each of them adapted and specifically designed to answer the most local specificities.

This does not exclude the possibility of learning from other experiences. In this sense, several authors recommend key points for a successful mobility hub. For example, RTP highlighted six components they find necessary for a successful mobility hub:

1. Multimodal transportation facilities and services
2. Economic activity
3. Intensified/concentrated land uses and urban density
4. Pedestrian facilities and accommodations
5. Embedded technology
6. A strong sense of place (RTP, 2022)

Therefore, to fulfill the above conditions and to go further, the implementation of a mobility hub will necessarily be preceded by a substantial preparation period. This phase of the project follows several phases, in particular the integration within the urban mobility policies of the city.

There are a variety of tools that can be adopted to support the implementation of mobility hubs. They include zoning regulations, a global parking strategy, and the identification of potential development in the catchment area. This involves the creation of a master plan for the mobility hub area. The goal is to help ensure that new transport installations are adapted to the various modes of transport, that they also encourage and support changes in modal split, and that they facilitate living and working possibilities (RTP, 2022).

There are several individual aspects that define and characterize mobility hubs and are well documented and studied in the available literature: optimal location, key characteristics and components, and leadership on their development. Within this framework, S. Aono, from Translink, followed three steps. She first reviewed common phases used for planning mobility hub implementation. Second, she outlined several *"partnerships and responsibilities involved in mobility hub creation, both internally and externally."* This is under the *"four main different topics of planning, services and elements, land development, and funding."* Then, she explored existing strategies *"to understand common approaches used by existing mobility hub studies."* Finally, she identified *"other key considerations and common challenges found in mobility hub implementation"* (Aono, 2019).

In the same vein, Go SEStran states that *"establishing new mobility hubs can take time and requires careful planning – working with multiple partners on a complex development may not happen fast or easily"* (GO SEStran et al., 2020). Like any other urban infrastructure, the implementation of one or more mobility hubs requires a series of steps such as planning, implementation, management, and maintenance or adjustment. It is worth pointing out here that it would be wise to involve at least future target users and local residents in these various steps. It would allow us to achieve a more consensus-based mobility hub that would better correspond to everyone's expectations. This kind of collective and inclusive planning will also strengthen local democracy and offer more transparency to citizens and users. We will develop here some major phases for the establishment of one or more mobility hubs: before, during, and after the implementation.

The idea of creating mobility hubs can be driven by a regional or national constraint in favor of more sustainable mobility. It may also express a more local desire. This desire arises from the awareness of elected representatives and/or citizens of the challenges of sustainable development and/or from the recognition of local problems. The latter are mainly related to the quality of urban space and mobility (i.e., congestion, noise or air pollution, accidents, etc.). It is important to point out here that convincing politicians is important during all steps of the project, especially where environmental policies are not yet considered a priority. It is also essential to stress that there are usually important negotiations with politicians, different departments of the territory, and other institutions concerned with mobility policy. This requires important and crucial coordination work.

When mobility hubs may appear as a suitable solution. Depending on the definition adopted, mobility hubs could take on different aspects and help to address issues that go beyond mobility (strengthening local life, inclusiveness, equity, etc.). Once the mobility hub option is chosen, the planning stage can begin. For this reason, communication between the city, residents, and users is already highly recommended for a more effective acceptance of potential future mobility hubs.

This first step can be considered a preplanning phase. Its goal is to develop *"a Vision and Framework for mobility hubs"* (GO SEStran et al., 2020).

While there may be political and public will, the implementation of mobility hubs or a network of mobility hubs requires a significant amount of time (1–2 years). The process includes several steps. We have highlighted seven of them and detailed the minimum measures included in each phase. In the following order:[1]

1. *Regulatory checking, feasibility, and integration*

   • Analysis of the regulatory context
   • Consider what funding is already available or can be made available
   • Consider economic aspect

2. *Urban analysis*

   • Analyze the needs and demands
   • Consider the urban environment in detail
   • Ensuring an adequate contribution to the various objectives of the city
   • Define clearer and more precise objectives

3. *Planning the (network of) mobility hubs*

- Adopt a global vision
- Prioritize the mobility hubs
- Identify one or more test areas
- The exact location of the mobility hubs can be precisely delineated
- The conception of the graphic documents could be launched

4. *Building the first mobility hubs*

- Identify the most appropriate stakeholders and partners
- Use public procurement
- Respect the principles of competition and foster innovation

5. *Impact measures and adjustment*

- Define specific indicators
- Collect a significant amount of data, negotiated in advance with the operators, continuously or regularly
- Measure indicators
- Adjustments may be necessary

6. *Generalization and wider implementation*

- Anticipate problems and be better prepared to deal with them
- Generalize the mobility hubs and thus create a more complete and more efficient network
- Choose new locations according to the results of the mobility hub tests

7. *Adaptation and permanent improvement*

- Maintain the attractiveness
- Reach increasingly ambitious targets
- Improvements (of vehicles, services, facilities, etc.) can continuously be considered

3.1 Choice of the Type of Mobility Supply

The choice of mobility modes to be provided in a mobility hub is often a delicate step, as it partly influences the success of the hub in achieving its objectives. It is important to be attentive to the needs and expectations of users and local residents. In some cases, it is necessary to take into consideration the goals of organizations covering large areas and their proper mobility plans (universities, businesses). However, this should not exclude the possibility of innovation. The involvement of associations and private partners is also important. It would be better to first focus on working with existing providers in the region. If there is none, it would be necessary to consider the need to launch services as a network (as in Bremen and Bergen). Isolated services may have more difficulties and work optimally only in very few cases. The partners, providers of shared modes, based on their experience, can use their own methods to evaluate the potential success (particularly in terms of use and profitability) of one or other modes in any given location.

In all cases, several parameters must be considered. We can cite as examples, without being exhaustive, land use, density, multimodal transportation network, transit density and service level, density of destinations, community demographics, individuals' ability to access transportation options, cost, efficiency, reliability, safety, and enjoyability of the options available, policy and programmatic structure already in place such as parking areas, cost of parking, shared mobility service areas, and similar (Crowther et al., 2020).

3.2 Choice of Partners/Providers, the Mix of Mobility Solutions

Like most urban operations, the implementation of a mobility hub often requires the mobilization of many and diverse participants and stakeholders. *"And the process of implementing mobility hubs is most successful when all responsible authorities for land use planning, urban design, transportation planning, and transportation engineering are all integrated into the design of a corridor"* (O'Berry, 2015). A mixture of public, private, political and associative stakeholders is not rare! Indeed, it is quite the opposite. We recall that the use of a pedagogical approach, a communication strategy, and participative and local democracy is also recommended. In this way, citizens and users are important partners. The several stakeholders involved in the success of a mobility hub should not work separately from each other. They should all be seen as committed partners, mobilizing their resources, knowledge, experience, and know-how for the success of the collective mobility hub project. *"As a concept that involves several public and private services, a key element in mobility hub implementation is partnerships"* (Aono, 2019).

Among the stakeholders involved in mobility hubs, we can highlight, for instance, the following: *"public transport operators, local community groups including residents and businesses, other government agencies and transport authorities, landowners and property developers, not-for-profit organizations including disability and other community groups, technology providers, major employment sites and other key trip generators, assets, infrastructure and utility companies, other established mobility hubs"* (GO SEStran et al., 2020). S. Aono states that *"the type of partnership and the stakeholders involved can vary across four main categories that are involved in mobility hub implementation"*: planning, services and elements, land development, and funding. Based on the work of S. Aono and discussions with European cities and private partners, Table 2 summarizes the roles of different

Table 2 Roles of different stakeholders in mobility hubs, based on the work of S. Aono and discussions with European cities and private partners (Hached, 2021)

Planning	Municipal government	Sets objectives for mobility hubs
		Adapt plans, regulations and urban planning documents
		Encourages local democracy by involving users and citizens in the different phases of the project
		Incorporating policies that promote mobility hubs and transit-oriented development in citywide plans
		Guide development around mobility hubs through different planning tools and incentives
		Encourage development while reducing processing times
		Initiate and develop a mobility hub plan
		Selects the operators in the mobility hub through public competitions or call for tenders while ensuring competition
		Regulates and controls the use of vehicles (provided in the mobility hub) in the public space
	Public transit agencies	Increasing service levels and improving transit infrastructure in a way that enhances customer service. This includes accessibility, safety, furniture, service, and information elements,
		Coordinate schedules both among different transit services and with the surrounding employers and institutions so transfers are made easily and match employee schedules
		Help address equity and accessibility by setting a precedent for subsidized fare programs
		Can reserve or create spaces in their station plans to lease for commercial and retail uses
		Can integrate mobility hubs into their stations or surroundings
		May consider mobility hubs as a way of improving access to public transport, especially in terms of the first and last mile
		Continuously assesses the impacts of the mobility hubs and their alignment with the objectives based on, among other things, anonymized data provided by the various partners. (If necessary, recommend improvements.)

(continued)

Table 2 (continued)

Services & Elements	On-demand ride-share agencies	Can form partnerships with transit agencies to encourage trips to and from major transit stations to ensure these services complement each other
		Utilize ride-hailing services to reduce parking demand
	Vehicle share services	Can encourage transit use through partnerships with the local government and/or transit agencies
		Integrate shared mobility with existing transit services is through an integrated access card
		Supply, maintain, and manage vehicles
		Opt for the least polluting vehicles
		Ensure that users of their vehicles comply with safety and local regulations
	Technology companies	Can help produce and operate mobile payment or trip planning app
		Obtain valuable travel data such as popular travel destinations and preferred travel modes
		Provide the necessary data for the city
		Promoting a variety of transportation modes
		Incorporate transportation options into their MaaS app
		Design and operate services that integrate payment and information technology for transportation services through their branch
		MaaS app provides information to the user regarding the available transit services near their location, while also acting as a mobile ticketing kiosk
	Wi-Fi providers	Critical to enhance user experience during their travel
		Direct sponsorships from advertising or technology companies
		Charging users for the Wi-Fi service
		Partnering with service providers
	Business Improvement Associations (BIA)	Can ensure transit plazas are utilized by holding public events and festivals that support local artists and community culture
		Help maintain the area as a clean and safe place and contribute to placemaking through initiatives that promote safety and active street uses
		Aligns with the common objectives of mobility hubs, where placemaking and safety are valuable elements for a successful mobility hub
		Potential for further partnerships with BIAs as mobility hubs present several elements that will help flourish local businesses such as shuttle services or on-demand ride-hailing
		There are opportunities to collaborate with BIAs to fund certain hub initiatives

(continued)

Table 2 (continued)

Land Development	Private developers	Help promote transit-oriented development by developing mixed-use buildings near mobility hubs
		By utilizing city incentives, developers can contribute to the incorporation of public art, public spaces, cycling or pedestrian amenities in new development
		Help develop buildings that accommodate both private and public sector agencies to achieve diversity in land use and achieve higher density
		Help connect public infrastructure with private buildings
Funding	Federal/provincial government	Help cost-share transportation investments and capital projects
		Recognizes the potential for senior levels of government to encourage development around mobility hubs by locating federal or provincial facilities near potential mobility hub locations.
	Sponsors	Help provides funding
		Sponsors provide payments to support the bike share system or components of the bike share program in exchange for branding on the bikes and stations

stakeholders in mobility hubs (Aono, 2019). Note that the stakeholders and roles may vary slightly from one project to another and from one country to another.

It is often the city, as manager of the urban space and project owner, who assumes the responsibility for the creation of mobility hubs. It has the leading role. The city initiates contact with various actors (mobility, energy, etc.) and brings them together around the same project. Although some cities could manage mobility hubs on their territory themselves, the majority outsource the service to private partners and providers.

The selection of private partners is often a delicate step, as the interests of private partners must be reconciled with the various objectives of the city. These include the social objectives of equity, safety, resilience, innovation, competition, etc. Cities frequently use a call for proposals to ensure competition. It also clearly specifies the conditions, obligations, and limits of the various future contractual parties and the objectives of the mobility hubs. The selection is therefore made on the basis of the best responses. Often, at least two (usually three) private partners are selected. The aim is therefore not only to ensure competition and innovation in the long term but also to foster resilience.

Once private partners are selected, this does not usually mean that they have a completely free hand. The city should still have the authority to control, adjust and adapt regulations. Some cities are very sensitive to the reactivity of private partners in solving problems that may arise and adapting to the requirements (temporary parking bans, speed limits in certain areas, provision of data, etc.). Although this concerns shared scooters, we can mention the Norwegian city of Bergen. It has developed internal software that allows communication in a fluid way with private partners. It allows them to locate each of the shared scooters in their territory and to display information on the identity of the private partner who manages them, the level of their battery charge, the last time they were used, etc. This system makes it possible to detect, for example, a large concentration of scooters in a particular area of the city and a lack of them elsewhere. The city then immediately informs the private partner, who is then required to dispatch the vehicles in a more harmonious manner. The partner then has limited time to meet this request, and if he fails to do so, he may be subject to financial sanctions (or even suspension of the partner's license and therefore a ban from operating). The particularity of the system developed by the city of Bergen is that it is collaborative. City agents and every citizen can report problems with shared scooters (such as parking problems) via a dedicated application. The priority of solving problems by private partners is obviously given first to the city and the city agents. Through this software, the city can specify on the map of the city the zones of traffic/parking of scooters that are allowed or not, set speed limits, open temporary parking places... These modifications are communicated in a fluid way to the private partner, which allows him to integrate and adapt them rapidly.

3.3 Difficulties During Mobility Hub Implementation

As with most actions in the urban space, planners will be confronted with difficulties and opposition when setting up mobility hubs. The difficulties encountered can take several forms (legal, territorial, economical, social, cultural), and the opposition can come from different profiles (inhabitants, local companies, associations, politicians). In this sense, some authors, such as S. Aono, list various difficulties commonly encountered. Among the possible challenges, we can mention, in a non-exhaustive way, the following examples: parking demand, land ownership, misalignment between transit and development, and equity considerations (cost of services, language and cultural barriers, accessibility) (Aono, 2019). To face these challenges, involving the different protagonists from the beginning of the project, taking into account their opinions and concerns, using pedagogy, diplomacy, and seeking consensus can, generally, be useful to solve many of these problems and improve the overall acceptability of the project.

In addition, when designing mobility hubs and selecting private mobility providers, cities focus on these social issues. In this case, engagement activity through local charities that already have strong community links can be an important tool (CoMoUK, 2021c). Some cities negotiate and/or condition the selection of private partners by an effective commitment to provide an equivalent service in all areas of the city. Private partners are often reluctant to consider the economic profitability of mobility hubs and the safety of their vehicles. However, some cities take care to minimize these risks by offering subsidies and more privileged locations.

4   Ways Forward: How Can Cities Advance

Most modern cities, especially large metropolises, offer undeniable advantages on the one hand but generate negative externalities on the other. The latter actually raises various and serious problems both on the environment (air pollution, water pollution, noise pollution, lighting) and on their inhabitants (deterioration of health, insomnia, stress, fatigue, security). Because of their urban forms, zoning (mainly induced by the Industrial Revolution and postwar urban planning), and the distribution of resources (work, shops, services, etc.), cities separate the places of demand and the places of supply. This inevitably generates greater or lesser distances to travel and thus greater or lesser mobility flows.

In addition to these negative effects of the urban planning of recent decades on humans and the environment, large cities create strong centralities and such a regional weight that it blurs neighboring towns and villages, gradually emptying them of their shops, services, etc., reducing their ability to retain or attract jobs, and thus emptying them of their inhabitants. This cycle of spatial specialization will contribute, on the one hand, to the sprawl of large cities and the accentuation of their negative externalities and the gradual decline of neighboring cities. In one way or another, the need to travel greater distances may increase regionally to reach the major city and within the major city itself.

The city (or urban space) is thus a very complex spatial object where different issues coexist (social, economic, environmental). The urban planner should always keep in mind a very broad and global vision when conducting any action on the territory. He should surround himself with a large panel of specialists and listen to citizens and various stakeholders to make the most balanced decisions possible. This is far from an easy task. Every intervention on the territory, whatever form it takes, can have positive effects in one sector and negative effects on the other. It is in this context that T. Saint Gérand's concept of spatial ergonomics takes on its full usefulness (Saint-Gérand, 2002).

To date, few cities can be proud to offer a sufficiently solid infrastructure capable of ensuring sustainable, peaceful, and equitable mobility. In this context, we can mention the detailed analysis of the ergonomics of access to resources in active modes in the Eurométropole de Strasbourg, which highlighted various disparities and their nature (Hached, 2019; Hached & Propeck-Zimmermann, 2020). Mobility hubs can therefore play an important role at the city and/or regional level that is not limited to more sustainable mobility. The objectives in implementing mobility hubs should go beyond mobility issues to look more broadly at the problems of the city and thus be part of a more global solution. To be part of this more integrative solution, different parameters should be taken into account when designing and implementing mobility hubs.

Following discussions with experts from the partner cities of the Mobi-Mix project, as nonexhaustive examples, we can consider some key parameters:[1]

- *Spatial distribution* that allows everyone to move throughout the whole studied territory without the need to own a car (a hierarchically structured network of mobility hubs) (Claasen, 2020; GO SEStran et al., 2020; Aono, 2019; Metrolinx, 2011; Waldron, 2007).
- *Multifunctionality*, which, in addition to providing mobility services, makes it possible to ensure additional services that are not available on-site. The aim is also to promote neighborhood life and activity while meeting local needs (GO SEStran et al., 2020; Aono, 2019; Queirós & González, 2019; SmartRail World, 2017; LA Urban Design Studio, 2016; 218 Consultants et al., 2015; O'Berry, 2015; Arup, 2014; Vahle, 2014; Midgley, 2009; Yeates & Jones, 1998).
- *Inclusiveness* in its broadest sense, for all users on an equitable basis regardless of age, physical ability, income… (GO SEStran et al., 2020; Aono, 2019; Metrolinx, 2011).
- *Security and safety* for the users themselves as well as third parties, whether physically, morally, or more virtually (such as personal data) (GO SEStran et al., 2020; Aono, 2019; LA Urban Design Studio, 2016; Metrolinx, 2011).
- *Comfort and ease of use* to make the use of mobility hub services as easy and comfortable as using one's own vehicle or even easier (RTP, 2022; GO SEStran et al., 2020; Aono, 2019; LA Urban Design Studio, 2016; Metrolinx, 2011; Waldron, 2007).
- *Reliability and resilience* by providing a stable and regular quality of service, in particular by supplying a sufficient number of vehicles so that the user can be sure of their availability when they need them. The aim is also to have a service that is sufficiently reliable and resilient (in financial and qualitative ways) to be part of the mobility policies of cities (ASQ, 2022; Raza, 2020; Aono, 2019; Géoconfluences, 2015; UNDRR, 2007).
- *Adaptation to technology habits and needs*, either at the level of vehicles or services (reservation, payment...) while allowing backward compatibility (GO SEStran et al., 2020; Aono, 2019).
- *Communication*, targeted and adapted to each interlocutor whoever they are, politician, user, opposition, neighbor, partner... The aim is to use pedagogical methods and to understand the needs, the concerns, or the problems and thus to find the essential compromises (Karbaumer, 2021c; Karbaumer & Metz, 2020).

All of this should occur within an urban setting designed for the way people and families would like to live, work and enjoy themselves. At the same time, the mobility hub is only one part of the equation. Because the transit system is the key connector to and between mobility hubs, the mix of land uses in the surrounding area is crucial to making it a destination conducive to transit choice. In other words, when developing the mobility hub concept […], we need a fundamental shift in thinking – away from land use patterns designed primarily for cars. That is why […] [the concept of] mobility hub is so important. They are the connection points in a transit-oriented metropolis – a concept very different from the car-based cities and towns we see today (CII – Kerala et al., 2022).

5   Further Research

Within the framework of the Mobi-Mix project, two European cities have chosen to set up two mobility hub demonstrators. These two cities are Norfolk and Valenciennes.

The shared mobility impact assessment in general and mobility hub impact assessment, in particular, is currently underway in the Mobi-Mix project. A sequential methodology has been developed to assess the impact of different shared mobility solutions deployed by Mobi-Mix partner cities. It allows to provide information at each phase of the project, adapting to the available data. The impact analysis is therefore carried out through three different approaches (exploratory, ex-ante, and ex-post evaluation). Surveys are conducted in each phase. The results and feedback of each phase make it possible to refine the estimates made in the previous phase and to minimize bias. In this way, they contribute to improving the precision of the method as a whole.

No empirical measurement of carbon emission changes has been included in this assessment, and all impacts are a result of measuring behavioral change. Converting these behavioral changes (changes in vehicle-km and passenger-km) into CO_2 savings will be achieved by applying standard emission factors per transport mode. Furthermore, it is important to highlight that the methodological approach is focused on the assessment of short-term impacts given the time frame of the proposed Mobi-Mix pilot projects.

The first estimations of the impacts of the mobility hub predict a positive impact on CO_2 reduction in the two partner cities, Norfolk and Valenciennes. The assessment is based mainly on estimates of change reported in mobility behavior surveys and on similar cities' experiences. The amount of CO_2 emission reduction depends on several parameters, such as the new shared modes being adopted (bicycle, scooter, car), the energy they use (fossil, electric, muscle), and the modes being abandoned (car, public transport, bicycle, walking…). Therefore, by sharing 50 bikes, 10 scooters, and 2 cars, Norfolk would replace 81,437 km done by private cars per year with less polluting vehicles and save approximately 23 tons per year in the short term. In the long term, following an estimated trend of behavioral change, the reduction in driving distances is estimated to reach 359,544 private car kilometers per year. Fifty-seven tons of CO_2 emissions per year, could be saved. With 100 shared bikes and 20 shared scooters, Valenciennes could lower CO_2 emissions from private cars by 67 tons per year in the short term and save 215,987 km done by private cars per year.

6   Conclusion

Within the framework of the Mobi-Mix project, we have taken a close look at the mobility hub concept. To do so, we carried out a literature review, attended specialized presentations, and discussed with experts from several cities. Two partner cities, Norfolk and Valenciennes, have chosen to implement mobility hub demonstrators and are studying their impact on CO_2 reduction and the adoption of more sustainable modes of transport.

Finally, what is a mobility hub? To the best of our knowledge, no author claims authorship of the term. The concept seems to have emerged from the reality of the field. However, various definitions exist. Some are more restrictive than others. They depend strongly on the project, the city, or the status of the person who defines it. This multitude of definitions could lead to confusion with other clear and well-established terms, such as multimodal hub. For this reason, we advocate for a discussion between stakeholders to find consensus for a definition that leaves a margin of maneuvers to the planners and offers flexibility of implementation. As the term "hub" expresses a centrality and thus a plurality of objects, we propose to define the mobility hub as a "place that regroups shared mobility modes." Although the presence of more than one shared mode suffices to define a mobility hub, it is recommended to integrate or to be connected to public transport. Other modes and facilities can also be integrated.

Thanks to its flexibility, the mobility hub concept can become a facility that allows the city/region to meet several objectives simultaneously. The primary objective is to enable more sustainable and less polluting mobility while reducing the use of private cars (especially private internal combustion engine vehicles). Depending on the location and design of the mobility hubs, other objectives that are part of the city/region's policy may be reflected in them. In particular, inclusiveness (for all, without depending on abilities, ages, genders...), equity (spatial and income equity...), safety (for users and others, data safety), etc. These parameters should be monitored to help make continuous adjustments to the mobility hub. A method of impact monitoring (focusing on CO_2 and taking into account the aforementioned parameters) is being developed within the framework of the Mobi-Mix project.

Each mobility hub is unique, but many of them share similar characteristics that allow them to be classified. Several typologies exist. However, most classifications consider the users for whom mobility hubs are intended (individuals, professionals, tourists, etc.), their temporality (temporary or permanent), their location (city center, suburban areas, etc.), their functions in the mobility network, their size and the vehicles they provide. The size of a mobility hub is often correlated with the surrounding density and the number of users. The type of vehicles provided often depends, among other factors, on the location of the mobility hub and the length of the expected trips. We believe that the classification of mobility hubs is relevant in the context of a network of mobility hubs and in contrasting mobility hubs from different cities or countries. Therefore, each city/region could adopt its own classification according to local specificities or objectives. However, the target users, the temporality, the geographical location, the size, and the type/number of vehicles and services offered all remain important parameters for defining a typology.

Although the mobility hub concept is flexible, the implementation of a mobility hub adapted to the needs and objectives can sometimes be complicated, as it requires several steps and may face difficulties at each step. Among these steps, we can mention foremost the emergence of the idea of creating a mobility hub and convincing both citizens and politicians of its usefulness. Then, we can mention the feasibility study and the verification of the correspondence to local regulations (if not, it will be necessary to plan the modification of these regulations). An analysis of the urban area enables a mobility hub network to be planned and adapted to meet the particular objectives of each city/region. The creation of first mobility hubs that serve as demonstrators may be necessary. Depending on the learning from these demonstrators, adjustments can be made and considered for future mobility hubs. Once the network of mobility hubs has been built, the process is not finished, and a process of continuous adjustments and modifications is recommended. The main difficulties that may arise are generally linked to the opposition of local residents, the choice of locations for mobility hubs, the choice of private partners if there are any (some cities can manage mobility hubs themselves, but the majority rely on private partners), and the modes of mobility to be provided.

Despite the possible challenges in implementing mobility hubs, the whole process could be worthwhile to allow cities/regions to meet several objectives at the same time. To support cities in this approach, we have proposed in this document several recommendations and guidelines for implementing better mobility hubs. First, it is necessary to create (or establish) features within the urban environment that support the implementation and functioning of mobility hubs (pedestrian and bicycle facilities, reduced traffic speeds, strict parking policies, pleasant urban environments, etc.). Second, it is important to consider mobility hubs as a network, where each node is adapted both to its function in the network and to local parameters. A mobility hub can be functional and provide additional services to meet the needs of local residents or users (e.g., cafés, snack bars, pick-ups of deliveries). The mobility hub should also be inclusive, helping everyone to meet their own mobility needs, regardless of their physical condition, age, or income. Safety and security within the mobility hub itself, and when using the vehicles it provides, are also important. In addition, to compete with private cars and be more attractive, comfort and ease of use are key considerations. When creating mobility hubs, cities/regions should ensure the reliability and resilience of the partners with whom they collaborate, as well as the flexibility of the infrastructure and its ease of adaptation to future technologies and compatibility with older technologies. Finally, the involvement of all stakeholders and communication is key when implementing mobility hubs. They should support all steps and be adapted to different stakeholders.

Within the framework of the Mobi-Mix project, two partner cities, Norfolk and Valenciennes, have been implementing mobility hubs. An evaluation method has been developed to monitor cities' objectives of reducing CO_2 and car use. Several indicators have been identified. A three-stage estimation and evolutionary method punctuated by surveys has been developed. It is based on the experiences of the other cities and takes into account the possible variations due to local contexts. Although there are still a number of reserves to be highlighted at this stage, the impact estimates of the mobility hub demonstrators in Norfolk and Valenciennes tend to confirm that it is possible to achieve the targeted objectives, in particular the reduction of CO_2, and even to go beyond them. In this regard, mobility hubs could be an important lever in future mobility policies by helping to reach more sustainable mobility.

Finally, the mobility hub concept seems to be an interesting, complex, and challenging topic worth investigating. The analysis of the impacts of this type of infrastructure is still to be fully achieved. It requires the collection of a multitude of data and their combination in a judicious way to establish sound measurements. It should also be remembered that mobility hubs are part of a more global context, which is relevant to consider. The question could also be asked as to whether this type of infrastructure could meet the needs of users at the lowest cost, harm, and risk. In this case, we are referring in particular to the research of T. Saint-Gérand on spatial ergonomics (Saint-Gérand, 2002) and that of W. Hached on the ergonomics of access (Hached, 2019).

Acknowledgment This chapter is produced within the framework of the Mobi-Mix project, funded by Interreg 2 Seas. We sincerely thank all the participants in this project for their participation in this work through their comments and relevant feedback. We also thank the members of the eHUBS project of Interreg Northwest Europe for sharing their experiences with us at the eHUBS International Academy in 2021. Special thanks to Ms. Saki Aono and Ms. Becky Lai from Translink for giving us their time and sharing their insights with us.

References

218 Consultants, University of California, Berkeley, Department of City and Regional Planning, Transportation Planning Studio, Anderson, K., Blanchard, S., Cheah, D., Koling, A., & Levitt, D. (2015). *City of Oakland: Mobility hub suitability analysis, technical report.*

Actionfigure. (2019). *How mobility hubs are changing city transit.* TransitScreen Actionfigure.

Aono, S. (2019). *Identifying best practices for mobility hubs* 72.

Arup. (2014). *Future of Rail 2050–Arup.* https://www.arup.com/perspectives/publications/research/section/future-of-rail-2050. Accessed 12.19.21.

ARUP. (2020). *Mobility hubs of the future, towards a new mobility behaviour.* https://www.ri.se/sites/default/files/2020-12/RISE-Arup_Mobility_hubs_report_FINAL.pdf. Accessed 1.21.22.

Asperges, T. (2020). *eHUBs and MOMENTUM: A match made in Leuven.* https://www.polisnetwork.eu/wp-content/uploads/2020/09/4_MOMOB_eHubs_Momentum-Leuven.pdf. Accessed 1.17.22.

ASQ. (2022). *What is reliability? Quality & reliability defined* | ASQ. asq.org. https://asq.org/quality-resources/reliability. Accessed 7.20.21.

Austin, J. (2021). *Mobility hubs–a transport planning concept whose time has.* https://www.transportxtra.com/publications/local-transport-today/news/69431/mobility-hubs%2D%2Da-transport-planning-concept-whose-time-has-come/. Accessed 1.21.22.

Bailey, G. (2020a). *A look at European mobility hubs.* https://www.metro-magazine.com/10122757/a-look-at-european-mobility-hubs. Accessed 1.21.22.

Bailey, G. (2020b). *A look at European mobility hubs.* https://www.metro-magazine.com/10122757/a-look-at-european-mobility-hubs. Accessed 1.18.22.

Basta, D. (2021). *BuurtHubs Amsterdam.* https://www.nweurope.eu/media/15319/ehubs-academy-presentations_compressed.pdf. Accessed 1.13.22.

Be. Brussels. (2022). *Accueil. be. brussels.* https://catalogue.be.brussels/fr/search-standalone/be.brussels. Accessed 1.18.22.

Belga. (2021). *COP26: le gouvernement flamand s'accorde in extremis sur son plan Climat*. RTBF. https://www.rtbf.be/article/cop26-le-gouvernement-flamand-s-accorde-in-extremis-sur-son-plan-climat-10873183. Accessed 1.18.22.

Børjesson, A. (2022a). *Smart mobility | Nordic Smart City Network*. https://nscn.eu/Bergen/SmartMobility. Accessed 1.4.22.

Børjesson, A. (2022b). *Mobility HUBS | Nordic Smart City Network*. https://nscn.eu/Citylabs/MobilityHUBS. Accessed 1.4.22.

Bremen. (2022a). *Cycling in Bremen*. https://www.bremen.eu/tourism/activities/cycling. Accessed 1.21.22.

Bremen. (2022b). *Bremen for cyclists – tips and info for your cycling holiday*. https://www.bremen.eu/tourism/bremen-for/bremen-for-cyclists. Accessed 1.21.22.

Bremen. (2022c). *Radfahren & Fahrradkultur in Bremen – BIKE IT!*. https://www.bremen.de/leben-in-bremen/fahrradstadt. Accessed 1.21.22.

Bremen. (2022d). *Bus und Straßenbahn in Bremen – Nachhaltig im Nahverkehr*. https://www.bremen.de/leben-in-bremen/mobilitaet-und-verkehr/bus-und-strassenbahn. Accessed 1.21.22.

Bremen. (2022e). *WK-Bike – Cycling in Bremen*. https://www.bremen.eu/wk-bike. Accessed 1.21.22.

Bremen. (2022f). *Fahrrad leihen in Bremen*. https://www.bremen.de/leben-in-bremen/fahrradstadt/fahrrad-leihen. Accessed 1.21.22.

Bremen. (2022g). *E-Scooter für Bremen – Verleih mit Voi und Tier*. https://www.bremen.de/leben-in-bremen/mobilitaet-und-verkehr/e-scooter-in-bremen. Accessed 1.21.22.

Bremen. (2022h). *Carsharing in Bremen – Anbieter, Infos, Standorte*. https://www.bremen.de/leben-in-bremen/mobilitaet-und-verkehr/carsharing. Accessed 1.21.22.

Bremer, T., Findeisen, S., Glotz-Richter, M., & City of Bremen. (2020). *SUNRISE-guidelines on "shared mobility"*.

Britaninica. (2009). *Bremen | History, facts, & points of interest | Britannica*. https://www.britannica.com/place/Bremen-Germany. Accessed 1.21.22.

Brussels Smart City. (2022). *Definition | Brussels Smart City*. https://smartcity.brussels/the-project-2-definition. Accessed 3.23.22.

Cerema. (2022). *Définition: qu'est-ce qu'une smart city?* Cerema. https://smart-city.cerema.fr/comprendre-smart-city/definition-smart-city. Accessed 10.12.20.

Chamberland, P., Markey-Crimp, D., & Pacheco, D. (2021). *Shoreline shared-use mobility study* 133.

CII- Kerala, Centre for Public Policy Research | Kumar Group. (2022). *Vytilla mobility hub: A gateway to Kerala*.

City of Amsterdam. (2020). *Operation plan Amsterdam DELIVERABLE 4.1*. https://www.nweurope.eu/media/12302/dt141-operational-plan-amsterdam.pdf. Accessed 1.13.22.

City of Amsterdam. (2022a). *Policy: Sustainability and energy*. Engl. Site. https://www.amsterdam.nl/en/policy/sustainability/. Accessed 1.13.22.

City of Amsterdam. (2022b). *Policy: Climate neutrality*. Engl. Site. https://www.amsterdam.nl/en/policy/sustainability/policy-climate-neutrality/. Accessed 1.13.22.

City of Amsterdam. (2022c). *Policy: Clean air*. Engl. Site. https://www.amsterdam.nl/en/policy/sustainability/clean-air/. Accessed 1.13.22.

City of Amsterdam. (2022d). *Policy: Traffic and transport*. Engl. Site. https://www.amsterdam.nl/en/policy/policy-traffic/. Accessed 1.13.22.

City of Amsterdam. (2022e). *eHUBS: mobiliteitshubs voor de buurt*. Innovatie. https://www.amsterdam.nl/innovatie/mobiliteit/ehubs-mobiliteitshubs-buurt/. Accessed 1.13.22.

City Ratings. (2022). *PeopleForBikes City Ratings | Every ride*. Every rider. Join us.. PeopleForBikes. https://cityratings.peopleforbikes.org/. Accessed 1.13.22.

Citypopulation. (2022). *Leuven (Leuven, Vlaams-Brabant, Belgium) – population statistics, charts, map, location, weather and web information*. https://www.citypopulation.de/en/belgium/vlaamsbrabant/leuven/24062__leuven/. Accessed 1.14.22.

Claasen, Y. (2020). *Potential effects of mobility hubs: Intention to use shared modes and the intention to reduce household car ownership*.

CoMoUK. (2021a). *Mobility hubs: The problem solving approach to congestion and parking.* https://como.org.uk/wp-content/uploads/2021/01/CoMoUK_Mobility-Hubs_Breman-Case-Study.pdf. Accessed 1.21.22.

CoMoUK. (2021b). *Mobility hubs toolkit.* https://como.org.uk/wp-content/uploads/2021/09/CoMoUK-Mobility-hubs-toolkit.pdf. Accessed 1.18.22.

CoMoUK. (2021c). *Bikes for all: A guide to setting up an equitable bike share scheme.* https://como.org.uk/wp-content/uploads/2021/04/Bikes-For-All-Report-2020.pdf. Accessed 2.15.22.

CoMoUK, Share North, Interreg North Sea Region. (2019). *UK mobility hubs guidance.*

Copenhagenize Index. (2019). *2019 Copenhagenize Index – Copenhagenize.* https://copenhagenizeindex.eu/. Accessed 1.13.22.

Coya. (2019). *Global Bicycle Cities Index 2019 | Coya.* https://www.coya.com/bike/index-2019. Accessed 1.13.22.

Crowther, J., Mangle, K., Abe, D., Maines, K., Hesse, E., Sherman, J., Hoyt-McBeth, S., Falbo, N., Berkow, M., Igarta, D., Hurley, P., & Lonsdale, S. (2020). *Mobility hub typology study.* Portland Bureau of Transportation (PBOT).

De Muelenaere, M. D. M. (2021). *Le plan climat flamand ne déclenche pas l'enthousiasme.* Le Soir. https://www.lesoir.be/404647/article/2021-11-04/le-plan-climat-flamand-ne-declenche-pas-lenthousiasme. Accessed 1.18.22.

Dirks Eskeland, I. (2021). *E-mobihubs in Stavanger and in the region Nord Jæren.*

e-MOPOLI, I.E. (2020). *[NEWS] Rogaland tests mobility hub.* Interreg Eur. https://www.interregeurope.eu/e-mopoli/news/news-article/10321/news-rogaland-tests-mobility-hub/. Accessed 1.11.22.

eHUBS. (2020a). *Belgium's Flanders Region will invest more than €100 mln on mobility hubs.* https://www.nweurope.eu/projects/project-search/ehubs-smart-shared-green-mobility-hubs/news/belgiums-flanders-region-will-invest-more-than-100-mln-on-mobility-hubs/. Accessed 1.18.22.

eHUBS. (2020b). *Leuven inaugurates its first eHUBS at the car-free day.* https://www.nweurope.eu/projects/project-search/ehubs-smart-shared-green-mobility-hubs/news/leuven-inaugurates-its-first-ehubs-at-the-car-free-day/. Accessed 1.17.22.

eHUBS. (2022). *eHUBS – smart shared green mobility hubs.* https://www.nweurope.eu/projects/project-search/ehubs-smart-shared-green-mobility-hubs/. Accessed 1.14.22.

ESPON. (2022). *Functional urban areas | ESPON FUORE.* https://fuore.espon.eu/. Accessed 4.22.22.

European Commission. (2019). *SHARE-North: Fostering shared mobility solutions for a low-carbon North Sea Region – Projects.* https://ec.europa.eu/regional_policy/en/projects/Germany/share-north-fostering-shared-mobility-solutions-for-a-low-carbon-north-sea-region. Accessed 1.21.22.

European Commission. (2022). *European green capital.* https://ec.europa.eu/environment/europeangreencapital/europeangreenleaf/egl-winning-cities/leuven/. Accessed 1.17.22.

Evenepoel, H. (2020). *Operational plan eHUBs Leuven DELIVERABLE 4.1.* https://www.nweurope.eu/media/12303/dt141-operational-plan-leuven.pdf. Accessed 1.17.22.

Evenepoel, H. (2021a). *The policy framework about eHUBS.*

Evenepoel, H. (2021b). *Planning process in Leuven.*

Fairfax County, Virginia. (2013). *Mobility hubs for Tysons corner metrorail stations, conceptual design plans.* https://www.mwcog.org/assets/1/6/Fairfax-Hubs.pdf. Accessed 1.21.22.

Flanders Environment Agency. (2018). *Air quality and emissions in the Flanders region.* Flanders Environ. Agency VMM. https://en.vmm.be/publications/annual-report-air-quality-in-the-flanders-region-2017. Accessed 1.18.22.

Flandre. (2021). *Toekomstvisie Vlaamse mobiliteit goedgekeurd.* www.vlaanderen.be. https://www.vlaanderen.be/departement-mobiliteit-en-openbare-werken/nieuwsberichten/toekomstvisie-vlaamse-mobiliteit-goedgekeurd. Accessed 1.18.22.

Flandre. (2022a). *Vlaamse mobiliteitsvisie 2040.* www.vlaanderen.be. https://www.vlaanderen.be/mobiliteit-en-openbare-werken/duurzame-mobiliteit/vlaamse-mobiliteitsvisie-2040. Accessed 1.18.22.

Flandre. (2022b). *Fietsbeleid.* www.vlaanderen.be. https://www.vlaanderen.be/departement-mobiliteit-en-openbare-werken/beleidsthemas/fietsbeleid. Accessed 1.18.22.

Flandre. (2022c). *Basisbereikbaarheid.* www.vlaanderen.be. https://www.vlaanderen.be/basis-bereikbaarheid. Accessed 1.18.22.

Flandre. (2022d). *Doelstellingen van basisbereikbaarheid.* www.vlaanderen.be. https://www.vlaanderen.be/basisbereikbaarheid/doelstellingen-van-basisbereikbaarheid. Accessed 1.18.22.

Frei Hansestadt Bremen. (2014). *SUMP Bremen 2025.*

FUB. (2022). *Les villes qui aiment le vélo en France et à l'étranger* | Fédération française des usagers de la bicyclette. https://www.fub.fr/velo-ville/villes-qui-aiment-velo/villes-qui-aiment-velo-france-etranger. Accessed 1.13.22.

Gemeente Amsterdam. (2019). *Clean air action plan.*

Gemeente Amsterdam. (2021). *Data en informatie.* https://data.amsterdam.nl/. Accessed 1.13.22.

Géoconfluences. (2015). *Résilience — Géoconfluences.* Geoconfluences. http://geoconfluences.ens-lyon.fr/glossaire/resilience. Accessed 7.20.21.

GO SEStran, steer, Transport Scotland. (2020). *Mobility hubs: A strategic study for the South-East of Scotland/SEStran region.*

Google. (2022). *Google Trends.* Google Trends. https://trends.google.com/trends/explore?date=2004-01-01%202022-06-01&q=%22mobility%20hub*%22. Accessed 6.01.22.

Gray, L. (2017). *Build your own mobility hub: 7 lessons for cities from Bremen, Germany.* Shar.-Use Mobil. Cent.

Hached, W. (2019). *Ergonomie d'accès aux ressources de la vie quotidienne en mobilité douce: Application à l'Eurométropole de Strasbourg* (PhD thesis). Université de Strasbourg, Strasbourg.

Hached, W., & Propeck-Zimmermann, É. (2020). Mobilité douce et disparités socio-spatiales: évaluation de l'ergonomie d'accès aux ressources du quotidien. *Territoire en mouvement Revue de géographie et aménagement. Territory in movement Journal of geography and planning.*

Henrik Haaland, N. (2022). *Stavanger — AI4Cities.*

I amsterdam. (2021). *Facts & figures* | I amsterdam. https://www.iamsterdam.com:443/en/about-amsterdam/amsterdam-information/facts-and-figures. Accessed 1.13.22.

IMS. (2019). *Moving forward with mobility hubs.* ImsInfo.

Intelligent Transport. (2021). *Scotland to introduce European style mobility hubs.* Intell. Transp. https://www.intelligenttransport.com/transport-news/117192/scotland-mobility-hubs/. Accessed 1.18.22.

Interreg NWE. (2019). *eHUBS – smart shared green mobility hubs.* https://www.nweurope.eu/projects/project-search/ehubs-smart-shared-green-mobility-hubs/. Accessed 12.19.21.

Intertraffic. (2021). *Mobility hubs* | *The multimodal stations at the centre of everything.* https://www.intertraffic.com/news/infrastructure/mobility-hubs-multimodal-stations-at-the-centre-of-everything/. Accessed 1.13.22.

ITDP. (2012). *Sustainable transport award finalist: Bremen, Germany.* Inst. Transp. Dev. Policy. https://www.itdp.org/2012/12/20/sustainable-transport-award-finalist-bremen-germany/. Accessed 1.21.22.

Karbaumer, R. (2018). *Bergen celebrates the grand opening of the city's first "Mobilpunkt".* Interreg VB North Sea Region Programme. https://northsearegion.eu/share-north/news/bergen-celebrates-the-grand-opening-of-the-city-s-first-mobilpunkt/. Accessed 1.10.22.

Karbaumer, R. (2020). *Reclaiming street space and place making with mobility hubs – Bremen's and Bergen's mobil. punkte.* https://www.polisnetwork.eu/wp-content/uploads/2020/12/7A_Rebecca-Karbaumer-City-of-Bremen.pdf. Accessed 1.21.22.

Karbaumer, R. (2021a). *Why shared mobility hubs rock – reclaiming street space and place making with car-sharing and mobility hubs in Bremen (and beyond).*

Karbaumer, R. (2021b). *Bremen's mobil. punkte – the planning process.*

Karbaumer, R. (2021c). *Bremen's mobil. punkte communication strategies for specific target groups.*

Karbaumer, R. (2022). *Engaging stakeholders in mobility hub planning: How we do it in Bremen.* https://www.duurzame-mobiliteit.be/sites/default/files/inline-files/2%20-%20Rebecca%20 Karbaumer%20%2C%20Stad%20Bremen%20-%20Engaging%20stakeholders%20in%20 mobility%20hub%20planning.pdf. Accessed 1.21.22.

Karbaumer, R., & Metz, F. (2020). *A planner's guide to the shared mobility galaxy* 252.

Kleiner, M. (2020). *Changing from oil city to smart city.* Nor. Am. https://www.norwegianameri-can.com/changing-from-oil-city-to-smart-city/. Accessed 1.11.22.

KU Leuven. (2022). *Leuven, a great city to live in.* https://lrd.kuleuven.be/en/hitech/leuven-a-great-city-to-live-in. Accessed 1.17.22.

LA Urban Design Studio. (2016). *Mobility hubs reader's guide.* http://www.urbandesignla.com/resources/MobilityHubsReadersGuide.php. Accessed 12.19.21.

Lanagarth. (2020). *Help shape future transport plans for Langarth.*

Leuven (2030). (2022). *Leuven 2030 – Roadmap 2025 · 2035 · 2050.* https://roadmap.leuven2030.be/intro. Accessed 1.17.22.

Leuven MindGate. (2019). *Leuven to install 50 "mobility hubs" to foster multimodality | Leuven MindGate.* https://www.leuvenmindgate.be/news/leuven-to-install-50-mobility-hubs-to-foster-multimodality. Accessed 1.17.22.

Liao, F., & de Almeida Correia, G. H. (2021). *How will people use eHUBS? Results from a survey in Amsterdam.*

Metrolinx. (2008). *Mobility hubs.* https://www.metrolinx.com/thebigmove/Docs/big_move/RTP_Backgrounder_Mobility_Hubs.pdf. Accessed 12.19.21.

Metrolinx. (2011). *Mobility hub guidelines draft for board approval: For the Greater Toronto and Hamilton Area.*

Midgley, P. (2009). *The role of smart bike-sharing systems in urban mobility* 9.

Miramontes, M., Pfertner, M., Rayaprolu, H. S., Schreiner, M., & Wulfhorst, G. (2017). Impacts of a multimodal mobility service on travel behavior and preferences: User insights from Munich's first Mobility Station. *Transportation, 44*(6), 1325–1342.

mobihub.com. (2022). *Mobility hubs in the UK – a short history.* MobiHub. https://www.mobihub.com/mobility-hubs-uk-history. Accessed 1.18.21.

Movmi. (2021). *Multimodal Mondays: Mobility hubs with Yuval Fogelson, Rebecca Karbaumer, Vlad Marica & Sandra Phillips – movmi.* https://movmi.net/blog/multimodal-mondays-mobility-hubs-2/. Accessed 1.21.22.

Mpact. (2022). *Mobihubs – your hub to mobility.* Mpact. https://www.mpact.be/en/project-event/mobipunt-your-hub-to-mobility/. Accessed 1.18.22.

NuMIDAS. (2022). *Leuven.* Numidas.

N-W Europe. (2022). *eHUBS – smart shared green mobility hubs.* https://www.nweurope.eu/projects/project-search/ehubs-smart-shared-green-mobility-hubs/. Accessed 1.13.22.

O'Berry, A. D. (2015). *Transportation engineering assimilated livability planning using micro-simulation models for South-East Florida.*

Ove Kvalbein, L. (2021). *E-hubs: The planning and design.*

Pais, R. R. (2019). *A tale of three cities.* Tale Three Cities 153.

Plymouth. (2022). *Mobility hubs | PLYMOUTH.GOV.UK.* https://www.plymouth.gov.uk/parkingandtravel/transportplansandprojects/transportplans/transformingcitiesfund/mobilityhubs. Accessed 1.18.22.

Queirós, A., & González, G. H. (2019). *Railway mobility hubs: A feature-based investment return analysis.*

Raza, M. (2020). Reliability vs availability: What's the difference? *BMC Blogs.* https://www.bmc.com/blogs/reliability-vs-availability/. Accessed 7.20.21.

Ripa, F. (2019). *Leuven to install 50 "mobility hubs" to foster multimodality | Eltis.* https://www.eltis.org/discover/news/leuven-install-50-mobility-hubs-foster-multimodality. Accessed 1.17.22.

Roelant, B. (2021). *EHubs in Flanders: A regional story.*

RTP. (2022). *Appendix 5A: Activity centers and regional mobility hubs.* https://static1.square-space.com/static/5bfc5ef3f93fd4e73b6c10fa/t/5c02bcec0e2e72190268a237/1543683308641/RTP-2035-Appendix-5A-Activity-Centers-and-Regional-Mobility-Hubs.pdf. Accessed 1.18.21.

Saelens, N. (2021). *La Flandre tient enfin son plan climat: que contient-il?* Bus. AM. https://fr.businessam.be/la-flandre-tient-enfin-son-plan-climat-que-contient-il/. Accessed 1.18.22.

Saint-Gérand, T. (2002). *S.I.G.: Structures conceptuelles pour l'analyse spatiale (HDR).* Université de Caen.

SANDAG. (2019). *5- Big-moves.* https://sdforward.com/mobility-planning/5-big-moves. Accessed 12.19.21.

SANDAG. (2022). *mobilityHubs.* https://sdforward.com/mobility-planning/mobilityhubs. Accessed 1.18.22.

Schmalholz, N. (2021). *That Leuven feeling.* POLIS Netw.

SHARE North. (2018a). *Results of impact analysis of car-sharing services and user behaviour delivers interesting results in Bremen–Share North.* https://share-north.eu/2018/05/results-of-impact-analysis-of-car-sharing-services-and-user-behaviour-delivers-interesting-results-in-bremen/. Accessed 1.21.22.

SHARE North. (2018b). *Analysis of the impact of car-sharing in Bremen 2018–Share North.* https://share-north.eu/2018/08/impact-analysis-of-car-sharing-in-bremen-english-report-published/analysis-of-the-impact-of-car-sharing-in-bremen-2018_team-red_final-report-english_compressed/. Accessed 1.21.22.

SHARE North. (2020). *Substantial funding for mobihubs in Flanders! —Share North.* https://share-north.eu/2020/08/substantial-funding-for-mobihubs-in-flanders/. Accessed 1.18.22.

SHARE North. (2022). *New concept in Flanders "Mobihubs".* Interreg VB North Sea Region Programme. https://northsearegion.eu/share-north/news/new-concept-in-flanders-mobihubs/. Accessed 1.18.22.

SHARE North, Ove Kvalbein, L., & Magerøy, M. (2019). *Bergen–A City dedicated to mobility hubs, emissions reduction and transnational learning–Share North.* https://share-north.eu/2019/07/bergen-a-city-dedicated-to-mobility-hubs-emissions-reduction-and-transnational-learning/. Accessed 1.4.22.

ShareNL. (2018). *1 2 M. concept deelhub door sharenl in opdracht van de Gemeente Utrecht – PDF Free Download.* https://docplayer.nl/105893320-1-2-m-concept-deelhub-door-sharenl-in-opdracht-van-de-gemeente-utrecht.html. Accessed 12.19.21.

SmartRail World. (2017). *Retails sales at train stations outstrip those on the High Street in the UK 8.*

Statistics Flanders. (2021). *Road casualties.* https://www.statistiekvlaanderen.be/en/road-casualties. Accessed 1.18.22.

Statistics Flanders. (2022a). *Population: Size and growth.* https://www.statistiekvlaanderen.be/en/population-size-and-growth-0. Accessed 1.18.22.

Statistics Flanders. (2022b). *Traffic jam severity.* https://www.statistiekvlaanderen.be/en/traffic-jam-severity. Accessed 1.18.22.

Stavanger Kommune. (2020). *Mobility hub* | City of Stavanger. https://www.stavanger.kommune.no/en/samfunnsutvikling/stavanger-smart-city/smart-city-projects/mobility-point/. Accessed 1.11.22.

Stavanger Kommune. (2021a). *Befolkning* | Stavanger Kommune. https://www.stavanger.kommune.no/om-stavanger-kommune/stavanger-statistikken/Befolkning/. Accessed 1.11.22.

Stavanger Kommune. (2021b). *Fakta om Stavanger* | Stavanger Kommune. https://www.stavanger.kommune.no/om-stavanger-kommune/fakta-om-stavanger/. Accessed 1.11.22.

Stavanger Kommune. (2021c). *Klima- og miljøplan 2018–2030* | Stavanger Kommune. https://www.stavanger.kommune.no/renovasjon-og-miljo/miljo-og-klima/klima%2D%2Dog-miljoplan-2018-2030/. Accessed 1.11.22.

Stavnes Hisdal, C. (2021). *Bergen kommune – Facts about Bergen.* Bergen Kommune. https://www.bergen.kommune.no/english/about-the-city-of-bergen/facts-about-bergen. Accessed 1.4.22.

Tague, N. (2021). *Place North West | Ancoats mobility hub advances in "UK first".* Place North West.

The Big Move. (2008). *Mobility hubs.* https://www.metrolinx.com/thebigmove/Docs/big_move/RTP_Backgrounder_Mobility_Hubs.pdf. Accessed 1.21.22.

The Explorer. (2020a). *Bergen leads the way for shared mobility in Norway.* https://www.the-explorer.no/stories/smart-cities2/bergen-leads-the-way-for-shared-mobility-in-norway/. Accessed 1.4.22.

The Explorer. (2020b). *Smart transportation essential for smart cities.* https://www.theexplorer.no/stories/smart-cities2/smart-transportation-essential-for-smart-cities/. Accessed 1.11.22.

Thorsnæs, G. (2020). *Stavanger–næringsliv.* Store Nor. Leks.

Times, T. B. (2021). *Air quality in Flanders improving, but health impact remains damaging.* https://www.brusselstimes.com/belgium/189236/air-quality-in-flanders-improving-but-health-impact-remains-damaging. Accessed 1.18.22.

Transit Forward. (2022). *Transit strategies, mobility hubs.* https://transitforwardri.com/pdf/Strategy%20Paper%2018%20Mobility%20Hubs.pdf. Accessed 1.21.22.

UNDRR. (2007). *Resilience.* U. N. Off. Disaster Risk Reduct. https://www.undrr.org/terminology/resilience. Accessed 7.20.21.

Universität Bremen. (2022). *Climate protection and mobility–Universität Bremen.* https://www.uni-bremen.de/en/umweltmanagement/referenzen/climate-protection-and-mobility. Accessed 1.21.22.

University of Stavanger. (2021). *Smart city–collaboration | University of Stavanger.* https://www.uis.no/en/smart-city-collaboration. Accessed 1.12.22.

Vahle, F. T. (2014). *Quo Vadis PRT? Review, update and outlook of an innovative mobility solution in the context of a changing urban mobility paradigm.*

VISITFLANDERS. (2022a). *Destinations à découvrir en Flandre | VISITFLANDERS.* https://www.visitflanders.com/fr/destinations/index.jsp. Accessed 1.18.22.

VISITFLANDERS. (2022b). *Leuven, Mecca of books and beer | VISITFLANDERS.* https://www.visitflanders.com/en/destinations/leuven/index.jsp. Accessed 1.17.22.

VisitLeuven. (2020). *Sustainable policy | VisitLeuven.* https://visitleuven.be/en/duurzame-gids. Accessed 1.17.22.

Waldron, L. (2007). *Mobility HUBs, Toronto, Ontario.* CRC Research. https://www.crcresearch.org/case-studies/case-studies-sustainable-infrastructure/transportation/mobility-hubs-toronto-ontario. Accessed 1.18.21.

Wegweiser Kommune. (2022a). *Bremen–Wegweiser Kommune.* https://www.wegweiser-kommune.de/kommunen/bremen. Accessed 1.21.22.

Wegweiser Kommune. (2022b). *Typisierung – Wegweiser Kommune.* https://www.wegweiser-kommune.de/demografietypen. Accessed 1.21.22.

Yeates, M., & Jones, K. (1998). Rapid transit and commuter rail-induced retail development. *Journal of Shopping Center Research, 5,* 7–38.

Zielinski, S. (2007). New mobility: The next generation of sustainable urban transportation. In *Frontiers of engineering: Reports on leading-edge engineering from the 2006 symposium (2007). Presented at the frontiers of engineering.* The National Academies of Sciences, Engineering, and Medicine, Michigan, p. 13.

Open Access This chapter is licensed under the terms of the Creative Commons Attribution 4.0 International License (http://creativecommons.org/licenses/by/4.0/), which permits use, sharing, adaptation, distribution and reproduction in any medium or format, as long as you give appropriate credit to the original author(s) and the source, provide a link to the Creative Commons license and indicate if changes were made.

The images or other third party material in this chapter are included in the chapter's Creative Commons license, unless indicated otherwise in a credit line to the material. If material is not included in the chapter's Creative Commons license and your intended use is not permitted by statutory regulation or exceeds the permitted use, you will need to obtain permission directly from the copyright holder.

Evaluating the Benefits of Promoting Intermodality and Active Modes in Urban Transportation: A Microsimulation Approach

Souhir Bennaya and Moez Kilani

Abstract The objective of this chapter is to show how microsimulation can be used to study urban transportation problems, in particular those issues related to sustainable transport and innovations. A theoretical, though representative, geometry of an urban area with a set of concentric and radial roads is considered for the analysis. Microsimulation, which provides a precise description of traffic flows, is used to draw a detailed accounting of emissions of pollutant gases and fuel consumption. In the base-case situation, the private car is the main transport mode. We then consider alternative scenarios where users are allowed to switch to public transportation or biking. A combination of walking, biking, and public transportation is also allowed. Under this intermodal setting, we find that congestion level, fuel consumption, and emissions of pollutant gases decrease significantly (up to 30%).

Keywords Urban transport · Microscopic simulation · Emissions and congestion · Active mobility · Public transport · Intermodal transport · Sumo traffic simulator

1 Introduction

Researchers from multidisciplinary perspectives, including economics, the environment, and transportation science, have been deeply investigating the paramount importance of reducing the emissions of greenhouse gases (GHG), the emissions of fine particles, and the consumption of fossil fuels. These objectives can be met only if ambitious reforms are identified and implemented. For example, to reduce road congestion, which is one of the main external costs generated by road transport, attractive alternatives to private cars should be proposed to users (Srisakda et al., 2022; de Palma et al., 2022).

S. Bennaya (✉)
LEM Lille Economie Management, UMR 9221, Lille Cedex, France

M. Kilani
University of Littoral, Opal Coast, Dunkerque, France
e-mail: moez.kilani@univ-littoral.fr

© The Author(s) 2024
F. Belaïd, A. Arora (eds.), *Smart Cities*, Studies in Energy, Resource and Environmental Economics, https://doi.org/10.1007/978-3-031-35664-3_15

Transport modeling and simulation have become valuable tools for planning the never-ending growth of the number of vehicles running in large urban areas. At the same time, the quality of simulation tools has improved significantly during the last decade. They became more accessible and reliable (Costeseque, 2013). The potential strength of traffic simulation has gained the attention of several researchers intending to study the dynamic adaptation of the existing infrastructure and its influence on traffic (Berdai, 2004). During the last 20 years, several transport simulations were conducted on several urban areas, mainly in developed countries. The macrosimulation approach was used by Leclercq (2002) to study traffic models. Appert and Santen (2002) emphasized road traffic modeling and employed a macroscopic simulation model by applying cellular automata. Guillotte et al. (2009) underlined the multimodal travel of the inhabitants of Quebec City simulation employing a mesoscopic simulation model. More recently, Kilani et al. (2022) utilized macroscopic simulation to develop a regional multimodal transport model in northern France to evaluate the impacts of free public transport and road pricing on congestion and pollution reduction.

Mobility currently accounts for extensive environmental, economic, and social concerns. Indeed, skyrocketing travel flows indicate infrastructure shortcomings, namely, roads. This also highlights the vital need for more sustainable mobility insights (Curiel-Esparza et al., 2016). The last decade witnessed a growing awareness of the substantial impact of intermodality and active mobility (AM), especially local walking and cycling practices in standing as relevant transportation modes in urban areas (Gebhardt et al., 2016).

These modes may serve as time and money savers, traffic attenuators, and overall carbon footprint reduction (Pini & Lavadinho, 2005). They also have health benefits and look convenient under particular situations, such as the COVID-19 pandemic (Loske, 2020). In fact, the modal shift from individual motorized transport to active modes was clearly observed in several countries, including North America (Shaheen et al., 2013), China (Shaheen et al., 2011), England (Midgley, 2011), and Australia (Share, 2011; Fishman et al., 2012, 2013).

AM (walking, cycling) is interconnected with intermodality insofar as these modes can efficiently complement public transportation. Related studies prove the effectiveness of improving and complementing public transport systems to make them more appealing (Weliwitiya et al., 2019). Thus, intermodal travel behavior plays a vital role in urban transportation systems. In other words, there could be feeders of the public transport network in underserved areas.

In this chapter, we are particularly interested in intermodal transport as the combination of public transport and active modes. This covers bike-sharing systems or bike parking systems located near bus stations. Despite the effectiveness and feasibility of this concept, studies related to intermodality remain very limited and generally follow a qualitative or descriptive analysis. On the other hand, studies based on analytical modeling or simulation are usually restricted to unimodal trips and do not consider possible combinations of two or several modes.

Recently, some studies were conducted on this line of research, but with a focus on freight transport (Willing et al., 2017; Gohari et al., 2022). Mode combinations in passenger transport were studied for some shared services. For example, Lorente et al. (2022), defined a method for intermodal assignment by combining public transport and carpooling. As a matter of fact, they have presented a new approach for a key component, with the mobility concept seen as a set of services, named MaaS (Mobility as a Service), hence integrating carpooling and public transport as components of intermodal trips. In this study, the authors proved that the whole intermodal system works effectively. Furthermore, Baum et al. (2019) discuss the variant of a bike-sharing service and a nonshared taxi as an auxiliary transportation mode that takes over public transportation networks, including them only as the first or last step of the trip (Last Mile Mobility). Their approach is to integrate public transportation along with auxiliary modes.

The results of these studies are very extensive and broad due to the complexity of this type of transport model. It is also worth mentioning that most of them integrate intramodality and are limited to macroscopic and mesoscopic simulations or theoretical studies. Therefore, our field of research focused on the study of effective methods to combine adequate public transportation systems and soft mobility.

This chapter will focus on studying the main stages of constructing a microsimulation transport model for a medium-sized city as a first step. In fact, the microsimulation model refers to a representation at the level of the agent with a fine description of the infrastructure and the interaction between the agents. For example, Fig. 1b below illustrates how the intersection details have been taken into account. On the one hand, simulation models at the urban scale often use macrosimulation and/or mesoscale simulation models to simplify (supply and demand) data processing and the implementation of the model (Axhausen et al., 2016). For example, intersections are oversimplified and (in general) are not explicitly taken into account. On the other hand, microsimulation requires too-refined and specific data, making it

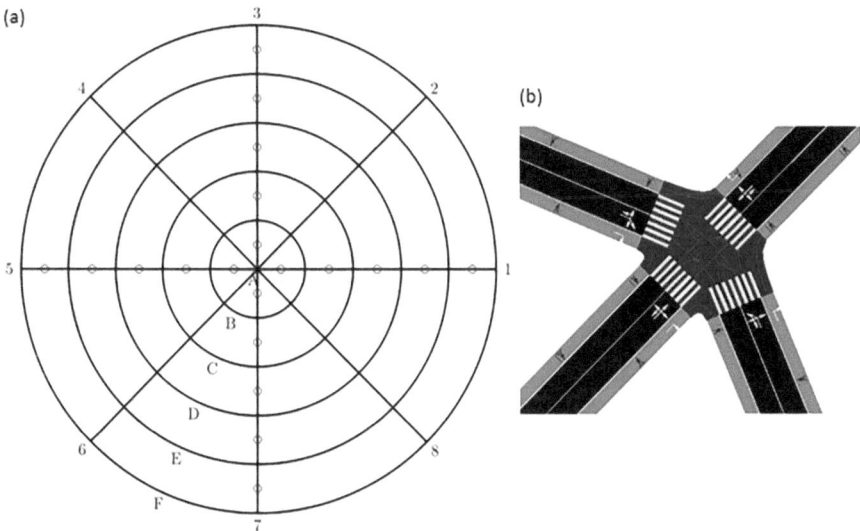

Fig. 1 Features of the network. (**a**) The network composed of several circumferential and radial roads. (**b**) An example of an intersection with its connections (SUMO framework)

difficult to generate demand and, more precisely, in the construction of origin-destination matrices (OD matrices). The benefit of this approach is that we produce a detailed description of all the movements of individuals, vehicles, and networks (positions, speeds, and accelerations).

After setting up the transport model, we will evaluate the effect of a modal shift toward sustainable mobility while integrating an intermodal system. We are particularly interested in the impacts on road congestion, pollution levels, and fuel consumption. Two scenarios of intermodal transport will be considered. In the first scenario, we model the combination of walking and public transport; in the second scenario, we consider the combination of cycling and public transport. For the simulation itself, we use the SUMO Traffic Simulator,[1] a microsimulation framework based on a car-following model (Lopez et al., 2018).

Comprehensive modeling of intermodal transport, where the chaining of transport mode is an endogenous choice, remains problematic for realistic models. This chapter reports these difficulties and proposes a modeling methodology to account for multimodal transport. The approach in this chapter can be replicated for other cases and can be conducted to develop realistic case studies.

The chapter is organized as follows. Section 2 describes the infrastructure and how demand has been produced. It also gives an overview of the modeling approach. Section 3 reports the simulation output for the base-case scenario and compares it with other scenarios where alternatives to the private car are made available. The last section concludes.

Note that all tables in this chapter report values produced by our simulations, and all the figures are also our own production.

2   The Transport Model

In this section, we describe the modeling framework and how the simulation framework is constructed. The used network is described in Sect. 2.1. The demand for trips is discussed in Sect. 2.2, and the workflow to conduct the simulation is described in Sect. 2.3.

2.1   Network

Recently, a number of empirical studies have stated that it is difficult to include the real dimension of a city, but we can work with simpler models that remain consistent with real-world cases.[2] A universal topology for all cities is difficult to provide,

---

[1] An open-source traffic simulation package developed at the Institute of Transportation Research at the German Aerospace Center https://www.eclipse.org/sumo/.

[2] https://blogs.worldbank.org/sustainablecities/how-do-we-define-cities-towns-and-rural areas.

but several cities in the world can be described as simple patterns based on road geometries. The Manhattan example, with a set of perpendicular and parallel roads, is usually considered a good representation of North American cities. However, this geometry does not reflect the structure of European cities, which were historically developed around rivers or main paths to facilitate trade and transportation. In this case, the city center is surrounded by a set of circumferential roads, which expand to the outskirt, and a set of radial roads that connect the suburban area to the city center. The model we will develop below is based on a methodology that can address any urban form, but to keep our discussion simple, we focus on circular cities.

The example we use is based on the structure of the city of Mons in Belgium. Its area is 146.6 km^2 with a radius of $r = 4.83$ km^2. The road network of this city is illustrated in Fig. 1a. The city center is denoted by A, and the five circumferential roads are labeled B, C, D, E, and F. There are eight (half) radial roads labeled 1 to 8, respectively. The circumferential roads are equidistant. With this labeling, it is straightforward to locate each node on the network as a combination of a circumferential and a radial road. For example, "B3" is the intersection between circumferential road B and radial road 3. Road links connect two nodes. For example, road link "C3D3" connects node "C3" to node "D3".

This network represents the main road axis in the city. Secondary roads are not directly considered. The network has been edited to add sidewalks, cycle paths, and crosswalks that are of first importance in modeling walker behavior. Indeed, "walk" is important to take into account public transport, which involves a transport step from home to the station of departure and another step from the arrival station to the destination. To conveniently model walk patterns, users should be able to change from a sidewalk to the opposite one, and this generally occurs at intersections. In practice, it adds to the complexity of the intersection.

Indeed, as shown in Fig. 1b several connections are required to allow each vehicle to leave the edge it is exiting and enter the next edge in its route. These connections add to crossings available for walkers, making traffic management particularly complex at the intersection level. At this step, we need to remove inconsistencies that may prevent walkers from performing realistic movements or may lead to collisions between vehicles and/or walkers.

This is one of the most challenging steps in setting up the network. If we have taken into account its initial state, this leads to unrealistic traffic, and teleportation problems will be created. To avoid teleportation during the simulation, we solved all connection-related inconsistencies. We proceed in two steps. In the first step, only private cars, buses, and walking are considered. In the second step, we add biking and update the main radial roads with dedicated cycle paths.

Access points (i.e., stations) should be added to the network to account for public transport. A public transport user should walk to the station's exact location, wait for the bus line, and ride when the bus arrives. He or she then leaves the bus when he or she arrives at a station near the destination and walks to the destination. The design of a bus transport system is an elaborate process that includes the design of the lines, the construction of timetables for each line, and the assignment of buses and drivers to perform the desired public transport service. Since we mainly focus on an illustration of these steps, we have implemented only two bus lines that run on two radial axes. The first ensures a round trip between the north and the south of the city (along edges 3 and 7 in Fig. 1a), and the second ensures a round trip between the city's west and east sides (along edges 1 and 5 in Fig. 1a). When using public transport, users in even-numbered edges need to walk through sidewalks and cross intersections to reach the station available on the nearest edge. This is also the case for users located on circumferential roads.

2.2   The Demand

For the construction of the OD matrices, we applied a gravitational-like model that we combined with a monocentric city shape. Activities are mainly located near the city center, while residential areas are in suburban areas. As we move away from the city center, the frequency of business activities decreases while the number of home locations increases. As a result, most traffic flows induced by home-to-work trips are realistically directed toward the city center (inward direction). The gravitational feature allows us to produce more trips on closer edges than far away ones.

The model involves three groups of trips. To easily refer to the trips, we used trip coding for the users. Trips with a simple commute to work are referred to as HWs. Trips with HW1 W2 codes refer to users with an intermediate trip, where W1 is the workplace of the first person and W2 is the workplace of the car driver. Home-School-Work (HSW) is specified for journeys having schools as an intermediate trip. Of course, several other trips can be considered. Specifically, multimember households with parents and children can have multiple trips involving chaining, for example, when parents need to take their children to school (HSW) or when couples want to drive one another to work (HW1 W2).

The transportation modes considered in this model are car, walking, bicycle, and public transport (bus). For cars, only two car types have been considered: diesel and gasoline. The European norm "6" has been chosen as an energetic class. For public transport, we designed four bus flows on the four main axes of the city. Each flow sticks to its corresponding line within a 4-h period and with a 15 min/20 min time frequency. Bus lines cover the north–south and east–west axes, axes 1, 3, 5 and 7 in Fig. 1a. Bus users may obtain the advantage of two types of routes: either one bus route for only one line user (e.g., origin at B1A1 and destination at E5F5) or two bus routes for two line users (e.g., origin at D5C5 and destination at D7E7).

Figure 2 depicts the multimodal transport with cars, buses, and pedestrians sharing the roads. The combination of these modes is also considered since some trips involve walking and public transport. At this station, some walkers board or alight from the bus.

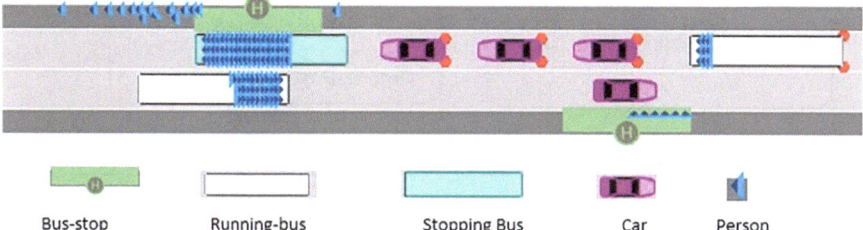

| Bus-stop | Running-bus | Stopping Bus | Car | Person |

Fig. 2 Dynamics of cars and buses driving. (Source: SUMO framework, Author's calculation)

2.3 Overview of the Simulation Workflow

Several steps are needed to end up with a comprehensive transport model. An overview of the corresponding workflow is given in Fig. 3. The construction of the simulation framework starts with data collection and organization and ends with examining policy scenarios.

The basis of the model is the network, including the construction of public transport services and demand for trips. The network needs to include the features required by all transport modes, not only cars. This is particularly important in microsimulation. With more aggregate approaches (macrosimulation), several of these details are ignored or oversimplified. To generate the OD matrices, random sampling households and work locations were used. As explained above, these samplings' distribution assumed that the work locations' frequency increases near the city center. In contrast, the frequency of home locations increases as we move away from the city center. Immediately after generating the demand and network, each user can select a departure time along with a route for his trip. Nevertheless, as the common goal for all network users is minimizing traveling costs, they may also have the choice of selecting the shortest distance routes. With a large number of users, this generally leads to hyper congestion (severe level of congestion)[3] in the edges located in the city center. As several users would gain the ability to improve and shorten their travel period by employing alternative routes, this does not stand as a traffic equilibrium. Indeed, traffic equilibrium is achieved when all users cannot improve their situation by unilaterally modifying their choices. Therefore, an equilibrium is approached when some users improve their situation by avoiding the city center links, which are given by the shortest routes. In this case, they may use the outside city lanes to reduce travel time.

Individuals overvaluing time may change routes to improve their satisfaction (increase utility or reduce generalized cost). Thus, they would accomplish a time-gaining trip by longer albeit less congested routes, therefore achieving balance. This

---

[3] The different levels of congestion are explained in the Fundamental Diagram of Traffic Flow (Li & Zhang, 2011).

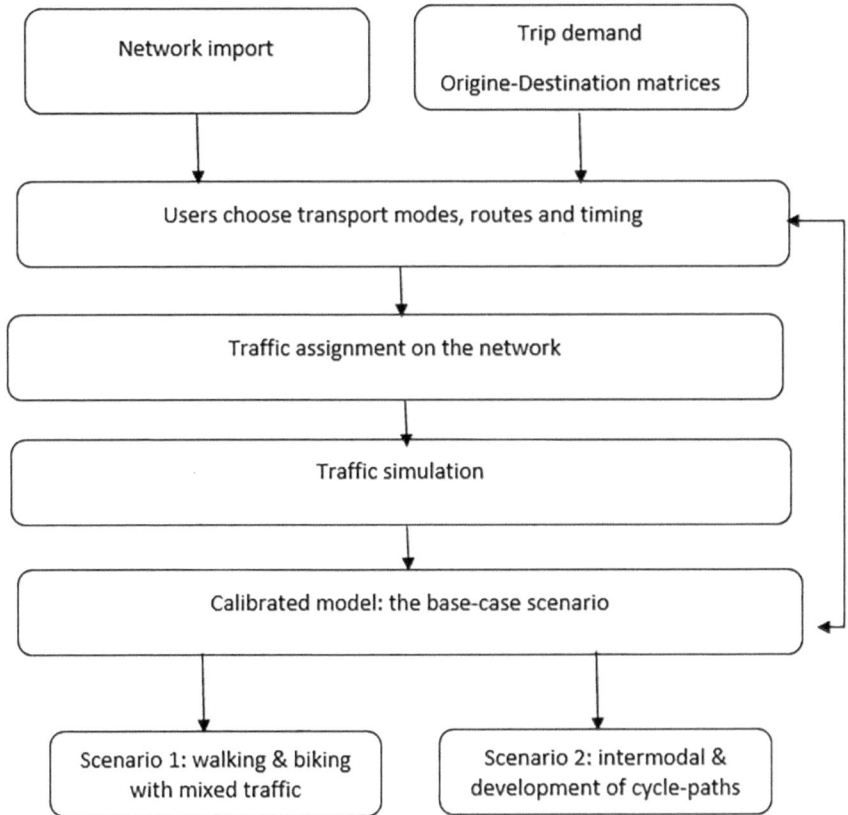

Fig. 3 The main steps in the development of the transport model

step should be repeated until the model reaches a stationary situation. The obtained model can then be used to evaluate transport policies.

This research will serve to measure the modal shift impact on active mobility as a first scenario and intermodality as a second scenario. We are interested in evaluating the impact of each scenario on congestion and pollution.

3   Transport Flows and Intermodality

In this section, a transport model for an average population size will be developed, and the amount of pollutant gas emissions and the level of congestion during the morning rush hour will be computed. Then, this initial situation will be compared with the results of a simulation where a modal choice that takes into account an intermodal choice (bike/walk and bus) is examined.

3.1   The Base-Case Scenario

Regarding our illustration, the first allocation hovers around investigating the congestion level, pollution, and energy consumption of 10,000 car trips over 4 h. Since we have a low traffic flow, network users will employ the shortest routes to minimize travel costs. In this simulation, the average travel speed is 36 km/h for the whole period.[4] Travel time consists of approximately 8 min/veh.

The simulation of daily traffic produces all statistics needed to compute fuel consumption for each car and the emissions of pollutant gases it produces. It then becomes straightforward to use models such as COPERT (COmputer Program to calculate Emission Road Transport) to evaluate environmental and energy impacts. After scaling the population size, we check that these values are consistent with those reported in the World Bank's data.[5] There are three GHGs (CO, CO_2, and HC) and two fine particle groups (PM and NO). These pollutants are measured in kilograms (kg), except for carbon dioxide (CO_2), which is produced in large quantities and measured in tons (t). Daily emissions of CO_2 are equal to 12 t, and emissions of carbon monoxide (CO) and hydrocarbon (HC) are measured in kilograms (their daily emissions are 184 kg and 3 kg, respectively). The emissions of fine particulates PMx are 0.51 kg, and those of NOx are 12 kg. The same model evaluates fuel consumption to approximately 5000 L. These values will be reported again in Tables 1 and 2, when we will be comparing the base-case situation with two alternative scenarios.

The distribution of trip distances produced by the calibrated model is given in Fig. 4. It is important to note that the share of shorter trips (less than 3 km) is higher than 15%, confirming that active modes (i.e., walking or biking) are attractive to many users, as reported in Schweizer (2008).

3.2   Public Transport Accessed by Walk

To reduce pollution and traffic density levels, the development of an intermodal system can be a good alternative. In our case, the intermodal system combines walking with public transport (buses). For each user performing such an intermodal trip, we need to find the station she uses to board the bus and the station she uses to alight.

We solve this problem by choosing the station near the home location (for boarding) and the station near the destination (for alighting). In a more general context, this choice may be influenced by crowding in vehicles and reliability, but to keep our analysis simple, we disregard these features in this model. To introduce intermodality, we assume that 30% of trips switch from cars to a combination of walking and public transport. For each of these trips, we find the corresponding bus line as

---

[4] https://www.statista.com/statistics/264703/average-speed-in-europes-15-most-congested-cities/
[5] https://donnees.banquemondiale.org/theme/environnement

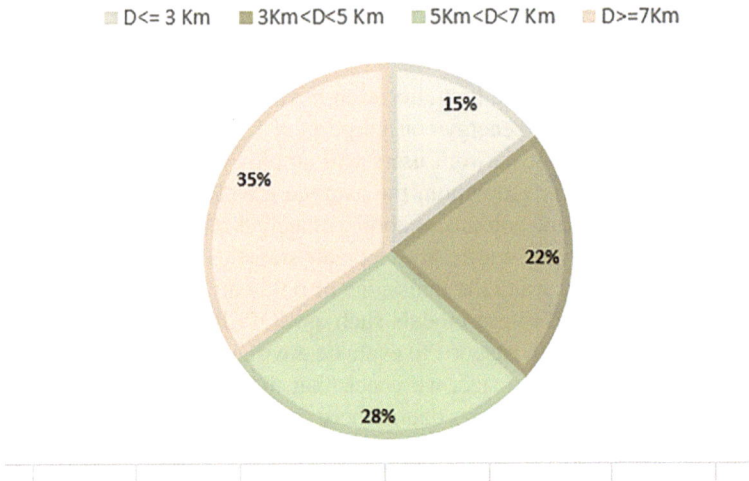

Fig. 4 Distribution of trips with respect to travel distances (produced by the model)

well as the boarding and alighting stations. AM (biking and walking) is particularly convenient for many urban trips when travel distance is short. AM reduces (monetary) user costs and produces much less congestion and pollution. To evaluate the impact of this change, we allow some users to switch to bikes or walk. We model this switch through a probabilistic model where walking and biking are more likely when the trip distance is smaller. Indeed, in the base-case scenario, 37% of trips are less than five kilometers long. AM are efficient alternatives for users making these trips. We adopt a simple choice model to evaluate the impact of a modal switch to bike or walk. Let us consider a total of n trips. We assume that the probability, denoted P_i, that trip $i = 1, ..., n$ becomes intermodal depends on the trip distance, denoted d_i, through the following logistic expression:

$$P_i = \frac{1}{1 + ae^{-\mu(d_i - d_0)}}$$

where d_0 is a parameter that corresponds to the threshold where switching to bike/walk becomes less probable, μ is a parameter that controls how P_i changes around d_0 and a is a positive parameter.

Note that as the value of d_i decreases, the value of P_i decreases quickly and vice versa. By adjusting parameters a and d_0, we can cover a large variety of possible cases to make intermodal transport more or less likely. A similar mode choice functional form will be adopted when intermodal trips involving biking are considered.

The simulation output shows that the average travel speed has decreased from 36 to 33 km/h. The average travel time decreases by 30%, from 8 to 5.6 min. This relief in congestion level is the consequence of the decrease in the network's running cars. From the values reported in Table 1, the quantity of GHG emissions, including CO_2,

Table 1 Emissions of pollutant gases and fuel consumption (output of the model)

Scenario	CO (kg)	CO_2 (tons)	HC (kg)	PMx (kg)	NOx (kg)	Fuel (kL)
Base-case	184	12	3	0.51	12	5
Walk-bus	126	8.4	1.6	0.45	10.74	3.6
Impacts in %	−32	−30	−47	−12	−10.5	−28

has witnessed a significant decrease and this impact is produced by modal shift. The magnitude of the decrease ranges from 30% (for CO_2) to 47% (for HC). This confirms that there is a clear benefit from the reduction in private car usage. For other pollutants (PMx and NOx), the impacts have the same sign with relatively smaller values. Fuel consumption, which is the source of all these emissions, decreases by 28%.

3.3 Biking as a Possible Alternative

We consider two situations. In the first one, traffic is mixed so that bikers share roads with cars and buses over the entire network. In the second one, cycle paths are developed along the main radial axes. This allows bikers to be safer but increases the complexity of the connections in the network. We use the same logistic expression as in the previous section to generate a modal switch to biking and walking.

3.3.1 Mixed Traffic

As discussed above, with mixed traffic, car usage for short journeys is expensive, and many users may prefer walking or biking. It is also a source of high external costs. In this simulation, we evaluate the impact of a modal shift toward active modes. As a result of the logit model we introduced above, some individuals with short trips will move to walking or cycling. We have also included the bicycle in the intermodal system. The user will likely switch to an intermodal trip involving walking when the distance between the home location and the nearest bus station is less than half a kilometer. With bikes, intermodal trips remain likely for longer distances between the home location and the nearest stations. Figure 5 illustrates a relatively complex morning commute trip. The user bikes to the nearest bus station, and this entails a short walk from the bike-sharing station to the bus station. She boards the bus to the station near the destination and then briefly walks to the bike-sharing station to take a bike to the office location (note that we may add a bike-sharing station near the office).

To evaluate the impacts of intermodal transport, we assume that a combination of cycling and buses replaces 30% of the trips initially made by private cars. We assume that 10% of trips, initially made by private cars, are made by a combination of walking and buses. Additionally, 8% of trips are made by bike and 5% by

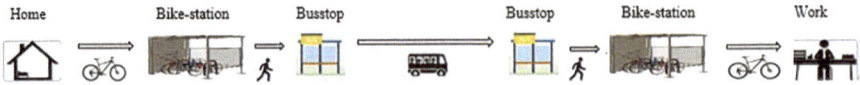

Fig. 5 Intermodality and bike sharing system

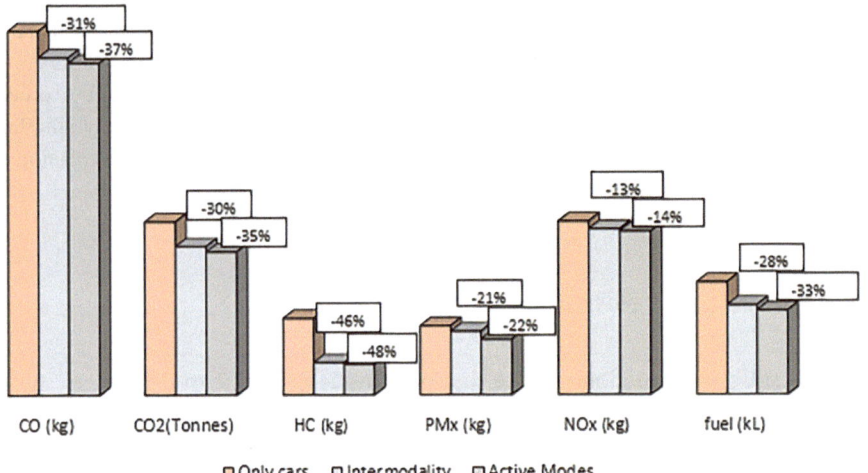

Only cars   Intermodality   Active Modes

Fig. 6 Emissions of pollutant gases and fuel consumption (output of the model)

walking. Walking is under-considered in this model. Indeed, it is so because walking does not produce (or almost) external costs and only needs a very limited infrastructure supply. Therefore, it is mainly added for illustrative purposes.

The output of the simulations shows that with more trips made by bike and walk, the levels of congestion and emissions of pollutant gases decrease. The average speed during the simulation period decreases from 36 to 28 km/h, but this is spelled out by the bicycle's low speed compared to the car and the moderate congestion level in the base-case scenario. This travel speed decrease leads to a slight increase in the average travel time from 8 to 10 min. At the same time, if we focus on cars only, the average travel time decreases from 8 to 5 min, which implies traffic congestion decline.

With respect to emissions, Fig. 6 shows that integrating cycling and walking into the daily commute has a positive effect on pollution. As depicted, most pollutant gases have decreased from approximately 30% to 35%. For example, the CO_2 emissions decreased from 12 to 7.8 t instead of 8.4 t, and the CO emissions decreased from 184 to 117 kg instead of 126 kg. Including active modes in daily trips and intermodality systems also has a significant impact on fuel consumption. We achieved a drop of over 5% compared to the second simulation (from 5000 to 3330 L instead of 3600 L), as indicated in Fig. 6.

3.3.2 Cycle Paths

We upgrade the road network to add cycle paths on the north–south and east–west axes. The purpose of these new lanes is to provide bikers with separate and safe travel conditions. In this case, the cyclists are not allowed to share links with cars in the presence of cycle tracks. As shown in Fig. 7, however, traffic remains mixed when no cycle tracks are available. Running the simulation with the new network and the same traffic flows as the last simulation, we have achieved a significant effect on road safety for cyclists and other network users. The results show that cycle paths also have positive effects on road congestion. The average travel time decreased from 5 min under mixed traffic to 4.6 min.

Moreover, Table 2 reports the results of the four simulations. As indicated in the Table 2, it was found that emissions have been reduced to approximately 45% instead of 35%. Likewise, a 41% reduction in fuel consumption. We can then conclude that AM and intermodal systems as well as bicycle lanes have a significantly positive effect on fossil energy consumption, pollution, and traffic congestion.

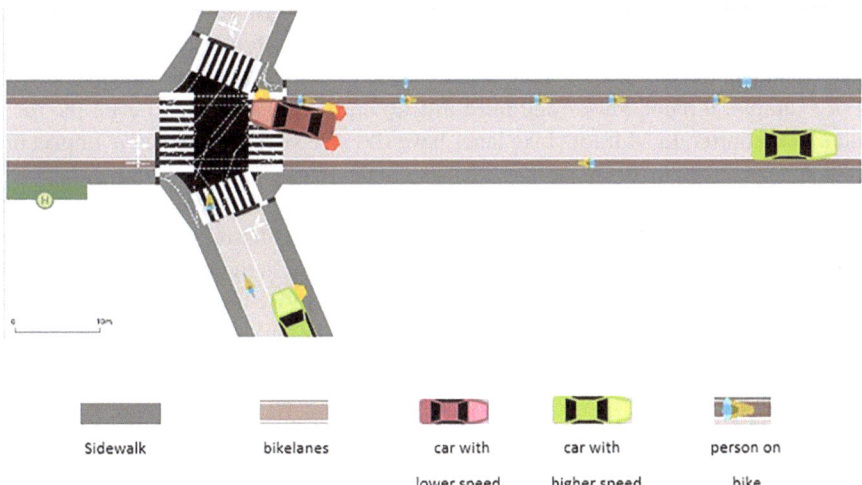

Fig. 7 Dynamic traffic with cycle paths

Table 2 Emissions of pollutant gases and fuel consumption (output of the model, update of Table 1)

Scenario	CO (kg)	CO$_2$ (tons)	HC (kg)	PMx (kg)	NOx (kg)	Fuel (kL)
Base-case	184	12	3	0.51	12	5
Intermodality impacts in %	−32	−30	−47	−12	−10.5	−28
Active mobility impacts in %	−36	−35	−48	−22	−13	−33.4
Cycle paths impacts in %	−37.5	−43.2	−55.6	−27.4	−15.4	−41.6

4   Conclusion

We have developed a microsimulation transport model for a representative city, and we have used it to discuss how the combination of several transport modes, including active modes, impacts energy consumption, traffic flows, and emissions of pollutant gases. We used the Sumo simulation engine to compute each vehicle's travel speed, acceleration, and location at each time t of the rush hour. The development of such a model has indisputable advantages for the planning of urban transport and the examination of distinct reforms of mobility. We have examined scenarios and focused on the issues of global warming and the limitation of fossil fuel energy consumption.

According to this study, the intermodality system was found to blend public transport with active modes. As a quantitative result, when we reduce the usage of private cars by 30%, through a modal switch toward a combination of walking and public transport, we obtain a significant decrease in the emissions of pollutant gases (at a comparable proportion to this modal share switch). The integration of another intermodal system that includes biking and buses, as well as cycling and walking as an alternative for short trips, has resulted in an increase of 35% instead of 30% in the decline of congestion. The results can be enhanced by 45% if we incorporate cycle paths in the city's main axes. We have also obtained another interesting result. For example, a reduction of one-third in CO_2 emissions, GHG, and even the rush hour gets shorter. In addition, bike lanes have shown a significant positive impact on road safety. It can be deduced that this has led to a reduction in external costs. Our methodology can be replicated for other case studies, including real cities of distinct geometries. All the modeling steps are transparent and based on data now made available for several cities. The use of microsimulation is the main contribution of our analysis. As shown in the output of the simulation, we trace back all the events at a very detailed level. This approach (microsimulation) is now limited to cities with less than 100,000 inhabitants, but more ambitious applications based on microsimulation will be possible in the near future.

This model may be improved in several directions. In particular, a more sophisticated combination of transport modes can be considered. These modes can include parking areas used by those who combine private cars and public transport modes. We may also extend the framework to consider user activities and how they are scheduled during the day and with respect to the dynamics of traffic flows. For example, one user may consider shopping activity in the morning (and not late in the afternoon) to avoid severe congestion. Updating the modeling in this direction requires an elaborate decision model describing how users choose among the several available alternatives and modes. We leave this task for future research.

At the same time, the model can be employed to examine other policy scenarios. Infrastructure upgrading can be considered through the development of dedicated cycle paths, for example. Additionally, it can be used to investigate the increase in the market share of electric vehicles. Electric vehicles (or bikes) can be examined through the deployment of solar panel barriers that provide clean energy sources.

Modeling the electric grid can be considered an extension module that connects vehicle consumption to electric production units to assess these solutions' overall energy and financial accounts.

References

Appert, C., & Santen, L. (2002). Modélisation du trafic routier par des automates cellulaires. *Actes INRETS, 100*, 1–18.

Axhausen, K. W., Horni, A., & Nagel, K. (2016). *The multiagent transport simulation MAT-Sim.* Ubiquity Press.

Baum, M., Buchhold, V., Sauer, J., Wagner, D., & Zündorf, T. (2019). Unlimited transfers for multimodal route planning: An efficient solution. *arXiv preprint arXiv:1906.04832*, 1–42.

Berdai, A. (2004). *Modélisation et simulation d'un réseau de transport public par une approche multiagents* [PhD thesis]. Besancon.

Costeseque, G. (2013). Modélisation et simulation dans le contexte du trafic routier. In F. Varenne & M. Silberstein (Eds.), *Modéliser et simuler. Epistémologies et pratiques de la modéélisation et de la simulation*. Editions Matériologiques.

Curiel-Esparza, J., Mazario-Diez, J. L., Canto-Perello, J., & Martin-Utrillas, M. (2016). Prioritization by consensus of enhancements for sustainable mobility in urban areas. *Environmental Science & Policy, 55*, 248–257.

de Palma, A., Stokkink, P., & Geroliminis, N. (2022). Influence of dynamic congestion with scheduling preferences on carpooling matching with heterogeneous users. *Transportation Research Part B: Methodological, 155*, 479–498.

Fishman, E., Washington, S., & Haworth, N. (2012). Barriers and facilitators to public bicycle scheme use: A qualitative approach. *Transportation Research Part F: Traffic Psychology and Behavior, 15*(6), 686–698.

Fishman, E., Washington, S., & Haworth, N. (2013). Bike share: A synthesis of the literature. *Transport Reviews, 33*(2), 148–165.

Gebhardt, L., Krajzewicz, D., Oostendorp, R., Goletz, M., Greger, K., Klötzke, M., Wagner, P., & Heinrichs, D. (2016). Intermodal urban mobility: Users, uses, and use cases. *Transportation Research Procedia, 14*, 1183–1192.

Gohari, A., Ahmad, A. B., Balasbaneh, A. T., Gohari, A., Hasan, R., & Sholagberu, A. T. (2022). Significance of intermodal freight modal choice criteria: MCDM-based decision support models and SP-based modal shift policies. *Transport Policy, 121*, 46–60.

Guillotte, K., Bédard, Y., Larrivée, S., & Badard, T. (2009). Conception et développement d'un outil de modification de la segmentation routière. *Geomatica, 63*(4), 365–381.

Kilani, M., Diop, N., & De Wolf, D. (2022). A multimodal transport model to evaluate transport policies in the north of France. *Sustainability, 14*(3), 1535.

Leclercq, L. (2002). *Modélisation dynamique du trafic et applications à l'estimation du bruit routier* [PhD thesis]. Lyon, INSA.

Li, J., & Zhang, H. M. (2011). Fundamental diagram of traffic flow: New identification scheme and further evidence from empirical data. *Transportation Research Record, 2260*(1), 50–59.

Lopez, P. A., Behrisch, M., Bieker-Walz, L., Erdmann, J., Flötteröd, Y.-P., Hilbrich, R., Lücken, L., Rummel, J., Wagner, P., & Wießner, E. (2018). Microscopic traffic simulation using sumo. In *2018 21st international conference on intelligent transportation systems (ITSC)* (pp. 2575–2582). IEEE.

Lorente, E., Barceló, J., Codina, E., & Noekel, K. (2022). An intermodal dispatcher for the assignment of public transport and ride pooling services. *Transportation Research Procedia, 62*, 450–458.

Loske, D. (2020). The impact of COVID-19 on transport volume and freight capacity dynamics: An empirical analysis in german food retail logistics. *Transportation Research Interdisciplinary Perspectives, 6*, 100165.

Midgley, P. (2011). Bicycle-sharing schemes: Enhancing sustainable mobility in urban areas. *United Nations, Department of Economic and Social Affairs, 8*, 1–12.

Pini, P., & Lavadinho, S. (2005). Développement durable, mobilité douce et santé en milieu urbain. In *Actes du colloque "Développent urbain durable, gestion des ressources et gouvernance"*. Université de Genève: Observatoire Universitaire de la Mobilité, Département de géographie, LEA, UNIGE.

Schweizer, P. (2008). *L'action "Bike to work": une voie vers la mobilité durable?: Les potentialités de l'événementiel dans la réalisation du transfert modal vers le vélo: apports et critères de réussite* [PhD thesis]. Université de Lausanne.

Shaheen, S. A., Zhang, H., Martin, E., & Guzman, S. (2011). China's Hangzhou public bicycle: Understanding early adoption and behavioral response to bikesharing. *Transportation Research Record, 2247*(1), 33–41.

Shaheen, S. A., Cohen, A. P., & Martin, E. W. (2013). Public bikesharing in North America: Early operator understanding and emerging trends. *Transportation Research Record, 2387*(1), 83–92.

Share, A. B. (2011). *Melbourne bike share survey*. Alta Bike Share.

Srisakda, N., Sumitsawan, P., Fukuda, A., Ishizaka, T., & Sangsrichan, C. (2022). Reduction of vehicle fuel consumption from adjustment of cycle length at a signalized intersection and promotional use of environmentally friendly vehicles. *Engineering and Applied Science Research, 49*(1), 18–28.

Weliwitiya, H., Rose, G., & Johnson, M. (2019). Bicycle train intermodality: Effects of demography, station characteristics and the built environment. *Journal of Transport Geography, 74*, 395–404.

Willing, C., Brandt, T., & Neumann, D. (2017). Intermodal mobility. *Business & Information Systems Engineering, 59*(3), 173–179.

Open Access This chapter is licensed under the terms of the Creative Commons Attribution 4.0 International License (http://creativecommons.org/licenses/by/4.0/), which permits use, sharing, adaptation, distribution and reproduction in any medium or format, as long as you give appropriate credit to the original author(s) and the source, provide a link to the Creative Commons license and indicate if changes were made.

The images or other third party material in this chapter are included in the chapter's Creative Commons license, unless indicated otherwise in a credit line to the material. If material is not included in the chapter's Creative Commons license and your intended use is not permitted by statutory regulation or exceeds the permitted use, you will need to obtain permission directly from the copyright holder.

Smart Cities Initiatives and Perspectives in the MENA Region and Saudi Arabia

Fateh Belaïd, Razan Amine, and Camille Massie

Abstract This chapter focuses on the evolution of smart cities in developing countries. It starts by mapping the definitions and evolutions of smart cities concepts. Then it outlines the progress and current practices that emerging economies, in general, have achieved in their transition towards smart cities and the significant key challenges and takeaways that can be acquired so far. It analyses what could have been done better and what factors are still missing in smoothening this transition. Furthermore, it zooms into the smart city initiatives in Saudi Arabia to better extract lessons learned and the way to move forward. The analysis suggests measures that each actor can take: the public, private and international sides to further smoothen the transition to address the rising challenges of urbanization in emerging economies.

Keywords Smart cities · Sustainability · Climate change · Energy transition · Saudi Arabia

1 Introduction

Urbanization of the world's population is increasing steadily. The world population approaches 8 billion, half of which lives in urban areas, and 85% of the world's GDP is generated in cities (UN, 2019). By 2050, the global population is projected

F. Belaïd (✉)
King Abdullah Petroleum Studies and Research Center, Riyadh, Saudi Arabia
e-mail: fateh.belaid@kapsarc.org

R. Amine
King Abdullah Petroleum Studies and Research Center, Riyadh, Saudi Arabia

Cambridge University, Faculty of Economics, Cambridge, UK

C. Massie
King Abdullah Petroleum Studies and Research Center, Riyadh, Saudi Arabia

Faculty of Management, Economics & Sciences, Lille Catholic University, UMR 9221-LEM, Lille, France

© The Author(s) 2024
F. Belaïd, A. Arora (eds.), *Smart Cities*, Studies in Energy, Resource and Environmental Economics, https://doi.org/10.1007/978-3-031-35664-3_16

to increase to approximately 9.8 billion, and more than two-thirds of the world's population is expected to live in urban areas (UN, 2019). Cities are particularly well positioned to play a leading role in tackling climate change and fostering the transition to a more sustainable world. This intensive urbanization already presents serious challenges to our society, including environmental quality degradation, increasing socioeconomic inequalities, energy security, intensive energy use, and increased natural and human-made disasters fueling climate change.

Smart cities promise to address these challenges and make cities that are more sustainable, resilient, eco-friendly, and livable. By integrating new digital technologies (such as the Internet of Things (IoT), artificial intelligence, 5G, cloud computing, and big data), communities, and policies, smart cities can potentially deliver well-being, competitiveness, transparency, and sustainability (Yigitcanlar et al., 2019). In recent years, the concept of smart cities has gained popularity in academia, industry, and public policies.

Although the concept of smart cities is widespread, research in this area is still in its infancy. According to Yigitcanlar et al. (2019), the notion is still ambiguous, with limited conceptualizations and practical frameworks that could assist policymakers in realizing their smart city initiatives.

One straightforward definition of smart cities is the use of different technologies, including the IoT, in urban areas to collect and share information and improve the operational performance of urban cities' services, such as mobility and energy. Conceptually, the notion of smart cities consists of six city-oriented elements: habitat, population, transportation, economy, environment, and government (Albino et al., 2015). An illustration is provided in Fig. 1.

The Middle East and North Africa (MENA) region will be a global urbanization hotspot over the next decade. Indeed, the region's urban population is expected to increase by one-quarter between 2020 and 2030 (UN, 2019). In this context, we

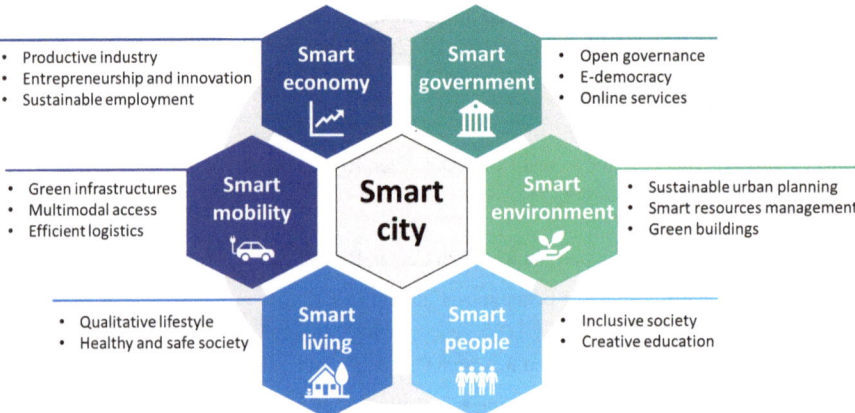

Fig. 1 The six pillars of smart cities. (Source: Authors)

consider it necessary to advance in this line by outlining the progress and current practices that emerging economies, in general, have achieved in their transition toward smart cities and the big key challenges and takeaways that can be acquired thus far.

Smart cities in the MENA region, particularly in Saudi Arabia, are pivotal to supporting the dynamic growth of population, diversifying economies, and showcasing the region's capabilities to the world. Building on this conjecture, this chapter aims to review the smart city concept and highlight insights into the policy implications of smart city development. It also aspires to contribute to the literature on the global governance of smart cities and their role in accelerating energy and ecological transition.

As many organizations and policymakers are under constant pressure to collect, process, and disclose detailed and accurate information on the considerable challenges posed by increased energy demand and urbanization, a systematic understanding of the complex nature of smart and sustainable cities becomes paramount. Especially in light of recent challenges facing urban developments (e.g., energy transition and consumption, improving air quality, adapting to climate change, improving interaction/integration between transportation and buildings, biodiversity preservation, etc.), aggressive urban agenda development becomes necessary to share information in real time, identify problems, anticipate risks and design solutions that enhance cooperation among stakeholders to improve growth, quality of life, and innovation in cities, and resolve societal challenges.

The remainder of this chapter proceeds as follows. Section 2 reviews the definitions and evolution of smart city concepts. Section 3 discusses the progress toward smart cities in MENA countries. Section 4 presents smart city initiatives in Saudi Arabia. Finally, Sect. 5 concludes the chapter and discusses policy implications and future work needed to build smart, sustainable cities in emerging economies.

2 Definitions and Evolution of the Smart Cities Concept

The smart cities notion was first introduced in the early 1990s (Orejon-Sanchez et al., 2022). In recent years, the "smart cities" concept has attracted increased interest in political, industrial, and academic circles (Orejon-Sanchez et al., 2022). However, even though the smart cities concept is popular, research in this area is still in its infancy. Furthermore, the notion is still vague and ambiguous, with limited conceptualizations and practical frameworks that could assist policymakers in realizing their smart city initiatives. The smart cities notion is an indication of a relationship between public and private sectors and cities where equipment is deployed on public utilities, using sensors to collect information. This information is used to manage resources, services, and assets efficiently. Smart city approaches offer municipal authorities and policymakers a new opportunity to improve municipal services and citizens' well-being. It is a concept that will shape the future of the evolution and transformation of the urban living environment.

Actually, the concept of smart cities is constantly evolving and remains under debate. The literature offers many definitions of smart cities. Caragliu et al. (2009) offer one of the most comprehensive definitions: a city is considered smart "...*when investments in human and social capital and traditional (transport) and modern communication infrastructure fuel sustainable economic growth and a high quality of life, with a wise management of natural resources, through participatory governance*" (Caragliu et al., 2009, p. 70).

A smart city is broadly considered by the International Organization for Standardization (ISO) to be "*a new model and concept, applying the next generation of Information and Communication Technology (ICT) to facilitate smart city planning, construction, management and services*" (ISO, 2014).

Alternatively, according to the EU, the smart cities concept is about using "*a smart city is a place where the traditional networks and services are made more efficient with the use of digital and telecommunication technologies, for the benefit of its inhabitants and businesses*" (European Commission, 2022).

The Organisation for Economic Co-operation and Development (OECD) defines smart cities as "*initiatives or approaches that effectively leverage digitalization to boost citizen well-being and deliver more efficient, sustainable and inclusive urban services and environments as part of a collaborative, multistakeholder process*" (OECD, 2019).

The smart city concept is often equated with other concepts, such as the knowledge city, sustainable city, intelligent city, ubiquitous city, digital city, and information city concepts. Essentially, all of these concepts focus on the application of information and communications technology (ICT) to urban management. These applications aim to improve the accountability, transparency, efficiency, and effectiveness of interactions between residents and local authorities. However, the concept of smart cities has shifted beyond a narrow focus on ICT diffusion. Instead, it addresses the needs and demands of individuals and communities holistically. Although ICTs are not the primary pillar of smart cities, they facilitate the establishment and development of smart communities (Pira, 2021).

The design of smart cities has traditionally focused on technology, smart devices, and urban infrastructure. However, in recent years, the concept has been expanded by several cities to incorporate socioeconomic aspects (Pira, 2021). The most relevant description of the concept as applied to urban projects is provided by Trencher (2019), who argues that smart city initiatives primarily focus on individuals, and technology is just a tool that is used mainly to serve citizens. Accordingly, this paradigm shift allows the smart cities approach to move beyond the techno-centric process and to expand its potential impacts on the economic, social, and environmental dimensions.

To date, the smart cities concept has been developed in three different phases. The first smart city generation, the so-called *Smart City 1.0*, is viewed as a technology-driven approach. This first generation of smart cities mainly concentrated on leveraging technology to enhance and facilitate urban activities, including the use of software, smart devices, and high-tech platforms in mobility, health, energy, and security domains. This has prompted investigation and research on the

Fig. 2 The three generations of smart cities. (Source: Authors)

commercial potential of digital technologies (Han & Hawken, 2018). From the early stage of smart city generation, six critical components have been identified as the core attribute of this concept: mobility, people, lifestyle, economy, environment, and governance (Albino et al., 2015), previously illustrated in Fig. 1.

While with the first generation of smart cities, the big IT companies led this movement in urban areas intending to provide municipalities with their products, *Smart City 2.0*, the second generation of the smart city, has been led by municipal authorities and decision-makers. The primary purpose was to improve services and enhance well-being and quality of life in urban areas by effectively harnessing the beneficial aspects of new technologies.

In recent years, a new phase of smart cities has emerged. Rather than adopting a technology-driven or city-driven model, large smart cities are moving to cocreation models involving citizens in developing the next generation of solutions. Accordingly, *Smart City 3.0* focuses on citizens' role and involvement in addressing community issues and helps municipality managers identify effective and reliable solutions for various city challenges, including social, economic, and environmental issues. Figure 2 illustrates the evolution of the smart city concept.

3   Progress Toward Smart Cities in Emerging Economies

Cities are at the origin of significant climate change activities and energy and mobility challenges; they must cooperate regionally and globally to identify and develop solutions. Most cities around the world today are embarking on important initiatives to make significant progress toward our societal and environmental objectives. This section seeks to understand how policies at the city level in emerging economies are developing the plans that country governments need to follow to achieve sustainable livelihoods for all.

In the context of the MENA region, an above-average population growth rate of 1.56% per year (vs. 1.1% globally) and a high speed of urbanization lead to the high importance of smart city solutions. While in 1960, less than 40% of the MENA population was living in cities, in 2020, this rose to above 60% and is expected to rise further significantly (World Bank, 2022). The MENA region also faces significant climate challenges impacted by changing precipitation patterns, rising sea levels, and water insecurity. This, in turn, hinders the development process through channels such as agriculture and the environment, which slows the economy's self-sufficiency (Sieghart & Betre, 2018).

This section reviews the progress toward smart cities in some MENA developing countries, namely, Algeria, Egypt, Jordan, Lebanon, Morocco, Saudi Arabia, and the United Arab Emirates (UAE). It then explores the critical elements and indicators of smart cities.

3.1 Overview of Progress in the MENA Region

The smart cities market and initiatives are proliferating in the MENA region. Many countries have planned to invest heavily in the sector.

Algeria has not set a clearly written (environmental) sustainable building policy; the focus is more on building and housing delivery. The country's regulatory framework ensures compliance with international standards but lacks specific links to sustainability performance improvement. The country recently developed a new smart city concept, Sidi Abdellah, which shows the government's wish to integrate sustainable development in the design and management of the city to create a liveable and sustainable environment for its residents.

Egypt has made ongoing efforts toward greening both the tertiary and residential sectors since 2009. A set of measures was adopted in 2010, reviewing sustainability, ecology, energy and water efficiency, resources and environmental quality, and technological innovation. The cities of Sharm El Sheikh, Kom Ombo, and Kuraymat are hybrid plant projects supported by the newly instituted New and Renewable Energy Authority (NREA), seeking to introduce and develop renewable energy technologies in the country.

Jordan faces severe issues of sustainability. Accordingly, the country has recently promoted the concept of smart cities and has achieved great work in transforming the construction market by instilling sustainable economic projects, products, and services. Amman, Sahab, and Irbid cities are great examples of the country's commitment to developing a national green economy based on renewable and sustainable energy sources.

The absence of legislation for green construction, energy conservation, water conservation, etc. in Lebanon, makes it lag behind its MENA neighbors in terms of economic commitments through the sustainability lens. Legal constraints and financial dependency hinder Lebanese municipalities in their duties and obligations. Promising programs have been launched to comply with international standards, such as the Country Energy Efficiency and Renewable Energy Demonstration

Project for the Recovery of Lebanon (CEDRO), which works on developing the energy market in effective and sustainable ways.

Morocco imports almost the totality of its energy needed to meet increasing demand. Thus, the country now seeks to sustainably improve its national energy production and has adopted the concept of sustainable development. To date, it has established several stepping stones to achieving a sustainable development vision that drives many reforms, including political, institutional, legal, and socioeconomic programs. Smart cities are at the heart of national research and innovation. Under the Sustainable Development Plan, which targets energy efficiency in energy-intensive sectors and promotes renewable energy, the city of Casablanca recently experienced a major transformation to improve its long-run livability.

Saudi Arabia has launched major smart city projects and is now positioned among the leaders in the MENA countries. Although sustainable development initiatives and programs continue to be slowly developed, they are only gradually being publicized, with the Saudi Green Initiative and Green Riyadh project being highlighted recently. Several mega projects are currently under development, including The Line, Oxagon, and Trojena in NEOM, a smart cities initiative that will be further discussed in this chapter.

The UAE aims to become the most sustainable country in the world (UAE, 2022) and has, before others, launched many smart city projects, such as Dubai and Masdar City. The country has addressed climate change by launching adaptation measures and policies and mitigation actions at many levels. It reflected its commitment toward sustainable development in its Vision 2021 and Green Economy Strategy for Sustainable Development. Abu Dhabi's Masdar city is often seen as a frontrunner for smart city development, reaching interconnectedness and minimal environmental pollution through large greenfield investments (Ringel, 2021). The country intends to promote the adoption and implementation of green strategies to accelerate the growth of the green construction sector.

These government-steered transitions to smart cities are typical for the MENA region's smart city approach (with dedicated initiatives of high prominence in Casablanca, Algiers, Cairo, Kuwait, Doha, Dubai, Abu Dhabi, and many more). The government's approach is to provide sustainable solutions that map across areas, including energy, mobility, and architecture. This is further based on integrating information technology and artificial intelligence. Table 1 presents the different commitments and progress toward smart cities in these countries.

Congestion, scarcity of resources, and waste management demand intelligently designed cities. Through first laying a foundation with aspirations (e.g., "The Line" in Saudi Arabia, or Tunis and Cairo's mostly conceptual stage in the smart cities transition), citizens, businesses, and investors are pooled toward jointly designing smart cities life. The second stage curtails the convergence of a city toward the outlined vision through a dedicated plan and investment efforts. Gulf Cooperation Council countries have pushed their "signature cities" into the convergence phase with successful implementations and large-scale investments. Finally, the transformation phase encompasses network-enabled utilities, security services integrations, and smart transport available to all inhabitants (Shokeir & Yahia, 2020). Ideally, a

Table 1 Progress toward smart cities in some MENA and GCC countries

Smart cities and buildings	Policies	Future actions
Algeria		
New city of Sidi Abdellah: shows the government's wish to integrate sustainable development in the design and management of the city, in order to create a livable and sustainable environment for its residents	Enforcement of building standards: inspired by Eurocode and supported by the National Centre for Studies and Integrated Research Building (CNERIB)	Develop procedures, promote concepts of sustainable planning and design
Egypt		
City of Sharm El Sheikh: signature of a cooperation agreement to transform it into a green city. Kom Ombo and Kuraymat cities: hybrid plant projects supported by the newly instituted New and Renewable Energy Authority (NREA)	The New and Renewable Energy Authority (NREA): expands efforts to develop and introduce renewable energy technologies in Egypt. The Green Pyramid Rating System (GPRS): promotes sustainable development, increased use of renewable energy, better management of natural resources, and efficient and effective use of resources	Enhance enforcement of building codes, develop and promote green labeling schemes, and launch major marketing strategies and media campaigns to raise awareness
Jordan		
City of Amman: embarked on developing a sustainable blueprint for development. Sahab and Irbid cities: have launched their own sustainability initiative, "Sahab a Green City" including renewable energy mega projects	The Renewable Energy and Energy Efficiency Law: sustains exploitation and development of renewable energy sources, contributes to the protection of the environment, and rationalizes and improves energy consumption efficiency	Understand stakeholder needs and added values to increase the interdisciplinary focus on the green economy
Lebanon		
The first sustainable "green" construction in Lebanon took place in 2009	Country Energy Efficiency and Renewable Energy Demonstration Project for the Recovery of Lebanon (CEDRO): funded by the European Union, objective of increasing energy use and efficiency opportunities in the country	Take constructive steps toward advancing the concept of sustainable development in the country to overcome legal constraints

(continued)

Table 1 (continued)

Smart cities and buildings	Policies	Future actions
Morocco		
City of Casablanca: transform the city into a more livable and greener city, by tackling water, waste, energy, and transportation practices	The Sustainable Development Plan: targets energy efficiency in the energy-intensive sectors and promotes renewable energy	Implement lower taxes programs for green buildings to foster their development, evaluate the performance of existing buildings, and guarantee their performance, savings, and benefits by establishing a financial mechanism
Saudi Arabia		
The Yanbu Industrial City, The Line, Oxagon, Trojena, in NEOM: several mega projects in the Kingdom which jointly pursue sustainable city development goals	Saudi Building Code, Energy Consumption: establishes minimum performance-related regulations for the design of energy-efficient buildings	Trigger public awareness, and encourage professionals and developers to implement sustainable practices at various levels: building, neighborhood, and city
United Arab Emirates		
Dubai is intended to be the most sustainable city in the world. Masdar Eco-city: applies new technologies and aims to promote the emergence of new industrial and commercial sectors in the UAE	Dubai's Green Building Regulations and Specifications (GBR&S): emphasize on sustainability and the development of green buildings	Foster awareness of the benefits of adopting and implementing green strategies to catalyze the growth of the green construction sector

Source: Authors, based on UNEP (2017)

transformation aligned to the needs of all stakeholders is thus reached: business, citizens, and nature. All stages have risks and challenges: aligning to a vision might be difficult with diverse and changing populations in cities across the MENA region. Successful convergence is dependent on significant governmental coordination and stable ground for investments, which cannot (yet) fully be found across the MENA region. In addition, Ringel (2021) finds that poor management is seen as the biggest barrier to successful convergence to smart cities. Last, the transformation is dependent on the continuous availability of sustainable energy (i.e., the ability to harness MENA's abundant renewable energy capacity).

3.2   Key Elements and Indicators of Smart Cities

We have assembled "indicators" or "pillars" of smart cities by combining various structures from the literature. Camargo et al. (2021) recommended that emerging economies follow a new structured model to adopt smart cities. They suggest that the model consists of four pillars (agile governance, urban planning, social cohesion, competitiveness, and growth). In a very similar fashion to the pillars of smart cities, Pira (2021) elaborates on the indicators of smart cities. We combine both frameworks to present a comprehensive framework that embraces five verticals of smart cities, also illustrated in Fig. 3.

Fig. 3 Verticals of smart cities. (Source: Authors)

The first vertical is agile governance. We dissect this into three main pillars: agile policies, advanced infrastructure, and technology. The first pillar of agile governance is agile policies, which suggests a more flexible approach to designing policies (economic policies, i.e., fiscal and monetary as well as other types of policies) to solve issues related to mobility, security, migration, and equity, among other issues. The second pillar is advanced infrastructure, which includes developed health and educational systems, safety and security platforms, and efficient and clean transportation. All of these elements ease the mobility of citizens (Pira, 2021). The third pillar is technology that supports this adaptability and agility of governance by embracing online sources, open data, and privacy maintenance simultaneously (Camargo et al., 2021). For example, designing smart tax policies as well as e-taxation platforms is an example that combines fiscal policy development, technology, and data privacy and sharing. E-taxation platforms allow the government to establish a better relationship with taxpayers and boost equity through income redistribution while at the same time collecting better structured data on filing and payment behavior that could be used for confidential research purposes. Lessons learned from research can then lead to policy recommendations on fine-tuning tax policies, increasing tax revenues from the "right" people, and maximizing public investment in this tax revenue.

The second vertical is the environment, which entails smart buildings that are sustainable, resource management that takes care of reducing one's carbon footprint and pollution, and sustainable urban planning that matches climate protection (Pira, 2021). Urban planning is crucial in the face of growing populations and is based on the efficient management of resources to match the rising demands of citizens (Camargo et al., 2021).

The third vertical is a comprehensive inclusive culture. This implies social cohesion, which is the effort to smooth the use and integration between a proper infrastructure, data, and people in a way that flows in favor of smarter cities (Camargo et al., 2021). A smart city also embraces a culture of education, positive spillovers, maximizing potential, and enhancing equity.

The fourth vertical is a well-established private sector that promotes economic growth, fosters a self-sufficient economy, and boosts trade. The first element is constructive competitiveness that promotes economic scales, product diversity, market entry requirements, and other economic factors (Camargo et al., 2021). The second element is innovation and creativity in producing according to comparative advantage, which enhances trade relations.

The fifth vertical involves international organizations and aid. This vertical relies on all other verticals because, first, receiving aid, and second, its proper utilization depends on the public and private sector's development and environmental status. We argue that a smart city with good governance and a self-contained private sector can smartly invest in the aid it receives without accumulating a large amount of debt. This is because investing in technology, infrastructure, and sustainable policies provides the private sector with a space to flourish and grow, which then generates an economic growth rate larger than the country's debt growth rate.

4 Smart City Initiatives in Saudi Arabia

This section focuses on the urban development projects of Saudi Arabia, which are on the trajectory of becoming a smart city. Why does this section focus on Saudi Arabia? Saudi Arabia is the largest country in the Middle East, both in terms of its land as well as the economy, and among the largest Middle Eastern countries in terms of its population. Namely, Saudi Arabia is inhabited by approximately 35.5 million people (2021 estimate), and its major economic indicators can be summarized by a real GDP growth of approximately 2.8% and its large oil reserves that constitute approximately 20% of the world's conventional oil reserves (2021 estimates). These are among the reasons why Saudi Arabia makes an interesting case and has the potential to set a stage for the development of smart cities in emerging economies, namely, the MENA region.

More specifically, Saudi Arabia has faced significant urbanization and population growth that have incentivized the need to establish smart cities. In the latter, Saudi Arabia's city populations almost tripled between 1980 and 2018, from approximately 9.32 million to 26.3 million, where most of the population is centered in Riyadh (AEC, 2018). This high population growth alongside migration from rural to urban areas induced urgency in devoting attention to urban development in Saudi Arabia. Such rapid urbanization has brought a suite of challenges, including resource exhaustion, such as water, road congestion, and pollution. This raised the need for the government to implement new solutions to address these challenges.

Since 1990, Saudi Arabia has been in the stage of developing structural reforms that boost its privatization, also known as "Saudization," encourage liberalization, and enhance investment regimes. In 2016, the country established the "Vision 2030 plan," which aims to expand the sustainability of resources for future generations while maximizing the well-being of citizens (Belaïd and Al-Sarihi, 2023; Saudi Vision, 2022). The vision adopts concepts of a smart city in regard to sustainability as well as the use and processing of knowledge. Aldusari (2015) illustrates how the Saudi government relies on knowledge-rooted technological advancement that would contribute to setting society on its path to sustainable development.

More directly related to the smart cities initiative, the government of Saudi Arabia has developed a vision to enhance its citizens' quality of life through smart reforms that begin by targeting five cities in Saudi Arabia: Makkah, Riyadh, Jeddah, Al-Madinah, and Al-Ahsa. Next, examples of existing and future smart initiatives in Saudi Arabia are given, namely, the cities of Yanbu and NEOM, respectively, followed by criteria for assessing smart cities.

4.1 Saudi Arabia's Yanbu and NEOM Smart City Concepts

The first smart city initiative in Saudi Arabia was the Yanbu Industrial Smart City. As the third-largest oil refinery center in the world, Yanbu has been more dependent on oil for its growth and development. This motivated the initiative that aims at diversifying the sectors that generate growth. The development of this initiative has been implemented through three stages. First, enhancing the public infrastructure, including roads, buildings, water, etc., as well as information infrastructure, including information networks and computing. The second stage consists of the smart usage of ICT applications. The third stage comprises the establishment of a smart community portal such as big data analytics to make city management as effective as possible (Doheim et al., 2019).

Recently, launched, NEOM is a smart city project focusing on three pillars: trade, innovation, and knowledge (NEOM, 2022a). First, it is a project that belongs to the Public Investment Fund. The location is also strategically chosen. The region is located in northwestern Saudi Arabia on the Red Sea, and it is expected to grow into a hub for business and creativity, raising hopes for it to be a smart city of the future. There are four main objectives for developing NEOM, represented in Fig. 4. First, it aims at diversifying Saudi Arabia's economic production of goods and services. This, in turn, enables the country to lead in world trade and thus boost its economic growth and connections with other countries. Second, the project aims at establishing a city that serves multiple purposes: it became a city that serves both residential and works purposes, in addition to including aspects of exploration and diversity that also serve the purpose of leisure for its residents. The third objective is to build robust sustainability for the city in terms of its urban, health, and environmental development. It aims to do so by first establishing high-quality standards for this sustainable development associated with measurable outcomes and, second, using technology to make the process more efficient and smarter. The fourth objective is complementary to the first three, which involves merging multiple communities into one: establishing research centers as well as leisure venues to supply the needs and multidimensional demands of citizens.

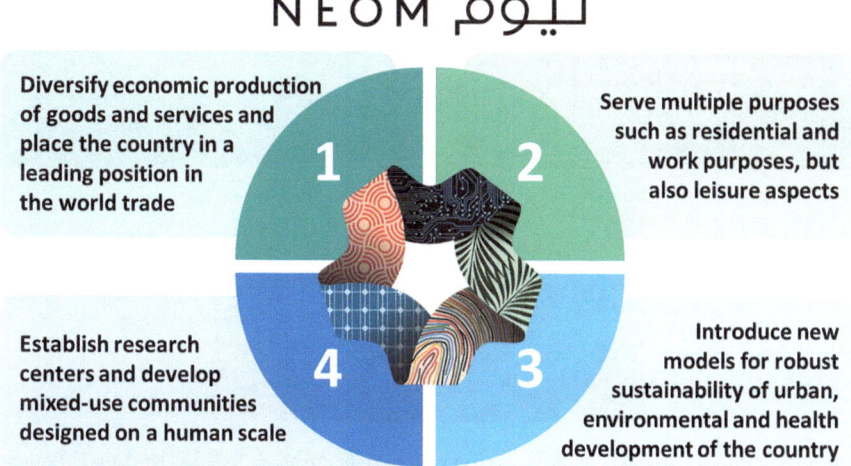

Fig. 4 The four objectives of NEOM. (Source: Authors)

What makes this project unique is its dependence on the comparative advantages of each of the countries it is implemented in: Egypt, Jordan, and Saudi Arabia (Doheim et al., 2019). NEOM offers a new perspective for building a smart city, as new land has been selected in the northwestern region of Saudi Arabia. The motivation is to build cities that are unique and different from the conventional aspects of a city, and one way to ensure this is to start from scratch in developing businesses, technology, artificial intelligence, and promoting skilled labor (Farag, 2019).

NEOM will host three new smart city projects: Oxagon, Trojena, and The Line (see Fig. 5; NEOM, 2022b). First, Oxagon is meant to become the region's financial and economic hub. Strategically located in the coastal part of NEOM, it will be the largest cruise terminal in the Red Sea, with 13% of global container traffic passing through the nearby Suez Canal. Up to 70,000 job creations and 90,000 inhabitants are expected by 2030. The city seeks to power "a fully integrated next-gen automated port and supply chain" with 100% clean energy, along with homes, research, industries, and business environments.

Second, Trojena seeks to adapt to the region's climate, located at a height of 2400 meters, and takes advantage of its large temperature range, from 0 to 30 degrees Celsius, by developing a premium luxury ski village, also including a freshwater lake, a thriving wildlife reserve, and residential areas. Driven by the principles of sustainability and cutting-edge technology and intended to become a year-round mountainous destination in NEOM and across the world, the city has very recently announced its bid to host the 2029 Asian Winter Games (NEOM Directory and News, 2022).

Fig. 5 NEOM's three smart city projects (from left to right: Oxagon, Trojena, The Line). (Source: NEOM, 2022b)

Third, The Line is envisioned to grasp multilevel benefits through its smartly planned infrastructure and even going beyond. It is by far the most publicized project of Saudi Arabia's new three cities in NEOM. As a new city, it is possible to design it on a line that is planned to optimize transportation efficiencies, minimizing the time needed for personal mobility and business logistics but also to minimize the footprint. Indeed, the recently announced designs of The Line provide its main characteristics: it will be 200 meters wide, 170 kilometers long, and 500 meters high, eventually accommodating 9 million residents, for a total and reduced footprint of 34 square kilometers. Led by the vertical city architectural concept, The Line redefines living, enabling inhabitants to attain a good work/life balance by ensuring close proximity to work, leisure, education, and health services. The city will be carbon neutral and 100% powered by renewable energy.

4.2 How Is a Smart City Assessed?

There are standard criteria that collect consensus on assessing a smart city. Doheim et al. (2019) use six criteria to assess how "smart" Saudi Arabia's cities are. First, a smart city is grounded in smart governance, that is, a government focused on citizens' well-being through democracy, transparency, and technology usage to bolster citizens' engagement (Kumar et al., 2016). This resonates with the "Ambitious Nation" theme of Saudi Arabia's vision, which relies on building an effective and transparent government (Doheim et al., 2019). The second pillar is building a smart economy that leverages the comparative advantage of workers and develops partnerships with other countries to enhance trade and international relations while also empowering innovation locally (Giffinger et al., 2007). Saudi Arabia plans to diversify and broaden its exports and income possibilities from its dependence on oil and gas, privatize its services, and invest in talent, which suggests that it is on the trajectory of adopting a smart economy (Doheim et al., 2019).

Third, smart cities rely on smart mobility by strengthening interconnections in the city through transport networks and better transmission of data and information (Giffinger et al., 2007). This dimension is not separable from the other dimensions of a smart city. Saudi Arabia's efforts are in line with the definition of this pillar, as it has eased transport and ensured its safety (Doheim et al., 2019). The fourth criterion of smart cities is establishing a smart environment that achieves a long-term goal of top quality of life for citizens by developing sustainable, environmentally friendly urban planning strategies (Giffinger et al., 2007). By focusing on preserving the environment, combating pollution, managing waste disposal, and other practices, Saudi Arabia is dedicated to developing a smart environment for the welfare maximization of its citizens (Saudi Vision, 2022; Belaïd and Massie, 2023). Fifth, what eventually makes the city smart is the people. Smart people are a notion that refers to the smart engagement of people in providing innovation and urban solutions to critical issues in the city. This indeed requires investment in the people, and the best form of investment in people is education (Giffinger et al., 2007). Saudi

Arabia's vision aligns with education and labor training by investing in people's talents by offering education through its top universities and spreading knowledge of its heritage and Islamic values among its people (Doheim et al., 2019). Relatedly, the last buttress of smart cities is smart living, which makes efficient use of all other pillars to satisfy people's needs and utilize their best potential to enhance the welfare of society. More practically, smart living involves smart management of public spaces and facilities and developing information infrastructure (Giffinger et al., 2007). "A Vibrant Society" is one of Saudi Arabia's vision themes that align with developing telecommunication and fostering an attractive environment to live in, including safe and good quality living spaces (Saudi Vision, 2022).

One could infer that the "smart cities" concept has already been incorporated into Saudi Arabia's vision and mission for one or two decades, without it being labeled with this exact terminology. While Saudi Arabia has yet to achieve all the pillars of a smart city, the assessment that was based on common criteria to evaluate smart cities hints toward the potential that Saudi Arabia's cities have to become smart.

Designing a framework for establishing a smart city relies on the city's history, challenges, and dynamic features. While some challenges or exogenous factors might affect many or all cities, such as the COVID-19 pandemic, its impacts are still diverse in different cities. In parallel, returning to our framework of verticals of smart cities, while we can generalize a few verticals and their corresponding pillars, the specificity and applicability of these verticals eventually depend on the complexity of the city itself. Even if located in the same region, not just in each country, each city still has a unique set of challenges and features that necessitate specific policy making. We can indeed learn lessons from one smart city to apply them to another. However, the dynamic evolution of the set of challenges faced by a city demands a dynamic evolution of how smart cities are being developed.

5   Conclusions and Policy Recommendations

In response to economic and environmental challenges, cities around the world are striving to develop multiple smart systems throughout their territory to foster efficiency, sustainability, technological transition, and improved quality of life. Smart city initiatives are recognized as a universal commitment to innovate, inspire and push cities to generate positive economic and social transitions. Ideologies such as smart territory, smart region, or nation are increasingly used by governments.

First, this chapter reviews the concept of smart cities. Then, it focuses on outlining the progress and current practices that emerging economies, in general, have achieved in their transition toward smart cities. Finally, it highlights the key challenges and opportunities that have been identified to date.

The analysis shows that even though the smart cities concept is widespread, there is still confusion about why we need smart cities and how we can transform existing cities and build new sustainable cities. The absence of a clear definition and the many unarticulated assumptions also obscure the smart cities concept. The analysis in Sect. 2 clearly shows an evolution of the purpose from the early 2000s. Initially, ICT infrastructure was given greater importance. Although environmental and economic sustainability were identified as major components early on, the concept of individual-centered and social capital development was recognized only recently.

Over the past three decades, the evolution of the smart city has followed a particular roadmap: it began with exigencies for Internet connectivity in the early 1990s, leveraging ICT for urban development in the early 2000s, and shifting to advancing innovation for urban sustainability in the early 2010s. Currently, the smart cities concept is more likely to be seen as an "advancement" turning cities into resilient, user-friendly, sustainable cities (Anthopoulos, 2017). Cities worldwide have evolved along with the smart cities concept, falling into several specific groups: (1) cities that focus on innovation and individuals to codesign their future; (2) cities providing specific innovative services to communities (e.g., lighting, parking, etc.); and (3) cities that are retrofitting or building entirely new neighborhoods (e.g., Sidewalk Toronto, Halifax, Songdo, etc.) or even entire cities that provide wide-scale smart services (e.g., NEOM, Toyota Woven city, Masdar, etc.

Smart city initiatives and projects in the MENA region and Saudi Arabia need to go beyond the TIC and urban planning dimensions and include other aspects such as well-being, sustainable economic development, and sound governance. Economic needs and the sustainability of society should be the cornerstone of the urban transformation agenda. Effective measures must be undertaken to reconcile social needs and urban development at different organizational levels.

To increase the socioeconomic outcomes and decrease the environmental impact of cities in emerging economies, we recommend the following strategies:

First, *existing data infrastructures should be built and strengthened.* The lack of data-driven planning and decision-making is one of the most difficult challenges facing many cities in emerging economies. Greater efforts are needed to strengthen data infrastructures and harness digital intelligence within existing urban services and systems to address this issue. The generated data will play a leading role in accelerating evidence-based city management and planning.

Second, *develop a framework to assess and monitor progress toward sustainable and resilient cities.* This can be achieved by framing key performance indicators (KPIs) that cover all aspects of smart cities. The performance indicators must incorporate both efficiency (process) and outcome metrics (sustainability, economic performance, life quality, inclusion, policy efficiency, etc.). KPIs need to be designed and approved at the earliest stage of any smart city project, as do their data collection, monitoring, and processing systems.

Third, *an appropriate governance framework* should be designed to help city managers select the most effective rules to help facilitate the economic, social, and operational sustainability of their integrated data exchange systems and, in tandem, their smart city initiatives and efforts.

Finally, *develop a collaborative planning and action model* involving all actors in every stage of the development, including the design, execution, reporting, and evaluation. Continuous commitment will guarantee respect for economic and social sustainability as a process to improve human outcomes.

References

Advanced Electronics Company. (2018). *Smart cities - The next stage of urbanization in KSA*. https://www.aecl.com/media/2371/urbane.pdf

Albino, V., Berardi, U., & Dangelico, R. M. (2015). Smart cities: Definitions, dimensions, performance, and initiatives. *Journal of Urban Technology, 22*(1), 3–21.

Aldusari, A. N. (2015). Smart city as urban innovation: A case of Riyadh North–West District. *Journal of Sustainable Development, 8*(8), 270.

Anthopoulos, L. G. (2017). The rise of the smart city. In *Understanding smart cities: A tool for smart government or an industrial trick?* (pp. 5–45). Springer International Publishing.

Belaïd, F., & Al-Sarihi, A. (2023). Saudi Arabia's energy transition in a post-paris agreement era: An analysis with a multi-level perspective approach. *Research in International Business and Finance*, p. 102086.

Belaïd, F., & Massié, C. (2023). The viability of energy efficiency in facilitating Saudi Arabia's journey toward net-zero emissions. *Energy Economics*, 106765.

Camargo, F., Montenegro-Marín, C. E., & González-Crespo, R. (2021). Toward a new model of smart cities in emerging countries. *Academy of Strategic Management Journal, 20*(6S), 1–20.

Caragliu, A., Del Bo, C., & Nijkamp, P. (2009). Smart cities in Europe. *Journal of Urban Technology, 18*(2), 65–82.

Doheim, R. M., Farag, A. A., & Badawi, S. (2019). Smart city vision and practices across the Kingdom of Saudi Arabia—A review. In *Smart cities: Issues and challenges, mapping political, social and economic risks and threats* (pp. 309–332). Elsevier.

European Commission. (2022). *Smart cities*. https://ec.europa.eu/info/eu-regional-and-urban-development/topics/cities-and-urban-development/city-initiatives/smart-cities_en

Farag, A. A. (2019). The story of NEOM City: Opportunities and challenges. In *New cities and community extensions in Egypt and the Middle East* (pp. 35–49). Springer.

Giffinger, R., Fertner, C., Kramar, H., Kalasek, R., Pichler-Milanović, N., & Meijers, E. (2007). *Smart cities: Ranking of European medium-sized cities*. http://www.smart-cities.eu/download/smart_cities_final_report.pdf

Han, H., & Hawken, S. (2018). Introduction: Innovation and identity in next-generation smart cities. *City, Culture and Society, 12*, 1–4.

International Standards Organization. (2014). *Smart cities preliminary report 2014*. http://www.iso.org/iso/smart_cities_report-jtc1.pdf

Kumar, H., Singh, M. K., & Gupta, M. P. (2016). Smart governance for smart cities: A conceptual framework from social media practices. In *Conference on e-Business, e-Services and e-Society* (pp. 628–634).

NEOM. (2022a). *Project objectives*. https://na.vision2030.gov.sa/v2030/v2030-projects/neom/

NEOM. (2022b). *Regions. Oxagon, Trojena, The Line*. https://www.neom.com/en-us/regions/oxagon; https://www.neom.com/en-us/regions/trojena; https://www.neom.com/en-us/regions/theline

NEOM Directory and News. (2022). *Saudi Arabia bid to host Asian Winter Games in TROJENA, NEOM*. Published: 04 August 2022. https://neom.directory/neom-news/saudi-arabia-bid-to-host-asian-winter-games-in-trojena-neom

OECD. (2019). *Enhancing the contribution of digitalisation to the smart cities of the future*. https://www.oecd.org/cfe/regionaldevelopment/Smart-Cities-FINAL.pdf

Orejon-Sanchez, R. D., Crespo-Garcia, D., Andres-Diaz, J. R., & Gago-Calderon, A. (2022). Smart cities' development in Spain: A comparison of technical and social indicators with reference to European cities. *Sustainable Cities and Society, 81*, 103828.

Pira, M. (2021). A novel taxonomy of smart, sustainable city indicators. *Humanities and Social Sciences Communications, 8*(1), 1–10.

Ringel, M. (2021). Smart city design differences: Insights from decision-makers in Germany and the Middle East/North-Africa region. *Sustainability, 13*(4), 2143.

Saudi Vision. (2022). *Vision 2030*. https://www.vision2030.gov.sa/

Shokeir, R. G., & Yahia, I. B. (2020). Moving toward smart cities: Insights from the MENA region. *International Journal of Web Based Communities, 16*(1), 92–108.

Sieghart, L. C., & Betre, M. (2018). *Climate change in MENA: Challenges and opportunities for the world's most water stressed region*. MENA knowledge and learning quick notes 164, World Bank.

Trencher, G. (2019). Toward the smart city 2.0: Empirical evidence of using smartness as a tool for tackling social challenges. *Technological Forecasting and Social Change, 142*, 117–128.

United Arab Emirates. (2022). *Green economy for sustainable development*. https://u.ae/en/about-the-uae/economy/green-economy-for-sustainable-development

United Nations. (2019). *World urbanization prospects: The 2018 Revision*. https://population.un.org/wup/Publications/Files/WUP2018-Report.pdf

United Nations Environment Programme. (2017). *State of play of sustainable cities and buildings in the Arab Region 2017*. https://dubailand.gov.ae/media/pjfjujk2/en_sscbar_un_2017.pdf

World Bank. (2022). *Population, total - Middle East & North Africa*. https://data.worldbank.org/indicator/SP.POP.TOTL?locations=ZQ

Yigitcanlar, T., Kamruzzaman, M., Foth, M., Sabatini-Marques, J., da Costa, E., & Ioppolo, G. (2019). Can cities become smart without being sustainable? A systematic review of the literature. *Sustainable Cities and Society, 45*, 348–365.

Open Access   This chapter is licensed under the terms of the Creative Commons Attribution 4.0 International License (http://creativecommons.org/licenses/by/4.0/), which permits use, sharing, adaptation, distribution and reproduction in any medium or format, as long as you give appropriate credit to the original author(s) and the source, provide a link to the Creative Commons license and indicate if changes were made.

The images or other third party material in this chapter are included in the chapter's Creative Commons license, unless indicated otherwise in a credit line to the material. If material is not included in the chapter's Creative Commons license and your intended use is not permitted by statutory regulation or exceeds the permitted use, you will need to obtain permission directly from the copyright holder.

Smart Transportation Systems in Smart Cities: Practices, Challenges, and Opportunities for Saudi Cities

A. H. M. Mehbub Anwar and Abu Toasin Oakil

Abstract Smart transportation is an approach that incorporates modern technologies into transportation systems to improve the efficiency of urban mobility. Cities worldwide call digital technologies to harness their development to address potential challenges and concerns, which provoke technology-driven practices in urban context. Big data and technologies now offer tools, techniques, and information that can improve how cities function. Consequently, urban process and practices are becoming highly responsive to a form of technology-driven urbanism, that is the key mode of production for smart urban development. This furnishes the prospect of building models of smart sustainable cities performing in real time from routinely available data. This in turn allows to monitor, understand, analyze, and plan such cities to improve their urban efficiency and promotes new urban intelligence functions as an advanced form of decision support. Although technology-driven approach to transport analysis and management is emerging as smart city principle, the application is limited in the Kingdom of Saudi Arabia (KSA). This chapter investigates the potentials and the role of technology-driven solutions in improving and advancing urban transport management in the context of smart cities. It also explores the relevant practices as well as potentials in smart urban development context for Saudi cities. Our approach of technology-driven urban management will envision cities as a complex social and technological ecosystem and build on lessons learned from the research at city level and conceptualizes actors and institutions in a technology-driven urban management for Saudi cities toward achieving liveable smart city.

Keywords Smart cities · Smart transport · GCC · Riyadh city · Urban mobility

A. H. M. M. Anwar (✉) · A. T. Oakil
Transport and Infrastructure Department, King Abdullah Petroleum Studies and Research Center (KAPSARC), Riyadh, Saudi Arabia
e-mail: mehbub.anwar@kapsarc.org

© The Author(s) 2024   315
F. Belaïd, A. Arora (eds.), *Smart Cities*, Studies in Energy, Resource and Environmental Economics, https://doi.org/10.1007/978-3-031-35664-3_17

1 Introduction

The United Nations (UN) predicts that 68% of the global population will live in cities by 2050 (UN, 2018), adding approximately 2.9 billion vehicles to the road (Djahel et al., 2018). This increasing demand will challenge the long-term sustainable goals of the present transportation system. In addition, people's choice regarding travel has been changing. Policies regarding sustainable transport and changing people's behavior make it harder to formulate solutions that protect individual well-being and societal interests. Traditional ways of depending only on automobiles or on public transport have come to an end (Sharmeen et al., 2020). The new generation of travel behavior suggests a more flexible way of traveling. Such a flexible lifestyle, together with a shared economy and the development of information technology, has encouraged the reconfiguration of transport systems (Sharmeen & Meurs, 2019). The system should serve the purpose without limiting travelers' intentions by means of transport. This requires having all possibilities open. For example, people may choose small cars when traveling to a place where parking can be difficult. However, people may prefer bigger cars when traveling to shopping centers or furniture shops. In terms of car ownership, it is difficult and costly to own cars for every purpose. Therefore, car-sharing options have become promising. People can choose whatever car they need at the time of travel. People may also choose to travel by public transport for many reasons, such as congestion charges and parking fees. Additionally, they may want to use an active transport mode such as walking or cycling. People may also find it beneficial to use multiple modes for one purpose. For example, people may take a car to the residential train station to catch a train and then take a cycle at the destination in a busy center to avoid congestion. Therefore, mobility should be considered as a single service, which may take different forms. Whereas this flexibility creates great satisfaction with travel at the individual level and benefits society by creating options for sustainability, such flexibility brings a high level of complexity in the transport system incorporating multiple stakeholders, clients, and actors in play, and therefore, an integrated transport system becomes difficult (Meurs et al., 2020). The more complex it gets, more sophisticated the system we need to manage our transport system. An intelligent/smart transport system (I/STS) is therefore coined. Although it is not necessarily a very new concept, the momentum that it achieved is recent (Debnath et al., 2014).

In general, the goal of a smart transport system is to solve direct and indirect transport-related problems in the most efficient way to maximize the well-being of individuals and society. Whereas the meaning of smart transport can be broad, ITSs add specific interest to the technology dependence of the system for the same goal. It creates a system that is more adaptive to changing circumstances. The information and communication technologies that connect diverse parts of the transport system seamlessly allow managing and solving issues in transport in a (near) real-time manner. It refers to an automatic, demand-responsive, and real-time system. According to the US Department of Transportation,

> Intelligent Transportation Systems (ITS) apply a variety of technologies to monitor, evaluate, and manage transportation systems to enhance efficiency and safety.

An ITS maximizes the use of existing infrastructure through a range of technological means, such as traffic signals, travel planners, smart ticketing, and cooperative systems (Bıyık et al., 2021; Bazzan & Klügl, 2013). ITSs often focus on smart technologies established in transport infrastructures and vehicles but are not limited to them. In general, smart transport refers to any technology used for efficient transport management. As per Shaheen and Finson (2013),

> Intelligent transportation system (ITS) is composed of a wide range of technologies—including electronics, information processing, wireless communications, and controls—aimed at improving safety, efficiency, and convenience of the overall surface transportation network.

In an introductory guide for ITSs, the Japan Society of Civil Engineers (JSCE, 2016) referred to ITSs as a new transportation system to resolve a variety of road traffic issues (such as traffic accidents and congestion) by linking people, roads, and vehicles in an information and communications network by cutting-edge technologies (Fig. 1).

With the increase in the Internet of Things (IoT), many devices are now connected through the internet, making information exchange among these devices fast and flexible. Whereas ITS started to develop infrastructure and vehicles with connected devices such as sensors and cameras, IoT has made it even easier through utilization of mobile devices such as mobile phones and navigation systems. Information coming from mobile devices is a convenient and inexpensive alternative to physically installed sensors within transport infrastructure (Alomari et al., 2021). Data and information are key to the efficient functioning of the transport system. Whereas data can be collected from different sources, transport sometimes requires real-time data to manage it, for instance, congestion control and traffic rerouting, public transport re/scheduling, accident response, etc. ITS harnesses the opportunities from the application of big data generated from different smart devices, either established in the infrastructure or carried by people such as mobile phones, navigation systems, or tablets. Not only data availability but also on-time availability is important. In addition, the ubiquitous nature of data provides more opportunities than ever before. Smart devices such as mobile phones have reached

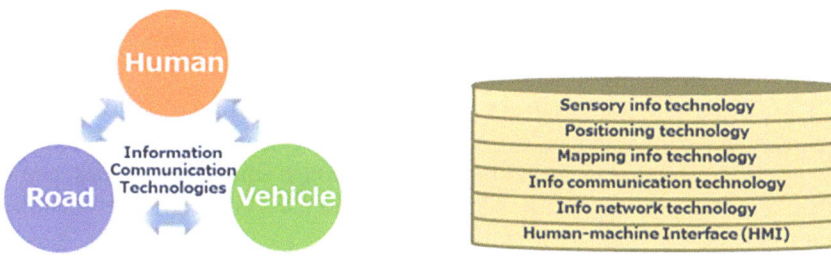

Fig. 1 Components of ITS. (Source: JSCE (2016))

many places in the world, reducing spatial disparity in functioning and management. Developed countries as well as developing countries are using smart devices to solve local problems and create efficiency. A key to problem-solving is to understand the problem, and information is a prerequisite in this regard. Although data ubiquity is the essence of modern-day smart devices, the application of data is not universal.

The remainder of this chapter is organized as follows: Sect. 2 provides a review of smart transport system (STS) applications around the world with a few examples. Section 3 focuses on STS practices in Saudi cities. Section 4 discusses smart cities from the perspective of Riyadh city. The final remarks, including the opportunities and challenges of STS, are put at the end of the paper.

2 Smart Transport System (STS): Applications

The goal of STS is many-fold—reducing carbon emissions and traffic congestion and increasing reliability, efficiency, and safety. To achieve these goals, transport systems must implement sustainable strategies in an efficient way, for which information and technology play a vital role. According to Nikitas et al. (2020), society needs to shift to a more sustainable techno-social paradigm to avoid the adverse repercussions of a resource-intensive and unthoughtfully opportunistic liveability philosophy. At the core of intelligent transport systems remains the efficiency, management, safety, and resilience of the transport system. The benefits of such efficient management could be real-time information, more efficient administration, the possibility for citizens to access the information system online via smartphones, and on-request traffic light control (Tomaszewska & Florea, 2018). According to Giannopoulos (2004), ITS achieves efficiency by improving the transport system in three main areas: (1) operation and management of transport system and networks, (2) information and guidance to the users of the system, and (3) operation and management of freight transport systems. It can be simplified into the following concepts for what makes up smart transportation: management, efficiency, and safety. In other words, smart transportation uses new and emerging technologies to make moving around a city more convenient, more cost-effective (for both the city and the individual), and safer. Thus, STS focuses on three dimensions (Yan et al., 2020): (i) smart cells, (ii) ICT (Information and Communication Technology), and (iii) development mechanisms. Smart cells should include smart cars, unmanned aerial vehicles (UAVs), smart infrastructure and devices, and smart base stations supported by ICT and enabled by various developmental mechanisms (Yan et al., 2020). In general, we can see the system as two interconnected segments—one is responsible for the operation and management of the system through smart technologies, and the other is the user of the system. STS creates a seamless connection between these two segments. STS does these in many ways.

2.1   Real-Time Traffic Updates and Management

There are two important components of STS: Vehicle to Vehicle (V2 V) and Vehicle to Infrastructure (V2I) solutions. V2 V solutions mainly increase travel safety by enabling advanced emergency braking systems. Similarly, V2I relies on different sensors installed at the road network to detect different traffic variables, such as speed, density, and waiting time (Mangiaracina et al., 2017). For instance, through V2 V applications, drivers will be alerted to imminent crashes, such as merging trucks, cars in the driver's blind spot, or when a vehicle ahead stops suddenly. This application includes forward collision warning, emergency electronic brake light, blind spot/lane change warning, not pass warning, intersection movement assist, and left or right turn assist. On the other hand, through V2I applications, for example, drivers will be alerted when they are entering a school zone, if workers are on the roadside, and if an upcoming traffic light is about to change. This application includes curve speed warning, red light violation warning, stop sign gap assist, smart roadside, and transit pedestrian warning.

Smith et al. (2013) analyzed the case of SURTRAC (Scalable Urban Traffic Control), a pilot implementation of an adaptive traffic signal control system installed for a nine-intersection road network in Pittsburgh, Pennsylvania (USA). The pilot test results demonstrated the effectiveness and potential of decentralized, adaptive traffic signal control in urban road networks. The SURTRAC system improved traffic flow efficiency by 25–40% and reduced emissions by over 20%.

2.2   Demand-Responsive Public Transport/ On-Demand Transport

There are over 40 cities around the world that are trialing on-demand public transport pilots and operational services. There is no universally accepted definition of on-demand public transport (sometimes referred to as demand-responsive transport or DRT). On-demand public transport (ODT) is a form of publicly subsidized transport that takes multiple passengers within a defined area from one place to another on a next-available or prebook basis. The ODT does not operate on fixed routes and fixed schedules. For the case in Australia, this ODT service becomes very popular initially in New South Wales (NSW) and later spreads over other parts of the country. This service picks passengers up from their home and takes them to their desired destination. For booking, the specified mobile apps must be installed; otherwise, computers can be used before the journey. Its flexible and nonregular route service encourages the use of public transport by providing mobility options for all in areas where daily demand is variable. On-demand services can connect passengers with other public transport hubs or direct to their destinations to enhance the mobility of the entire community.

2.3 Smart Ticketing

Smart ticketing is a ticketing system where a travel ticket is electronically stored on a smart card or a smartphone, eliminating the need for traditional paper tickets and enabling users to simply tap their smart card or device on a gate or validator to access travel. This system allows the user to skip the ticket counter at the station. Users can buy tickets for a specific period in advance or load credit onto their account. For example, in London, back in the 1990s, the Transport for London (TFL) began introducing the Oyster travel card, a contactless card to speed up access for people traveling on the London Underground. Over 60 million Oyster cards were recorded until 2013, with 85% of all rail and bus travel paid for using this system.[1] Contemporarily, various types of smart tickets, such as mobile tickets and smart cards, are seen, and they use significantly different technologies. Some widely used contactless smart cards include Melbourne's myki, Sydney's Opal Card, South Korea's T-money, Hong Kong's Octopus card, Stockholm's Access card, Japan's Suica and Pasmo cards, Singapore's NETS FlashPay and EZ-Link cards, Manila's Beep cards, Paris's Calypso/Navigo, the Dutch OV-Chipkaart, Greater Toronto's (as well as Hamilton and Ottawa) Presto card and Lisbon's LisboaViva card, including many in the USA, such as Go-To card in Minneapolis–Saint Paul, SEPTA Key in Philadelphia, CharmCard in Maryland, etc.

2.4 Ridesharing

Ridesharing refers to a mode of transportation in which individual travelers share a vehicle for a trip and travel costs such as gas, toll, and parking fees with others that have similar itineraries and time schedules. Thus, ridesharing appears to be an advantageous approach to society and the environment that includes saving travel costs, reducing travel time, mitigating traffic congestion, conserving fuel, and reducing air pollution (Ferguson, 1997; Kelly, 2007; Morency, 2007; Chan & Shaheen, 2012). Mitropoulos et al. identified approximately 29 rideshare platforms in their paper and noted that most of the ridesharing platforms were found to operate in the EU (48%), with 27% of them operating in Italy (Mitropoulos et al., 2021). In addition, ridesharing platforms operate in the USA, Asia, and Latin America remarkably. The majority of ridesharing platforms (93%) started their operation in 2005 or after, while 62% were found to start operations in or after 2010, which might be explained by the rapid development of mobile applications and the spread of smartphones. Smartphone annual sales doubled between 2007 and 2010 (i.e., 122.32 vs. 296.65 million units) and increased by a factor of 4.2 between 2010 and 2014 (i.e., 296.65 vs. 969.72 million units), reaching 1540.66 million sold units in

---

[1] Pablo Vinuesa. 2021. Smart ticketing: the revolution for the tickets of the future or just an anti-covid measure? URL: https://tomorrow.city/a/smart-ticketing accessed on May 26, 2022.

2019 (Mitropoulos et al., 2021), meaning that the increasing rate of smartphone users has made ridesharing convenient for travelers.

2.5   Smart Parking System

Sustainable strategies target many aspects of transport systems to encourage less carbon-intensive travel as well as to increase efficiency, comfort, and safety. For instance, transport authorities consider car parking charges in busy areas to avoid congestion, encourage public transport, discourage car use, and save space for productive or recreational uses (Anwar & Oakil, 2021; Calthrop et al., 2000; Buehler et al., 2017). However, the outcome might not be positive if we cannot implement such car parking policies efficiently. Reducing the availability of car parking in a busy urban center may discourage car use, but it may also create difficult situations for people who need to use cars. Finding a parking space is time-consuming and requires unnecessary driving around the city burning fuels (Awaisi et al., 2019; Durga Devi et al., 2017). This may also aggravate traffic congestion in the inner city. For instance, cars searching for parking spaces generate up to 40% of the total traffic on the street (Kazi et al., 2018).

A smart parking system can acquire information regarding the availability of parking spaces and help reserve a space needed. It can also reduce the stress of maintaining the time on the parking meter because the billing system can be adjusted for use. It utilizes smart sensor technologies that can be installed on the street to collect real-time data and update a web-based portal to monitor the parking spaces and feedback the drivers using a mobile app that can enable the user to reserve the space and make online payments as well as navigate to the nearest vacant space (Durga Devi et al., 2017; Mainetti et al., 2015). Furthermore, smart parking systems can reduce parking pressure in busy areas by sharing parking spaces (Atif et al., 2016).

Many parking apps have been developed, such as Parkmobile, BestParking, Parker, SpotHero, ParkMe, Parking Mate, Parking Panda, ParkWhiz, and VoicePark. These apps help you to

- Locate available parking spots in garages and lots as well as local street parking
- Quickly compare the price, distance, hours availability
- Navigate to the nearest open/vacant spot
- Run a timer, so there is no need to worry about parking tickets
- Reserve a spot
- App may own dedicated unsold spots from parking facilities in an area.
- Provide a discounted rate.
- Record the history of your parking (location, time, price)
- Notify about parking expiration based on parking regulations such as limited hour zone

- Collaborate with attraction points such as stadiums, family attractions, and other events to ensure attraction.

2.6 Public Transport Management

Whereas smart parking management leads to better outcomes, the acceptance of parking policy can vary and may lead to inequitable implications for society (Hamer et al., 2012). We need to create opportunities for travelers. If cars have been discouraged from encouraging public transport use, then public transport should be attractive, acceptable, and satisfactory to users. Often, public transport frequencies affect user experience. Waiting for a train or bus, not finding a better route, and the absence of appropriate transfer/connection information may discourage public transport use. A smart public transport system intends to make not only the system efficient and flexible but also easily accessible for users. For instance, smart ticketing and real-time journey planning are two important features of smart public transport. Smart ticketing allows travelers to use one card for any combination of travel, including hiring a taxi or bicycle or renting a car. In addition, multiple service plans are provided, such as pay-as-you-go, monthly billing afterward, seasonal tickets, or just using your own bank card. Similarly, real-time journey planners are no longer for one travel mode; they combine bus, tram, train, metro, and other available public transport. Weng et al. (2016) estimated the real-time bus travel speed and the location of the bus. Combining the real-time bus positioning system data with the bus geographical information system map, the study accurately matched the bus location onto the map so that it could determine the bus position of the given route and estimate the bus arrival time, which might increase the wait time performance for the passengers.

2.7 Mobility as a Service (MaaS)

The increasing number of transport services offered in cities and advancements in technology and ICT have introduced an innovative concept called mobility as a service (MaaS), which makes it easier for citizens to access and utilize several complementary mobility services. MaaS combines different available transport modes to offer a customized mobility package and includes other attractive complementary services in comparison to privately owned vehicles, such as trip planning, reservation, and payments, through a single interface (Hietanen, 2014). Such mobility bundles promote a shift from private transport to smart access-based transport. MaaS-like services have been trailed and are planned to be implemented in upcoming years, including in Europe, North America, Asia, and Oceania. Smith and Hensher (2020) pointed out numerous benefits that we can obtain from this service, such as a smaller carbon footprint from personal mobility and reduced congestion, which in turn could lead to higher productivity, improved air quality, fewer traffic accidents, and reduced social exclusion.

3   Smart Transport System Practices in the Kingdom of Saudi Arabia (KSA)

3.1   Past to Present

The Ministry of Interior signed a contract with Doxiadis Associates (Consultants on Development and Ekistics of Athens, Greece) for the formulation of a master plan to guide the development of Riyadh city up to the year 2000, and Doxiadis submitted a Master plan in 1971. In terms of mobility, the master plan was designed in a way that promoted automobiles as the only mode of transport (Al-Hathloul, 2017). Later, this plan became obsolete due to predicting the size of urban growth inadequately; therefore, it revealed the necessity of further adjustments to the master plan to accommodate the subsequent city's rapid growth. However, the ADA (Arriyadh Development Authority) established a Comprehensive Riyadh Strategic Plan called the MEDSTAR (Metropolitan Development Strategy Arriyadh) project in 2003 to plan, manage and control city growth by setting up urban growth boundaries. Under the MEDSTAR project, the Riyadh Public Transport Network (RPTN) has been proposed to achieve reliable, efficient, and affordable public transport for residents. The network is composed of metro lines with Bus Rapid Transit (BRT), community bus lines, feeder buses, and park and ride facilities. The project is underway and expected to be in operation soon. As a result of implementing the RPTN, the Royal Commission for Riyadh City (RCRC) is currently undertaking a preliminary study of transit-oriented development (TOD) within Riyadh city, which symbolizes the TOD strategy for future Riyadh, intensifies mixed-use activities and increases the density around metro stations. Cervero and Murakami (2008) argued that TOD embodies physical features of high density and mixed land uses accommodating enough population living within the acceptable distance that encourages walking, cycling, and public transport that can also contribute to improving social cohesion.

3.2   Traffic Condition and Accessibility Analysis

Accessibility is the most important measure in transport systems. Evaluating and assessing how a transport system might and will perform depends on how the system makes different activities or destinations accessible. However, the measurement of accessibility is difficult. A high level of accessibility does not come with only a high level of mobility (El-Geneidy & Levinson, 2006). We can achieve high mobility without achieving a high level of accessibility. Mobility comes with high speed, and accessibility comes with a combination of time, speed, and destination. Therefore, density, population, network and connectivity, travel cost, time, and speed, among others, are combined in accessibility measures. A rigorous collection of data is required to accurately assess accessibility.

KSA's planning principle is often toward mobility. Accessibility is often ignored in the planning process. Therefore, Saudi cities represent urban sprawl. Whereas you can achieve high speed on the road using automobiles with wide roads, you cannot reach a destination in a short time. Moreover, sprawl, meaning a low density of people leading to the overall accessibility of a city, is reduced. Since the city has been built without considering accessibility, a detailed description of built environment characteristics, walk time to destinations, and concentration of services are often not known. A physical as well as social survey is necessary to collect such data. This is often time-consuming and costly and cannot cope with the rapidly changing environment of Riyadh. Ubiquitous data generated from different smart devices can sometimes be used to proxy such large-scale data collection.

Al-Hosain and Alhussaini (2021) evaluated the potential accessibility to Riyadh's planned public transport system using geospatial analysis. The study showed that approximately 34% of Riyadh's population could be served if 5-minute driving to the metro station is assumed, while this percentage would increase to 74% if 10-minute driving is considered. Smaller percentages of the population were found within an acceptable walking distance to the stations. Specifically, 5% and 14% of the population could walk to a metro station within 5 and 10 minutes, respectively.

3.3 Freight Management and Analysis

The KAPSARC (King Abdullah Petroleum and Studies and Research Center) estimated freight transport activities in China, India, and Saudi Arabia using Nighttime Lights Satellite Data sourced by the National Ocean and Atmospheric Administration. This paper focuses on using night satellite radiometry to estimate total freight transportation activity in any city, region, or country based on the intensity of nighttime lights and their relationship with economic and human activity on the ground (Lopez-Ruiz et al., 2019).

3.4 Walk Behavior Analysis

Walking is a mode of contributing positive impact to health and social benefits. Almahmood et al. (2017) studied this component in the Riyadh context using a method that is a combination of movement tracking data using GPS technology and map-based workshops where participants can reflect on their walking behavior and spatial preferences. The results map the city by exploring gender-specific walk-scapes. According to Riyadh respondents, women enjoy shopping malls as a function of urban shelters, indicating spaces for walking, while young men mainly walk in urban streets. Thus, streets are perceived as men's walkscapes, where women's presence is limited. The key message of this study is that walkscapes are determined not only by the quality of the space nut but also by sociocultural settings.

3.5   Traffic Safety and Event Analysis

Roadside events are sometimes crucial to detect either a mechanical breakdown or an accident. A mechanical breakdown can cause traffic congestion and delays, leading to economic and time losses if not solved in a timely manner. Accidents can be life-threatening, and immediate detection of the event can save lives directly. Notification can reduce traffic congestion and delays due to any accident since on-time measures can be taken to divert traffic to another situation. However, diversion is also needed to provide support to the accident itself and may also reduce chances of other accidents. Deaths in road transport account for 1.25 million deaths globally (Aqib et al., 2019). The car accident rate in the KSA is very high, making the KSA one of the world's highest rates of death and casualties (Al-Mosaind, 2018). Information boards are not well designed. Updates are not coming on the road. People rely on social media, which has increasingly become a convenient and inexpensive alternative for physical sensors, such as in the transport sector in smart societies (Alomari et al., 2021).

4   Riyadh from the Perspective of Smart Cities

Due to urbanization challenges, the concept of "smart cities" has become a popular tale in the modern urban planning era. The extant research indicates that the latest technological advancements are fully capable of offering solutions for those challenges (Kumar et al., 2020). These challenges are not different in Saudi Arabian cities, particularly in Riyadh. Saudi Vision 2030 (SV2030) is committed to revitalizing economic cities, establishing special zones, and deregulating the energy market to make it more competitive. The Ministry of Municipal and Rural Affairs (MOMRA) launched the "application of smart cities" initiative as one of the municipal transformation projects emanating from the National Transformation Program 2020 (NTP2020) and SV2030 to enhance urban development in the Kingdom through smart city initiatives. The Ministry conducted a field study in 2017 to determine the readiness and feasibility of Saudi cities to be transformed into smart cities. This study covered 17 major cities that make up nearly 72% of the Kingdom's population and indicated that there could be disparity in the Kingdom's cities' preparedness to shift to smart cities, with Makkah coming first, followed by Riyadh, Jeddah,

Madinah, and Asha.[2] Moreover, in the First Saudi Conference for Smart Cities at Riyadh, the Minister of MOMRA unveiled plans to establish 10 smart cities in the Kingdom, and Riyadh is one of them.

4.1 A Concept of Smart City

Substantial infrastructural development can direct positive outcomes for citizens as well as adverse effects and challenges on the smooth functioning of the city. Some of the challenges of current cities are as follows (not limited to):

1. Urban sprawl (Artmann et al., 2019; Kovács et al., 2019; Mahmoud & Divigalpitiya, 2019)
2. Urban pollution (Luo et al., 2019; Munksgaard et al., 2019; Munoz-Pandiella et al., 2018)
3. Inadequate citizen participation in planning and management process (Glaas et al., 2020; Slaev et al., 2019; Sou, 2019)
4. Technological infrastructure (Appio et al., 2019; Pham & Phan, 2018; Juwet & Ryckewaert, 2018)
5. Waste management (Zabara & Ahmad, 2020; Das et al., 2019; Zhang et al., 2019)
6. Poverty: inequality of wealth distribution (Bradshaw et al., 2019; Siwar et al., 2016; Muktiali, 2018)
7. Urban mobility (de Oliveira et al., 2019; Kijewska et al., 2019; Goetz & Alexander, 2019)
8. Aging population (Cheng et al., 2019; Heffner et al., 2019; Jayantha et al., 2018)

Due to identified challenges and complex issues, there is a growing need for innovative and dynamic solutions to address these issues in urban life. In the last couple of decades, the smart city concept, which enables better housing, transport, energy, and other infrastructure needs and is treated as a key strategy to combat poverty and inequality, unemployment, and energy management, has gained considerable attention in the urban development domain. The diverse scope, concept, and empirical approaches incorporated in the smart city debate have contributed to the lack of an absolute definition of smart cities. However, Lee et al. (2014) suggest that effective and sustainable smart cities emerge as a result of dynamic processes in which the players of the public and private sectors coordinate their activities and resources on an open innovation platform. Ahvenniemi et al. (2017) conceptualized smart cities as an "urban development model" geared to the utilization of human, collective, and technological capital within urban agglomerations. In a common sense, the concept of smart cities is embedded within the advances of ICT and their effective use and application in the context of cities and urban spaces (Visvizi & Lytras, 2019). Thus,

---

[2] Arab News, 2017. Smart City Initiative Launched. http://www.arabnews.com/node/1087402/saudi-arabia accessed on May 26, 2022.

the smart city is an aggregate concept of connectedness of the various aspects of a city that plays a role in its function. Doheim et al. (2019) emphasize the main concept of a smart city, which focuses on the interconnection of all aspects of the city, such as social, urban, economic, and environmental institutions. and reflects a holistic approach to handle urban problems by taking advantage of the new technologies of ICT. This means that a balance of technological, economic, environmental, and social factors involved in an urban system is a prerequisite to defining a smart city. Technology, people, and institutions are the key strategic principles of smart cities, such as the integration of infrastructures and technology-mediated services, social learning for strengthening human infrastructure, and governance for institutional improvement and citizen engagement (Nam & Pardo, 2011). Other research has identified two domains (Doheim et al., 2019) to define smart cities: *hard domains* such as buildings, energy grids, natural resources, water management, waste management, and mobility (Neirotti et al., 2014), where ICT can play a vital role in the functions of the systems, and *soft domains* such as education, culture, policy innovations, social inclusion, and government, where the applications of ICT are not usually decisive (Albino et al., 2015). Accordingly, it is difficult to obtain an absolute definition of smart cities. However, Giffinger et al. (2007) defined a smart city by assessing six basic dimensions: (i) smart governance, (ii) smart economy, (iii) smart mobility, (iv) smart environment, (v) smart people, and (vi) smart living (Fig. 2) as pillars of the smart city that make it the most practical classification of the smart city concept.

Smart governance can be expressed by the intelligent use of ICT to improve decision-making through better collaboration among different stakeholders, including government and citizens (Pereira et al., 2018). Smart governance also has an important role in smart city initiatives, which require multilayered interactions between governments, citizens, and stakeholders. *The smart economy* is understood as an economy based on innovation, entrepreneurship, high productivity, flexibility in the labor market, international and interregional embeddedness, and the ability to transform (Giffinger et al., 2007). *Smart mobility* is largely permeated by ICT, used

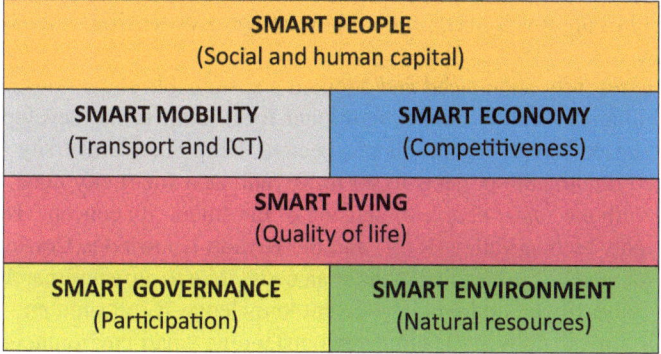

Fig. 2 Six main pillars (dimensions) of smart cities. (Adopted from Giffinger et al. (2007))

in both backward and forward applications, to support the optimization of traffic fluxes but also to collect citizens' opinions about liveability in cities or the quality of local public transport services (Benevolo et al., 2016). *The smart environment* is measured by maintaining the attractiveness of the natural environment, pollution levels, environmental protection activities, and resource management (Winkowska et al., 2019). *The smart people* concept focuses on the sensible participation of citizens in urban life and their ability to adjust to new solutions through adequate qualification, and motivation for lifelong learning is a critical requirement for a smart city (Doheim et al., 2019). *Smart living* is understood as an attempt to achieve a high quality of life by enhancing health conditions and individual safety; combining different values, cultures, and styles harmoniously; building opportunities for social cohesion using public space; and having sustainable and environmentally friendly buildings (Giffinger et al., 2007).

Based on the above discussion, Doheim et al. (2019) categorized the smart city concept into three classes:

1. *The digital city* (technology-based city) refers to a city where ICT connects and facilitates services interactively. The dimensions of *smart governance, smart economy,* and *smart mobility* may fall in this class.
2. The *intelligent city* (knowledge-based city) refers to a city where people are well educated, knowledgeable, and curious to learn lifelong. The smart city dimension, called *smart people,* may fall in this class.
3. The *environmental city* (community-based city) refers to a city where establishments promote the opportunity to build social cohesion/capital among citizens for a better quality of life and well-being. It is a community-driven city that ensures citizens' participation in decision-making processes at each level. The practices of *smart living* and *smart environment* dimensions may fall in this class.

4.2   Smart City Practices in Riyadh City

Riyadh city is the capital and financial hub of the Kingdom. It is economically vibrant and has approved many initiatives relating to smart cities, which are elucidated below.

KSA has recently announced and adopted a sustainable strategy of urban planning that inflames its shift from a traditional focal point to the development of a knowledge-based society. From this perspective, King Saud University (KSU) has initiated two remarkable projects to set up Riyadh as a smart city node, which are associated with the *smart people* dimension of the smart city concept. The projects are the Riyadh Techno Valley (RTV) and the Riyadh Knowledge Corridor (RKC). RTV is developed as the establishment of an ecosystem to attract research and business for a competitive and highly proficient knowledge-based economy (Aldusari, 2015). RTV aims to deliver a leading and outstanding smart environmental campus to enhance research and development for efficient operation, maintenance, and service delivery. The RKC is designed to promote an interlinked group of innovation outlets and pioneering knowledge that is planned to develop a knowledge-based society. In this context, Riyadh implements a range of education and economic entities to achieve such societies, proposed by KSU, known as RKC, which is one of the greatest initiatives to develop and enhance knowledge, intellectual, economic, research, and educational entities (Aldusari, 2015). Moreover, another objective of the RKC project is to improve and enhance the communication and interaction among all these entities in a smart manner. The RTV and RKC projects can also represent the *smart economy* dimension of the smart city concept, as both are designed to be the gateway of the knowledge-based economy. KSU has built a partnership with RTV to facilitate the Saudi economy to a knowledge-based economy.[3] The term "knowledge-based economy" can be understood as production and services based on knowledge-intensive activities that contribute to an accelerated pace of technological and scientific advancement. Moreover, it includes a greater reliance on intellectual capabilities than on physical inputs or natural resources, combined with efforts to integrate improvements in every stage of the production process, from the research and development lab to the industry to the interface with customers (Powell & Snellman, 2004). There are several projects relating to the *smart economy* of the smart city concept, such as the King Abdullah Financial District, the City of Communication, and Information Technology, the first and second industrial cities, and other centers that are all emerging, helping to provide growth and development opportunities in the city (Doheim et al., 2019). Riyadh is set to be the Arab world's first digital capital and leads the region in the promotion of ICT as a catalyst for increased economy.[4]

Regarding *smart governance* of smart city vision, it relies on the good governance of having open (i.e., transparent), accountable, collaborative (i.e., involving all stakeholders) and participatory (i.e., citizens' participation) principles, and on Electronic Government (e-Government) (Lopes, 2017). In Riyadh, establishments facilitate the ease of the administrative process and increase transparency. For example, Al Helal and Mokhtar (2018) proposed the "Riyadh Wiki Information and Complaining System" for citizen engagement in Riyadh. It is an initiative to make Riyadh smarter and presents a hybrid model of citizens' engagement where they can involve themselves in making their city smart by publishing issues relating to Riyadh and data regarding different sectors such as health and education. Additionally, citizens are allowed to add new features, functionalities, and even new sectors to the system to contribute to and support the government in making the city smart. This tool is embedded with two approaches: "codesign," which allows citizens to be involved in the planning and decision-making process to build new services, and "cloud sourcing," which allows citizens to act as sources of data to

---

[3] KSU News, 2018. Riyadh Techno Valley, King Saud University's Gateway for the Knowledge Economy. King Saud University. https://news.ksu.edu.sa/en/node/101739

[4] Business Wire, 2019. Riyadh Set to Become the Arab World's First Digital Capital in 2020 https://www.businesswire.com/news/home/20191219005518/en/Riyadh-Set-Arab-World%E2%80%99s-Digital-Capital-2020

cooperate with the government and eventually improve their city. This type of inter-action between citizens and the government can enhance the transparency and trust between them, which improves administrative governance by listening to the concerns of citizens; thus, the level of satisfaction is increased among city dwellers. This approach can strengthen the *smart people* concept of smart city vision because of its wide range of peoples' participation in urban life as well as ability to cope with new services.

In Riyadh, a huge investment project has been initiated to facilitate *smart mobility* in developing urban public transport systems and efficient traffic management. For instance, the MEDSTAR project was initiated in 2003 to manage the city smartly. Under the MEDSTAR project, ongoing metro service development is the discussed component. The metro will run automatically (without a driver) and be operated and monitored by a central control room. The required electric power will be optimized by providing electricity from independent sources. Metro cars and stations will be under continuous advanced surveillance systems.[5] This new system of public transport provides effective solutions for citizens to move across Riyadh efficiently and easily through comfortable train cars equipped with the latest technological advancements. Moreover, the video stream cameras are placed along the Riyadh Road network that has been integrated into a video-monitoring system to oversee major traffic "corridors" of the city (Doheim et al., 2019). The public transport system that Riyadh has introduced enables the enhancement of the quality of life of people, and it can also be labeled *smart living* due to its opportunity for efficient, safe, and comfortable mobility, which also brings about the opening of active transport to connect with open/public spaces and public transit stops.

In terms of the *smart environment* dimension, Riyadh city has initiated a few air-quality monitoring networks to control air pollution, which is composed of dust and other toxic chemicals. The initial monitoring network was found to be insufficient; therefore, the Natural Resources and Environmental Research Institute (NRERI) at King Abdulaziz City for Science & Technology (KACST) took the initiative to construct such a network in Riyadh (Alharbi et al., 2014). The first monitoring station was constructed in 1999 at the KACST campus, and then another four stations were established across the city by 2002. Another example is the HYDRUS smart water meters and Hydrometer Automatic Meter Reading (AMR) system that has been introduced in some parts of the Riyadh region. Smart water systems use ultrasound measuring techniques to measure the amount of water and computers to collect, analyze, and control data for efficient water use.[6] Diehl Metering and Abunayyan Trading have worked together to achieve smart solutions in the metering segment in Saudi Arabia since 2008. In the last 10 years, in partnership with Abunayyan Trading, Diehl Metering has provided an innovative solution to cope with the environmental and technical challenges in the water system. Recently, almost 20 fixed

---

[5] RCRC 2020, King Abdulaziz Project for Riyadh Public Transport https://www.rcrc.gov.sa/ accessed on 27th Feb. 2020.

[6] Abunayyan Trading, 2016. Technical Workshop on HYDRUS Smart Water Meters & AMR Systems for GDOW-Riyadh. Abunayyan Trading.

networks have been installed, and one million HYDRUS are connected around Saudi Arabia.[7]

5   Conclusions: Opportunities and Challenges

Saudi Arabia's Public Investment Fund (PIF) and STC Group (STC) have announced the signing of a joint venture (JV) agreement to establish a new company specializing in the Internet of Things (IoT). It will contribute toward boosting IoT adoption by being a technology-agnostic service provider with offerings in the smart industrial manufacturing sector, smart logistics transportation sector, and smart cities. Additionally, the company will also help to create an ecosystem by providing consulting, implementation, and training support as well as facilitating funding models to support businesses in their adoption of IoT. According to local market studies, there is vast growth in the size of the Kingdom's IoT market to potentially reach SAR10.8 bn by 2025, with an annual growth rate of 12.8%.[8]

Recently, the Saudi Ministry of Transport and Huawei, a provider of ICT infrastructure and smart devices, have signed a memorandum of understanding (MoU)[9] to enhance future transport and technology adoption in the transport and logistics sector by exploring the prospects of utilizing advanced technologies such as 5G, artificial intelligence (AI) and big data. According to the MoU, the Kingdom is set to receive urban solutions in the fields of automation, big data, and digitization. In addition, it will also contribute to the provision of solutions for shared mobility, sustainability, and the use of disruptive technology in the development of the logistics sector and smart transport systems in the Kingdom. The MoU ensures the utilization of the latest technology in the Saudi transport sector to enhance its performance in transporting freight and people, which is part of the Vision 2030 objectives. The adoption of advanced technologies such as AI and IoT improves the logistical performance of the Kingdom's transport systems, hence directly boosting Saudi Arabia's rank on the Logistics and Trading Across Borders performance indexes, further establishing the Kingdom as a global logistics hub connecting three continents: Europe, Africa, and Asia. The agreement also contributes to enhancing the quality of life across the country through the adoption of advanced smart

---

[7] Diehl Metering, 2018. One million HYDRUS ultrasonic water meters installed in Saudi Arabia. https://www.diehl.com/metering/en/news-events/news/1-million-hydrus-ultrasonic-water-meters-installed-in-saudi-arabia/ accessed on 27th Feb. 2020.

[8] Zainab Mansoor, 2022. Saudi's PIF and STC Group sign agreement to establish a new IoT company. URL: https://www.msn.com/en-ae/money/news/saudi-s-pif-stc-group-sign-agreement-to-establish-new-iot-company/ar-AAVPfYt?ocid=entnewsntp&cvid=d5ba2bb4872b45d288dbc3854e0e5bf4 accessed on May 26, 2022.

[9] Manda Banda, 2021. SMT and Huawei sign MoU to enhance mobility adoption in the transport sector. URL: https://www.intelligentcio.com/me/2021/05/05/smt-and-huawei-sign-mou-to-enhance-mobility-adoption-in-transport-sector/ accessed on May 26, 2022.

transport systems as well as improving the services provided to citizens, residents, and visitors across all transport facilities, including airports and railway stations. It focuses on finding ways to implement the latest automation and IoT practices in operations in addition to increasing multimodal integration to enhance transportation inside and across cities and reduce travel time. The Ministry of Transport targets boosting the adoption and utilization of technology in the field of transportation and logistics through collaboration with leading ICT organizations to enhance the services it provides and realize the goals of SV2030. NEOM, a brand-new high-tech city in progress, is a good example in this regard. The fundamental purpose of this city is to create a sustainable smart city that incorporates cutting-edge technologies and heals the environment from the damaging emissions it has experienced in recent years. NEOM aspires to remain committed to sustainability and putting in place mobility systems that are fueled entirely by renewable energy. Zero-emission vehicles will be deployed across all urban and regional mobility modes. In addition, NEOM plans to introduce shared mobility services, driverless shuttles, electric boats, delivery drones, and solar-powered mobility hubs. To align with this, the Kingdom has collaborated with the German aircraft manufacturer Volocopter, who will propel the city of NEOM to new heights of technical growth. Air taxis will be fully integrated into the city with the aim of carrying passengers and transporting logistics.

Furthermore, Saudi Arabia's Royal Commission for AlUla (RCU) launched an autonomous pod vehicle service in AlUla early in 2022,[10] which will experience residents in the future of sustainable, zero-emission mobility on their doorstep. It is planned to extend further to new locations, including Dadan, Hegra, and Al Jadidah, later in 2022. This service is part of the RCU's comprehensive Journey Through Time (JTT) master plan to develop a range of fully integrated, accessible, and environmentally friendly public transport options. The launch of this pilot scheme for the autonomous pod public transport service is the first step toward RCU's goal of providing the AlUla community with access to the very latest clean, safe, and energy-efficient mobility solutions. As outlined in the JTT master plan, sustainability is the driving force behind the ambitions for the future of AlUla and the goal to establish the wider northwest Arabian region as a hub for innovation, business, and tourism. The pilot project of this service monitors energy consumption, connectivity, and practicality carefully before rolling out the autonomous service across specially chosen sites in Al Ula. The position, progress, and performance of the pod are monitored by smart transport systems in the RCU called the Smart County Control Platform, while cameras can be added to enhance passenger safety and capacity planning.

Finally, the pattern of mobility, the way of city planning and transporting people and goods will experience drastic changes in the coming years. Smart transportation systems are aligned with the concept of smart cities and sustainable development

---

[10] Intelligent Transport, 2022. Saudi Arabia's RCU launches autonomous pod vehicle service in AlUla.   URL: https://www.intelligenttransport.com/transport-news/132935/saudi-arabia-rcu-autonomous-pod-alula/ accessed on May 26, 2022.

goals (SDGs), as defined by the United Nations, considering the 17 SDGs for 2030. The increasing population will add vehicles to roads; therefore, authorities need to act to reduce traffic congestion and plan new route optimizations with a reduced ecological footprint, which are essential factors of smart transportation systems. Emerging technologies such as AI, IoT, blockchain, and big data technology will serve as the main entry points and fundamental pillars to promote new innovative solutions that will change the current paradigm of cities and their citizens. The growth in cities would translate into the need for new route optimization algorithms for vehicles and people, traffic management to reduce congestion, and more significant optimization in processes. With the integration of smart technologies and mobility, we will see several changes in the coming years, such as mobility-as-a-service (MaaS) traffic flow optimization, which will transition existing cities into smart cities. This is how smart transportation will be a part of a smart city ecosystem.

References

Ahvenniemi, H., et al. (2017). What are the differences between sustainable and smart cities? *Cities, 60*, 234–245.

Al Helal, E., & Mokhtar, H. (2018). Towards smart Riyadh: Riyadh Wiki information and complaining system. *International Journal of Managing Information Technology, 10*(2), 95–106.

Albino, V., Berardi, U., & Dangelico, R. M. (2015). Smart cities: Definitions, dimensions, performance, and initiatives. *Journal of Urban Technology, 22*(1), 3–21.

Aldusari, A. N. (2015). Smart city as urban innovation: A case of Riyadh north-west district. *Journal of Sustainable Development, 8*(8), 270.

Alharbi, B., Pasha, M., & Tapper, N. (2014). Assessment of ambient air quality in Riyadh City, Saudi Arabia. *Current World Environment, 9*(2), 227.

Al-Hathloul, S. (2017). Riyadh development plans in the past fifty years (1967–2016). *Current Urban Studies, 5*(01), 97.

Al-Hosain, N., & Alhussaini, A. (2021). *Evaluating access to Riyadh's planned public transport system using geospatial analysis*. KAPSARC Discussion Paper. https://doi.org/10.30573/KS%2D%2D2021-DP10

Almahmood, M., et al. (2017). Mapping the gendered city: Investigating the socio-cultural influence on the practice of walking and the meaning of walkscapes among young Saudi adults in Riyadh. *Journal of Urban Design, 22*(2), 229–248.

Al-Mosaind, M. (2018). Applying complete streets concept in Riyadh, Saudi Arabia: Opportunities and challenges. *Urban, Planning and Transport Research, 6*(1), 129–147.

Alomari, E., et al. (2021). Iktishaf+: A big data tool with automatic labeling for road traffic social sensing and event detection using distributed machine learning. *Sensors, 21*(9), 2993.

Anwar, A. M., & Oakil, A. T. (2021). *Can parking fees influence people to use the metro instead of cars in Riyadh City?* KAPSARC Commentary.

Appio, F. P., Lima, M., & Paroutis, S. (2019). Understanding Smart Cities: Innovation ecosystems, technological advancements, and societal challenges. *Technological Forecasting and Social Change, 142*, 1–14.

Aqib, M., et al. (2019). Rapid transit systems: Smarter urban planning using big data, in-memory computing, deep learning, and GPUs. *Sustainability, 11*(10), 2736.

Artmann, M., Inostroza, L., & Fan, P. (2019). *Urban sprawl, compact urban development and green cities. How much do we know, how much do we agree?* (pp. 3–9). Elsevier.

Atif, Y., Ding, J., & Jeusfeld, M. A. (2016). Internet of things approach to cloud-based smart car parking. *Procedia Computer Science, 98*, 193–198.

Awaisi, K. S., et al. (2019). Towards a fog enabled efficient car parking architecture. *IEEE Access, 7*, 159100–159111.

Bazzan, A. L., & Klügl, F. (2013). Introduction to intelligent systems in traffic and transportation. *Synthesis Lectures on Artificial Intelligence and Machine Learning, 7*(3), 1–137.

Benevolo, C., Dameri, R. P., & D'auria, B. (2016). Smart mobility in smart city. In *Empowering organizations* (pp. 13–28). Springer.

Bıyık, C., et al. (2021). Smart mobility adoption: A review of the literature. *Journal of Open Innovation: Technology, Market, and Complexity, 7*(2), 146.

Bradshaw, S., Chant, S., & Linneker, B. (2019). Challenges and changes in gendered poverty: The feminization, de-feminization, and re-feminization of poverty in Latin America. *Feminist Economics, 25*(1), 119–144.

Buehler, R., et al. (2017). Reducing car dependence in the heart of Europe: Lessons from Germany, Austria, and Switzerland. *Transport Reviews, 37*(1), 4–28.

Calthrop, E., Proost, S., & Van Dender, K. (2000). Parking policies and road pricing. *Urban Studies, 37*(1), 63–76.

Cervero, R., & Murakami, J. (2008). *Rail+ property development: A model of sustainable transit finance and urbanism*. Institute of Transportation Studies, University of California.

Chan, N. D., & Shaheen, S. A. (2012). Ridesharing in North America: Past, present, and future. *Transport Reviews, 32*(1), 93–112.

Cheng, Y., et al. (2019). Understanding the spatial disparities and vulnerability of population aging in China. *Asia & the Pacific Policy Studies, 6*(1), 73–89.

Das, S., et al. (2019). Solid waste management: Scope and the challenge of sustainability. *Journal of Cleaner Production, 228*, 658–678.

de Oliveira, L. K., et al. (2019). Challenges to urban freight transport in historical cities: A case study for Sabará (Brazil). *Transportation Research Procedia, 39*, 370–380.

Debnath, A. K., et al. (2014). A methodological framework for benchmarking smart transport cities. *Cities, 37*, 47–56.

Djahel, S., Sommer, C., & Marconi, A. (2018). Guest editorial: Introduction to the special issue on advances in smart and green transportation for smart cities. *IEEE Transactions on Intelligent Transportation Systems, 19*(7), 2152–2155.

Doheim, R. M., Farag, A. A., & Badawi, S. (2019). Smart city vision and practices across the Kingdom of Saudi Arabia—A review. In *Smart Cities: Issues and challenges mapping political, social and economic risks and threats* (pp. 309–332). Elsevier.

Durga Devi, T. J. B., Subramani, A., & Solanki, V. K. (2017). Smart city: IOT based prototype for parking monitoring and management system commanded by mobile app. *Annals of Computer Science and Information Systems, 10*, 341–343.

El-Geneidy, A. M., & Levinson, D. M. (2006). *Access to destinations: Development of accessibility measures*. University of Minnesota Center for Transportation Studies.

Ferguson, E. (1997). The rise and fall of the American carpool: 1970–1990. *Transportation, 24*(4), 349–376.

Giannopoulos, G. A. (2004). The application of information and communication technologies in transport. *European Journal of Operational Research, 152*(2), 302–320.

Giffinger, R., et al. (2007). *City-ranking of European medium-sized cities* (Vol. 9, pp. 1–12). Centre of Regional Science.

Glaas, E., et al. (2020). Visualization for citizen participation: User perceptions on a mainstreamed online participatory tool and its usefulness for climate change planning. *Sustainability, 12*(2), 705.

Goetz, A. R., & Alexander, S. (2019). *Urban goods movement and local climate action plans: Assessing strategies to reduce greenhouse gas emissions from urban freight transportation*. Mineta Transportation Institute Publications, San Jose State University.

Hamer, P., Currie, G., & Young, W. (2012). Equity implications of parking taxes. *Transportation Research Record, 2319*(1), 21–29.

Heffner, K., Klemens, B., & Solga, B. (2019). Challenges of regional development in the context of population ageing. Analysis based on the example of Opolskie Voivodeship. *Sustainability, 11*(19), 5207.

Hietanen, S. (2014). Mobility as a Service. The new transport model. *ITS & Transport Management Supplement. Eurotransport, 12*(2), 2–4.

Jayantha, W. M., Qian, Q. K., & Yi, C. O. (2018). Applicability of 'Aging in Place' in redeveloped public rental housing estates in Hong Kong. *Cities, 83*, 140–151.

JSCE. (2016). *Intelligent transport systems (ITS): Introduction guide*. Japan Society of Civil Engineers.

Juwet, G., & Ryckewaert, M. (2018). Energy transition in the nebular city: Connecting transition thinking, metabolism studies, and urban design. *Sustainability, 10*(4), 955.

Kazi, S., et al. (2018). Smart parking system to reduce traffic congestion. In *2018 international conference on Smart City and emerging technology (ICSCET)*. IEEE.

Kelly, K. L. (2007). Casual carpooling-enhanced. *Journal of Public Transportation, 10*(4), 6.

Kijewska, K., Iwan, S., & Korczak, J. (2019). Challenges to increase the sustainable urban freight transport in South Baltic Region–LCL project. *Transportation Research Procedia, 39*, 170–179.

Kovács, Z., et al. (2019). Urban sprawl and land conversion in post-socialist cities: The case of metropolitan Budapest. *Cities, 92*, 71–81.

Kumar, H., et al. (2020). Moving towards smart cities: Solutions that lead to the Smart City Transformation Framework. *Technological Forecasting and Social Change, 153*, 119281.

Lee, J. H., Hancock, M. G., & Hu, M.-C. (2014). Towards an effective framework for building smart cities: Lessons from Seoul and San Francisco. *Technological Forecasting and Social Change, 89*, 80–99.

Lopes, N. V. (2017). Smart governance: A key factor for smart cities implementation. In *2017 IEEE international conference on smart grid and smart cities (ICSGSC)*. IEEE.

Lopez-Ruiz, H. G., et al. (2019). *Estimating freight transport activity using nighttime lights satellite data in China, India and Saudi Arabia*. King Abdullah Petroleum Studies and Research Center.

Luo, X., et al. (2019). *Challenges and adaptation to urban climate change in China: A viewpoint of urban climate and urban planning* (pp. 1157–1161). SAGE Publications Sage UK.

Mahmoud, H., & Divigalpitiya, P. (2019). Spatiotemporal variation analysis of urban land expansion in the establishment of new communities in Upper Egypt: A case study of New Asyut city. *The Egyptian Journal of Remote Sensing and Space Science, 22*(1), 59–66.

Mainetti, L., et al. (2015). A smart parking system based on IoT protocols and emerging enabling technologies. In *2015 IEEE 2nd world forum on internet of things (WF-IoT)*. IEEE.

Mangiaracina, R., et al. (2017). A comprehensive view of intelligent transport systems for urban smart mobility. *International Journal of Logistics Research and Applications, 20*(1), 39–52.

Meurs, H., et al. (2020). Organizing integrated services in mobility-as-a-service systems: Principles of alliance formation applied to a MaaS-pilot in the Netherlands. *Transportation Research Part A: Policy and Practice, 131*, 178–195.

Mitropoulos, L., Kortsari, A., & Ayfantopoulou, G. (2021). A systematic literature review of ride-sharing platforms, user factors and barriers. *European Transport Research Review, 13*(1), 1–22.

Morency, C. (2007). The ambivalence of ridesharing. *Transportation, 34*(2), 239–253.

Muktiali, M. (2018). Policy analysis of poverty alleviation in Semarang city using spatial and sectoral approach. *IOP Conference Series: Earth and Environmental Science, 123*. https://doi.org/10.1088/1755-1315/123/1/012046. IOP Publishing.

Munksgaard, N. C., et al. (2019). Environmental challenges in a near-pristine mangrove estuary facing rapid urban and industrial development: Darwin Harbour, Northern Australia. *Regional Studies in Marine Science, 25*, 100438.

Munoz-Pandiella, I., et al. (2018). Urban weathering: Interactive rendering of polluted cities. *IEEE Transactions on Visualization and Computer Graphics, 24*(12), 3239–3252.

Nam, T., & Pardo, T. A. (2011). Conceptualizing smart city with dimensions of technology, people, and institutions. In *Proceedings of the 12th annual international digital government research conference: Digital government innovation in challenging times.*

Neirotti, P., et al. (2014). Current trends in Smart City initiatives: Some stylised facts. *Cities, 38*, 25–36.

Nikitas, A., et al. (2020). Artificial intelligence, transport and the smart city: Definitions and dimensions of a new mobility era. *Sustainability, 12*(7), 2789.

Pereira, G. V., et al. (2018). Smart governance in the context of smart cities: A literature review. *Information Polity, 23*(2), 143–162.

Pham, T. T., & Phan, C. T. (2018). Risk management: Awareness, identification and mitigation in public private partnerships of technical infrastructure projects in da nang. In *2018 4th international conference on green technology and sustainable development (GTSD).* IEEE.

Powell, W. W., & Snellman, K. (2004). The knowledge economy. *Annual Review of Sociology, 30*, 199–220.

Shaheen, S., & Finson, R. (2013). Intelligent transportation systems. In *Reference module in earth systems and environmental sciences.* Elsevier.

Sharmeen, F., & Meurs, H. (2019). The governance of demand-responsive transit systems—A multi-level perspective. In *The governance of smart transportation systems* (pp. 207–227). Springer.

Sharmeen, F., Drost, D., & Meurs, H. (2020). A business model perspective to understand intra-firm transitions: From traditional to flexible public transport services. *Research in Transportation Economics, 83*, 100959.

Siwar, C., et al. (2016). Urbanization and urban poverty in Malaysia: Consequences and vulnerability. *Journal of Applied Sciences, 16*(4), 154–160.

Slaev, A. D., et al. (2019). Overcoming the failures of citizen participation: The relevance of the liberal approach in planning. *Planning Theory, 18*(4), 448–469.

Smith, G., & Hensher, D. A. (2020). Towards a framework for Mobility-as-a-Service policies. *Transport Policy, 89*, 54–65.

Smith, S. F., et al. (2013). *Surtrac: Scalable urban traffic control.* Carnegie Mellon University.

Sou, G. (2019). Household self-blame for disasters: Responsibilisation and (un) accountability in decentralised participatory risk governance. *Disasters, 43*(2), 289–310.

Tomaszewska, E. J., & Florea, A. (2018). Urban smart mobility in the scientific literature—bibliometric analysis. *Engineering Management in Production and Services, 10*(2), 41–56.

UN (United Nations). (2018). https://www.un.org/development/desa/en/news/population/2018-revision-of-worldurbanization-prospects.html. Accessed on July 5, 2023.

Visvizi, A., & Lytras, M. (2019). *Smart cities: Issues and challenges: Mapping political, social and economic risks and threats.* Elsevier.

Weng, J., et al. (2016). Real-time bus travel speed estimation model based on bus GPS data. *Advances in Mechanical Engineering, 8*(11), 1687814016678162.

Winkowska, J., Szpilko, D., & Pejić, S. (2019). Smart city concept in the light of the literature review. *Engineering Management in Production and Services, 11*(2), 70–86.

Yan, J., Liu, J., & Tseng, F.-M. (2020). An evaluation system based on the self-organizing system framework of smart cities: A case study of smart transportation systems in China. *Technological Forecasting and Social Change, 153*, 119371.

Zabara, B., & Ahmad, A. A. (2020). Biomass waste in Yemen: Management and challenges. In *Waste management in MENA regions* (pp. 313–336). Springer.

Zhang, A., et al. (2019). Barriers to smart waste management for a circular economy in China. *Journal of Cleaner Production, 240*, 118198.

Open Access   This chapter is licensed under the terms of the Creative Commons Attribution 4.0 International License (http://creativecommons.org/licenses/by/4.0/), which permits use, sharing, adaptation, distribution and reproduction in any medium or format, as long as you give appropriate credit to the original author(s) and the source, provide a link to the Creative Commons license and indicate if changes were made.

The images or other third party material in this chapter are included in the chapter's Creative Commons license, unless indicated otherwise in a credit line to the material. If material is not included in the chapter's Creative Commons license and your intended use is not permitted by statutory regulation or exceeds the permitted use, you will need to obtain permission directly from the copyright holder.

Smart Cities: GCC and Kuwait Experience

Mohammad I. Elian and Khalid M. Kisswani

Abstract This study has two main objectives. First, spread out attention toward modern urban modeling and smart cities development in the GCC countries in general and Kuwait in particular. Second, expand existing scholarly concern on digital urbanization notion beyond the scope of the worldwide developed economies. There is a need to include the GCC countries to the scholarly urban literature since such countries are now highly urbanized. Also, the knowledge-based economy is applied through adopting the digitalization processing toward offering public and private services to their communities. The study concludes that smart cities have become very much significant through offering preeminent solutions to several environmental, economic, and societal problems in urban regions. For the GCC, it is crucial to identify the concept and structure of the targeted smart city initiative which should be linked to the country's national strategic vision. Additionally, what proxies are used to judge on how smart is a city? Not less important, adopting the smart cities format would require overcoming significant limitations in order to overcome urban pressures through introducing the smart cities initiatives.

Keywords Smart City · Urban Development · Informational urbanism · GCC Urban Modeling

M. I. Elian
Department of Economics and Finance, Gulf University for Science and Technology, Hawally, Kuwait
e-mail: Elian.M@gust.edu.kw

K. M. Kisswani (✉)
Department of Economics and Finance, Gulf University for Science and Technology, Hawally, Kuwait

Center for Sustainable Energy and Economic Development (SEED), Gulf University for Science and Technology, Hawally, Kuwait
e-mail: kisswani.k@gust.edu.kw

© The Author(s) 2024
F. Belaïd, A. Arora (eds.), *Smart Cities*, Studies in Energy, Resource and Environmental Economics, https://doi.org/10.1007/978-3-031-35664-3_18

1   Introduction

Urban cities are growing across the world as a result of the rapid growth in urban populations. This unprecedented progression initiates the need to introduce an effective urban city design to accommodate the urbanization confronts in urban areas (i.e., population detonation pressure, increase in cities' consumption, variability in people's standards of living, expensive cost of living, environment pollution, insufficient infrastructure competences, mainly traffic congestion, etc.). Thus far, insightful design is becoming enormously crucial for official city planners and practitioners as the form of a smart city is introduced. Under the smart city format, it is expected that cities would be more resilient, environmentally sustainable, and efficient in providing services that would lead to better quality of life (QoL) for residents.

In fact, future urbanization statistics boost the smart city's prominence as a relevant solution to urban population pressure. These statistics signify the introduction of the *informational urbanism* notion to most urban regions.[1] For example, while approximately 55% of the world's population is currently living in cities, the United Nations expects that an average of 70% of the world population will live in urban areas by the year 2050. Additionally, it is estimated that more than 80% of the entire population of Europe and 50% of the population in the Middle East and North Africa (MENA) will live in urban cities (see Constro, 2021; Ringel, 2021; United Nation, 2018). When considering the level of income, the 2050 statistics estimate that the urban population would rise to 88.4% in high-income countries in general and 90% in the countries of the Gulf Cooperation Council (GCC) in particular. This estimated increase is a clear indication of the interdependence between the urbanization rate and quality of life. The lower the income is, the lower the level of urbanization (Asmyatullin et al., 2020; Bufetova, 2013).[2]

Similar expansion is mirrored crosswise in the GCC region, where urbanization is expanding, and many challenges are encountered.[3] Some of these challenges include but are not limited to upward regional growth rates of urbanization, limited diversified economies (which are restricted to oil products), variability in sources of income (since oil prices are volatile), infrastructure issues (particularly those related to education, healthcare, and transportation), climate concerns, upgrading quality of public services provided to residents, stimulating sustainable economic development, inspiring quality of residents' life, etc. To overcome these challenges and

---

[1] "Information urbanism is an interdisciplinary endeavor, which incorporates computer and information sciences as well as urban studies, city planning, architecture, city economics, and city sociology" (Barth et al., 2017).

[2] For instance, the urban population pressure in Europe is very much clear, especially in Germany where the population in Frankfurt and Munich grew by 7% between 2008 to 2016. Similar urban pressure exists in the MENA population (Ringel, 2021; Griffiths, 2017).

[3] The average urbanization rate for the GCC countries will be 95.1% by the year 2050, especially Bahrain 93.2%, Kuwait 100%, Oman 94.9%, Qatar 99.7%, Saudi Arabia 90.4%, and UAE 92.4% (see, Asmyatullin et al., 2020).

others alike, GCC countries are now considering initiating smart cities and adopting digitalized economies as urgent vision-oriented initiatives. The initiative designates an informational-technological perspective of urban management processing. Hence, developing and adopting smart appliances has been significantly endorsed in the future national strategies of the GCC countries.

In this chapter, we argue that while there are ideal industrial cities (industrial societies) and ideal services cities (services societies), there will be an ideal digital or smart city (digital or smart societies) very soon. Apparently, the ideal digital or smart cities will keep making headway as the world is changing rapidly toward the *informational urbanism* notion, which will be the norm in the very near future. We argue that the *informational* notion will be achievable by many societies in the world, and the typical knowledgeable cities will be the dynamic that would categorize societies futurewise. Many cities are currently evolving their digital information abilities by centering their interest on knowledge-based urban expansion in areas of higher education, health, infrastructure, telecommunication, transportation, science labs, ingenuity of human capital, and so on (see Barth et al., 2017; Carillo et al., 2014; Madanipour, 2011). This is quite true since cities are competing in constructing and updating their information and communication technology (ICT) infrastructure, which qualifies many aspects of knowledge and information to be converted to digital applications rather than physical functioning.

This study aims to increase attention to smart city development in GCC countries in general and Kuwait in particular. It seeks to expand existing scholarly investigations on urban modeling and digital urbanization beyond the scope of developed economies worldwide. In fact, little attention has been given to the notion of smart city development in the Middle East and North Africa (MENA) and the GCC countries, although most of their populations are now urbanized, and technology infrastructure is based on global best practices and mutual learning. Additionally, digitalization is adopted in many public and private services and deliberated as an option to create a knowledge-based economy. For Kuwait, the country is moving ahead toward modeling new urban cities to overcome the pressure caused by the urban population.

Three reasons motivated the work in our chapter. First, smart city initiatives are still in their go-forward stages since the impact of those initiatives toward overcoming the urban population challenges in the GCC countries has not yet been clearly achieved. Second, the smart city format is supposed to propound solutions to the environmental, economic, and social challenges in urban zones, but this is not yet confirmed by the related GCC literature. Finally, most smart cities are characterized by their capability to achieve integration between people, technology, and information. The objective is to generate efficient, sustainable, and resilient infrastructure and quality of services that would boost the quality of life (QoL) of residents in terms of education, health, transportation, public and private services, public safety, and the like. This is not approved thus far as per the current status of the GCC countries.

The study proceeds as follows: Sect. 2 presents the concept of smart cities as per the literature. Section 3 focuses on the necessity of smart cities. In Sect. 4, the different structuring formats of a smart city are outlined. How smart is a city? is outlined in Sect. 5, while the GCC experience is discussed in Sect. 6. The case of Kuwait is detailed in Sect. 7, which shows the country's strategic concern of smart city initiatives and their needs to the country. The *Saad Al-Abdullah* smart city initiative is summarized as a smart city format in Kuwait. Section 8 concludes by reviewing some of the overall limitations confronting smart city initiatives, particularly those in the context of Kuwait.

2 What Is a Smart City?

The term smart city or smart city was first introduced in the 1990s, where the focus was on the significance of information communication technology (ICT) to update infrastructure within cities (see Harrison & Donnelly, 2011; Albino et al., 2015). During the last two decades, smart cities have been advanced and labeled informational cities, *digital cities, ubiquitous cities, knowledge cities,* or *creative cities.*

As per the literature, there is still no unified concept or coherent definition of a "smart city" or "informational city." Most definitions evidently overlap and are not mutually exclusive. Narrowly, some definitions of a smart city are classified by its concern. Specific definitions emphasize green and sustainable cities that focus on environmental traits. Such definitions consider a smart city as an environmental front and hence count urban cities as secure and environmentally green. The environmental front definition is sharply interrelated to natural resources and energy, transport and mobility, and living population conditions (Hall et al., 2000; Martin et al., 2018; Barth et al., 2017; Chourabi et al., 2012).

In a broader sense, many other definitions detect a smart city from an informational perspective, hence talking about the *informational city*. Giffinger et al. (2007) define smart cities by a list of crucial features, namely, smart economy, people, governance, mobility, environment, and living. Foth et al. (2011) differentiate between urban informatics cities *informativeness* and informational urbanism cities. They show that urban informatics underlines the responsibilities of people, place, information, and ICT with a focus on cities. Informational urbanism includes not only ICT but also all categories of information and knowledge (tacit knowledge as well as explicit knowledge). Likewise, (Chourabi et al., 2012) define smartness with eight analytical factors, including management and organization, technology, governance, policy context, people and communities, economy, built infrastructure, and natural environment.

While Albino et al. (2015) conclude many definitions of a smart city, Cheu et al. (2015) summarize us the following major definitions as used by foremost industrial associations and standards authorities. The Smart City Council defines a smart city as *"digital technology, and intelligent design have been harnessed to create smart, sustainable cities with high-quality living and high-quality jobs"* (Smart Cities Council 2014 as cited by Cheu et al., 2015). The British Standards Institute (BSI) defines a smart city as the *"effective integration of physical, digital and human systems in the built environment to deliver a sustainable, prosperous and inclusive future for its citizens*" (BSI 2014 as cited by Cheu et al. (2015)).

The International Standard Organization (ISO) describes a smart city as "*a new concept and a new model, which applies the new generation of information technologies, such as the internet of things, cloud computing, big data and space/geographical information integration, to facilitate the planning, construction, management and smart services of cities*" (ISO 2015 as cited by Cheu et al., 2015). The IEEE identifies that a smart city is the one that "*brings together technology, government and society to enable the following characteristics: smart economy, smart mobility, smart environment, smart people, smart living, and smart governance*" (IEEE 2015 as cited by Cheu et al., 2015). The European Commission states that "*smart cities are characterized and defined by a number of factors including sustainability, economic development and high quality of life. Enhancing these factors can be achieved through infrastructure (physical capital), human capital, social capital and/or ICT infrastructure*" (EC 2015 as cited by Cheu et al., 2015).

Cheu et al. (2015) proposed their own definition in which "*a smart city is characterized by its ability to integrate people, technology and information to create an efficient, sustainable and resilient infrastructure that provides high quality services while improving the quality of life of its residents.*" Their definition of a smart city emphasizes two crucial integrated terms (people and technology) with targeted objectives (efficiency of operations and residents' quality of life). Mohanty (2016) identify a smart city as "*a city connecting the physical infrastructure, the information-technology infrastructure, the social infrastructure, and the business infrastructure to leverage the collective intelligence of the city.*" Mohanty (2016) report another comprehensive definition of a smart city as "*a smart sustainable city is an innovative city that uses information and communication technologies (ICTs) and other means to improve quality of life, the efficiency of urban operations and services, and competitiveness while ensuring that it meets the needs of present and future generations with respect to economic, social and environmental aspects.*"

Other definitions differentiate between an "intelligent" city and a "digital" city. The former focuses on competences and knowledge (Meijer & Bolívar, 2016), and the latter focuses on ICT components and their interconnectivity. Barth et al. (2017) identify that most definitions of a smart city reflect the interface between two major domains. First, the city's control system and its subsystems of infrastructures, economics, politics and administration, spaces, and location factors. Second, the information behavior of the city's stakeholders, namely, residents, companies, administrations, visitors, and alike. Saxena and Al-Tamimi (2018) report that "*smart cities (are) those innovative urban systems that strategically invest in new technologies and human capital, seeking to improve services effectiveness, quality of life, economic competitiveness, environmental sustainability, and participatory gover*nance" (see Fernandez-Guell et al., 2016).

The World Bank defines smart cities as "*a technology intensive city that delivers 'intelligent' energy and mobility solutions in cooperation with its citizens*" (Muente-Kunigami & Mulas, 2019). The European Commission defines smart cities as "*cities using technological solutions to improve the management and efficiency of the urban environment*" (European Commission, 2021). Others add the dimension of *sustainability* to the World Bank definition, implying both the minimization of resource streams and environmental impacts as well as adaptation to a changing global climate (Ringel, 2021; Al-Thani et al., 2018; Berg, 2017). Constro (2021) considers the smart city an urban area that utilizes numerous electronic techniques and sensors to collect data that can be used to enhance operations across the city by efficiently managing the available assets, resources, and services.

In conclusion, smart city definitions emphasize the amalgamation of many factors, including people, technology and information, enhanced services, and sustainable and resilient systems of infrastructure, where all factors stimulate residents' quality of life (QoL). Most of the above concepts of a smart city indicate significant technological progress during the past two decades. This incorporates the hardware and software designs that cause the development of ICTs worldwide. In fact, the above conceptual development of a smart city indicates the transfer toward the *informational* city framework. This may justify why most broader concepts of a smart city highlight the *informational urbanism* notion, which has theoretical applications on the one hand and computer applications on the other hand.

3   What Makes Smart Cities a Necessity?

Smart cities have become very significant by offering preeminent solutions to many environmental, economic, and societal problems in urban groups. The environmental contribution of smart cities is very obvious, as smart city format can avoid the negative impact of energy consumption on the environment. Specifically, the environmental solution concludes the smart cities' impact toward reducing energy consumption, stabilizing water utilization, controlling carbon emissions, and easing transportation supplies. In this regard, the redesign of energy supply, use of renewable energies, energy efficiency, and transport infrastructure would reduce energy consumption and alleviate global warming. This is one of the most preeminent solutions introduced by smart cities since cities are responsible for 70% of total energy consumption and approximately 75% of greenhouse gas production in Europe (Halleux, 2016; Balta-Ozkan et al., 2014). In other words, cities are consuming 75% of the world's resources and energy, which generates 80% of greenhouse gases. The same is mirrored in the MENA cities. Qatar is an example of one of the most energy- and carbon-intensive countries worldwide (Ringel, 2021; Charfeddine et al., 2018).

Smart cities also lead to economic optimization through energy savings, raw material consumption, and stabilization of resource flow in urban areas. Economic optimization is expedited by using information and communication technology (ICT), which can improve living conditions, facilitate urban management, or mediate energy and climate problems. Additionally, smart economies support the education, entrepreneurial spirit, innovation, and productivity of city populations (see Constro, 2021; Letnik et al., 2018; Sicilia et al., 2017; Monfaredzadeh & Krueger, 2015; Abellá-García et al., 2015; Jedlin'ski, 2014).

Smart cities and their technology would allow city formers to interact immediately with both community and city infrastructure, where many social problems can be solved within urban groups. It seems this is quite true since smart cities are basically initiated to get used to urban growth and help ease people's intensification and to effectively address urban issues such as traffic congestion, low housing affordability, and social inequity. The social contribution from a smart city would also offer better quality of life and a higher standard of living for its residents. An additional social benefit is by establishing new economical districts for financial, tourism, and entertainment (Tok et al., 2015).

4 Smart City Structure

A smart city structure is indicated by some smart core constituents, various attributes, core themes, and infrastructure. Those dimensions can be considered smart city features to be distinguished from a conventional city. Cheu et al. (2015) summarize the main features used by foremost industrial associations and standards authorities. The IEEE (2015) refers to six features of a smart city: smart economy, smart mobility, smart environment, smart people, smart living, and smart governance. The European Commission (2015) characterizes smart cities by three features: sustainability, economic development, and quality of life (QoL). ISO (2015) outlines the characteristics that are required for smart cities: *"The city will be instrumented; The data from different sources will be available to be easily aggregated; The data will be easily visualized and accessible; Detailed, measurable, real-time knowledge will be available at every level; Analytics and decision-making systems will be used; The city will be automated; The city will have a network of collaborative spaces; and The decision making processes are to be much more open and inclusive."* The Smart Cities Council (https://www.smartcitiescouncil.com/) provides a list of responsibility areas and a list of enablers rather than explicitly providing a list of the features of smart cities.[4]

Mohanty (2016) identifies that a smart city would have nine constituents, four attributes, four core themes, and three infrastructures. The following is a brief overview of the conclusions of Mohanty (2016). The nine constituents of a smart city include: (1) smart citizens, (2) smart infrastructure (e.g., physical, ICT and services), (3) smart building (e.g., different hardware, software, sensors and smart appliances for different automated operations), (4) smart transportation (e.g.,

---

[4] *"The responsibility areas (which correspond to services provided by the cities to its residents) are: Built environment; Economic development; Energy; Health and human services; Payments (electronic payment of fees); Public safety and security; Telecommunications; Transportation; Waste management; and Water and wastewater. The list of enablers (technologies and strategies) are: Analytics; Citizen engagement; Computing resources; Connectivity; Data management; Finance and procurement; Instrumentation and control; Interoperability; Policy and leadership; and Security and privacy"* (Cheu et al., 2015).

Intelligent Transport Systems (ITS) as communication and navigation systems between vehicles (car-to-car) and between vehicles and fixed locations (car-to-infrastructure), in addition to the ITS for the rail, water, and air transport systems and their interactions), (5) smart energy (e.g., smart power generation, smart power grids, smart storage, and smart consumption, reflecting the integration of decentralized sustainable energy sources, efficient distribution, and optimized power consumption), (6) smart healthcare (e.g., calls for traditional healthcare to be intelligent, efficient, and sustainable), (7) smart technology (e.g., infrastructure, buildings, physical structures, electrical infrastructure, electronics, communication infrastructure, information technology infrastructure, and software), (8) smart governance, and (9) smart education.

The four attributes of smart cities include sustainability (infrastructure and governance, energy and climate change, pollution and waste, social issues, economics, and health) and quality of life (QoL) in terms of the emotional and financial well-being of citizens. Urbanization (through multiple aspects and indicators such as technology, infrastructure, governance, and economics) and smartness (there are various aspects of city smartness that include the smart economy, smart people, smart governance, smart mobility, and smart living.). The four core themes for a smart city are society (signifies that the city is for its inhabitants or citizens), economy (signifies that the city is able to thrive with continuous job growth and economic growth), environment (indicates that the city will be able to sustain its function and remain in operation for current and future generations), and governance (suggests that the city is robust in its ability to administer policies and combine together the other elements). The three infrastructures of a smart city are indicated by its physical (buildings, roads, railway tracks, power supply lines, and water supply system), electrical, and digital infrastructure ICT infrastructure. Saxena and Al-Tamimi (2018) summarize the four basics of a smart city, including technology, sustainability, human and social capital, and governance.[5] A smart city structure can be shown by the below-proposed design.

---

[5] "Technology (to collect and manage new data sources, conduct analyses and use networked tools and technologies to manage cities), sustainability (adoption of a diversity of urban growth management policies where the dimensions of environmental, economic and social sustainability are being emphasized), human and social capital (knowledge exchange, creativity, and innovation through collaboration among stakeholders to create 'smart solutions') and governance (institutional preparation and community governance)" (see, Chauhan et al., 2016; de Wijs et al., 2016; Fernandez-Guell et al., 2016; Monfaredzadeh & Berardi, 2015; Tranos & Gertner, 2012 as cited by Saxena and Al-Tamimi (2018)).

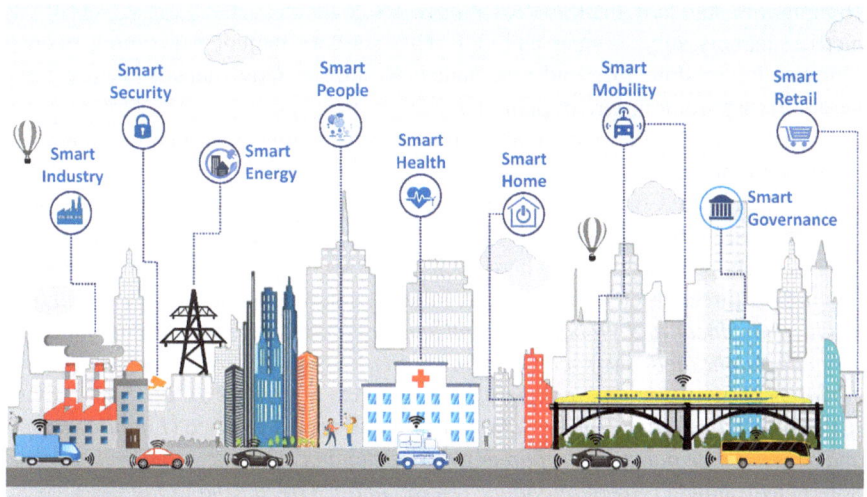

Source: Constro Facilitator (2021)

5 The Smartness of a City

How smart is a city? The answer to this question may differentiate between an urban city and a smart city per se. As shown in the previous section, smart city structures aim to improve the QoL of residents to several extents, namely, smart ICT devices that improve public services, smart mobility of society, smart environment, smart living, and smart governance. The concern is now how someone can assess the level of a city's smartness.

Thus far, the overall smartness of a city can be indexed by measurable indicators that have to show the resident's partialities, in line with the status of the city's advancement and other country-specific constraints. The city smartness indicators look to examine to what extent a city is close to smart city status, test for the city's performance results, explore growth and expansion trends, and discover a city's strengths and weaknesses grounds (see Vázquez-Castañeda and Estrada-Guzman 2014 as cited by Cheu et al., 2015).

Cheu et al. (2015) present most smart city indicators as proposed by several industries and professional interested parties, mainly the BSI, ISO, and the Smart City Council. The BSI indicators of smart cities include "Broadband connectivity," including "*GPS, Wi-Fi and satellite availability; Knowledge workforce; Digital inclusion; Innovation; and Marketing and advocacy.*" ISO (2015) has introduced indicators related to life status, including "*education, health, recreation, safety, transportation, water, finance, etc.*" and environmental sustainability issues such as energy, water, and transport. Throughout its sustainable development of communities standard, the ISO (2015) provides 20 themes related to two sets of city metrics, namely, city services and QoL. City services include Education; Energy; Finance; Recreation; Fire and emergency; Response; Governance; Health; Solid waste; Transportation; Urban planning; Wastewater; and Water. The QoL includes civic engagement; culture; economy; environment; shelter; social equity; and technology and innovation.

The Smart City Council (2014) indicators followed the Cohen (2012) indicators of a smart city. The indicators are organized into six dimensions that are similar to the European categorization, where each of the six dimensions is further divided into almost three targeted indicators as follows: smart economy (*entrepreneurship, productivity, local and global interconnectedness*); smart people (*twenty-first century education, inclusive society, embrace creativity*); and smart mobility (*mixed-modal access, clean and nonmotorized options, integrated ICT*). Smart living (*Culturally vibrant and Happy, Safe, Healthy*); Smart governance (*Enabling supply and demand-side policy, Transparency and open data, ICT and e-government*); and Smart environment (*Green buildings, Green energy, Green urban planning*).

Part of the literature has emphasized indicators that enhance the QoL of residents since this is the ultimate objective of smart city initiatives. Mercer (2014) suggested many QoL index tests for targeted related categories, namely, "*Political and social environment; Economic environment; Sociocultural environment; Medical and health considerations; Schools and education; Public services and transport; Recreation; Consumer goods; Housing; and Natural environment.*" Pribyl and Horak (2015) focused on individuals' insights into QoL and concluded that there are discrepancies among individual preferences on what composes QoL.

In conclusion, we argue that there is a crucial need to establish smart city indicators. Not less important is the impact the indicators would have on the QoL of residents. QoL indicators are critical and challenging factors given differences among countries in terms of the history and size of the population, population growth rates, geographical location, availability of domestic resources and economic growth rates, residents' QoL priorities, and cultural heterogeneity. Additionally, there is a need to ask for city stakeholders' transparent engagement while structuring smart city initiatives and also while deriving key performance indicators of the city. Stakeholders include government, industry, practitioners, residents, and any other related parties.

Overall, having clear smart city indicators is relevant for the GCC experience. Those countries share similar features in terms of their population culture, geographical location, natural resources, sources of income, development growth, and so on. However, the impact of smart city initiatives on residents' QoL and economic growth is not very quantifiable among countries. Additionally, special concern has to be directed to the integration among smart city ingredients, namely, industry, IT security, energy, people, health, building, retail, and governance. In fact, having clear smart city indicators may overcome many such concerns and other structural smart city issues.

6 The GCC Experience

The GCC smart city initiatives are already well documented in their national strate-gic plans, as introduced by Saxena and Al-Tamimi (2018) and Asmyatullin et al. (2020). The proceeding section concludes their findings. For Bahrain, the country lodged the e-government initiative in 2007, which was a strategic segment as per their Vision 2030. The economic vision 2030 focuses on attaining advanced classes of infrastructure and innovation, improving the quality of public services and reduc-ing expenses, creating a safe and secure environment, and "adopting the updated technologies and alike." Bahrain has a plan of setting up ten smart cities. Information technology is applied for the online transformation of public services that can be acquired by the public through e-government networks, including website portal, open data portal, call centers, booths, smart cards, and mobile applications. The e-government applied the "Cloud First" policy where government affairs are trans-ferred to the cloud.

For Oman, acquiring digital skills, digital literacy, and new technologies was introduced in 2017 as part of the Digital Oman Vision 2030. The e-Oman strategy emphasizes three main key dimensions, namely, IT industry development, society, and e-government services. Among many related looking forward objectives, Digital Oman Vision 2030 provides special emphasis on information technology training and digital literacy, free and open-source software usage endeavors, state online application enhancements, Internet law (eLaws) developments, and the expansion of mobile access to public services. The smart city initiative is verified through the Duqm Special Economic Zone in Oman, which has its main concern toward utilities, tourism, security, smart port solutions, traffic lights, road lighting, smart building management, and waste management. The country is achieving digi-tal transformation by deploying ICT and other enablers of "technological progress." This is indicated by many initiatives of the Information Technology Authority (ITA), i.e., the introduction of smart government, "Digital Oman Strategy"; "e-transformation plans"; "e-payment gateway"; "National Unified Addressing System"; "e-Health portal"; "Educational portal"; "Open Data initiatives"; and "Oman National Spatial Data Infrastructure (ONSDI)."

Qatar had e-Government 2020 aims to upgrade the efficiency of services pro-vided by the government through several initiatives, including mobile applications, establishing digital applications and cloud infrastructure, etc. The Lusail Smart City is an example of a smart city format where advanced services are provided to resi-dents and visitors by tapping a high-technology environment in addition to handling crisis and disaster situations. As an indication, the building's centralized cooling system is established, an exterior smart pipe network that transmits waste to recy-cling plants is there, wastewater is reprocessed to dampen the city's green spaces, and a driverless and fully automated advanced metro is being erected in Doha to ease the 2020 World Cup activities. The metro transport network acts as a connec-tion nexus between Doha and Lusail Smart City. Smart city formation is expected to continue, given the country's national vision 2030. The National Vision, 2030 focuses on residents' QoL requirements, broader technological infrastructure, and how to achieve further progress toward having a "digital economy." Examples of smart city initiatives include the establishment of Qatar Digital Oasis and Msheireb Downtown Doha and Qatar Rail Development Programs. To accelerate the Qatar Smart Program, the country has manpower training programs that are associated with sophisticated software and hardware applications. The Qatar "smart city vision" is expected to improve people's standards of living and lifestyles by empowering businesses through the use of an integrated ICT infrastructure.

For Saudi Arabia, there is a deep desire to convert most Saudi cities to smart city format with high-quality services, sophisticated digital infrastructure adaptation, and a developing "digital economy" platform. This is one of the main objectives of the country's National Transformation Program and the country's Vision 2030, where many initiatives are considered, such as the digital traffic control system and applying new technologies to infrastructure. Smart city programs are applied to 17 cities in which 72% of the country's population lives. A very excellent example of a smart city is the city of Neom, with an estimated cost of $500 billion. Some of the looking forward smart projects that are under the concern of the Neom smart city are energy and water supply, travel, digital technology, food processing, biotechnology, advanced production, entertainment, tourism, education, healthcare, and robots. Digital transport systems, solar, and wind energy. Another example is the King Abdullah Economic City with its four areas of concern, i.e., home automation platforms, smart lighting, security, and intelligent energy consumption. The government is interested in having partnerships with the private sector to build up a "smart city" innovation facility that would provide the necessary training sessions to participants who are involved in designing and constructing smart city initiatives.

For the United Arab Emirates (UAE), there is no doubt that the UAE is leading the GCC region in terms of smart city initiatives and smart services. Dubai's smart government strategy is one of the key examples in this regard. The Smart Dubai initiative was launched in 2014, aiming to deliver and promote an efficient, seamless, safe, and impactful city experience for people. Dubai has an advanced e-government formation, a very highbrow network for solar energy and hybrid gas stations. Areas of concern as per the smart Dubai 2021 strategy include sustainability, advanced technological economy, digital accessibility of social services, digital transportation, clean environment, and ensuring collaboration with the private sector. The Smart Dubai Index is an example of the progress achieved toward assessing smart performance and attaining the economic diversification strategy of the country. Many artificial intelligence infrastructure projects have been developed to meet smart city needs. Dubai Silicon Oasis (DSO) is an example of a smart city application that aims to install a charging station for electric vehicles. Dubai Telecommunication is using its "Wi-Fi UAE" initiative to provide Wi-Fi services in public places across the country. The same can be said about Abu Dhabi city. Masdar City is an example of smart city formation, as indicated by several smart initiatives, such as solar energy systems, green building infrastructure, and digital applications. The Zayed smart city project, which was initiated in 2018, is an additional example of smart city formation in the UAE.

In conclusion, all GCC countries share comparable socioeconomic concerns expected from initiating smart cities in the region. They are addressing their urban population pressure issues and forming a knowledge-based economy to meet the progress of the digital world. Keeping in mind that there is variation among the GCC countries in terms of the level of development of digitalization and smart city development. Digital applications are being substantially applied to provide enhanced public services to people and boost urban initiatives in the region. However, it seems that the GCC concern is most likely focusing on IT-led and ICT smart city initiatives (e-government and e-services applications), while other smart city dimensions are not efficiently considered. Additionally, stakeholders (private sector and residents) are not clearly involved in smart city development, leading to doubt about ensuring sustainable results. Table 1 summarizes smart city initiatives in the GCC region.

Table 1 Smart city initiatives in the GCC region

Country	National strategy	National strategy	Initiatives and programs	Country vision focus
Bahrain	Bahrain Economic Vision 2030	National Development Strategy	National eGovernment Strategy	Infrastructure, innovation, quality of public services, business environment, and e-government networks
Kuwait	New Kuwait 2035	Kuwait National Development Plan	eGovernment Program	Advanced IT infrastructure, human capital, effective public services, sustainable diversified economy
Oman	Vision 2020; Vision 2040	National Program for Enhancing Economic Diversification	Oman	IT industry development, society, and e-government services
Qatar	National Vision 2030	National Development Strategy	Qatar e-Government 2020; Lusail Smart City Vision	Efficiency of public services, QoL requirements, empowering businesses, and digital economy
Saudi Arabia	Vision 2030	National Transformation Program	Smart Cities Program	Quality of public services, advanced digital infrastructure, and adopting a digital economy platform
United Arab Emirates (UAE)	UAE Vision 2021	UAE National Agenda 2021	National Innovation Strategy; Dubai Smart Government; Smart Dubai 2021; Abu Dhabi Economic Vision 2030	People's QoL standards, advanced e-government formation, energy, high tech-sustainable economy, digital transportation, and environment

Source: Asmyatullin et al. (2020), Saxena and Al-Tamimi (2018)

7 The Case of Kuwait

7.1 Smart Cities as Urban Strategic Concerns

Kuwait's interest in developing smart cities is clearly documented through the New Kuwait Vision 2035, which has seven areas of concern, including *"global positioning, developed infrastructure, creative human capital, effective public administration, high-quality health care, sustainable diversified economy, sustainable living environment"* (Saxena & Al-Tamimi, 2018). The overall vision aim is to recognize Kuwait as a regional financial, commercial and cultural center. As per Vision 2035, the government creates initiatives to modernize the collection and processing of data and support the development of an *information society*. Additionally, to date, many e-government services have been introduced, including the online transformation of public services for education, health, infrastructure, etc. Other initiatives include intellectual mobility (traffic management and infrastructure monitoring) and installing smart electricity meters and water meters that would have immediate data transformation.

7.2 Need for Smart Cities

Kuwait, like other GCC countries, is confronting rapid urban growth issues, mainly those related to traffic congestion and housing affordability. Alghais and Pullar (2017) identify that urban modeling would overcome urban growth issues. Otherwise, traffic congestion and housing availability issues will be intensified in the country. Other urban growth issues are also critical for the country, such as alternatives to public transportation, which currently is limited to busses, excessive car reliance usage, and government housing policies. (see Alshalfan, 2013; Al-Nakib, 2014; Dakkak, 2016 cited by Alghais and Pullar (2018)).

As per the literature, Alkandari and Alshailhi (2012) identify that there are three major issues under concern, namely, pollution (sea, groundwater, air, and soil), traffic, and slow on services. The issues can be resolved by having smart cities with smart infrastructure with information and communication technology (ICT). AlEnezi et al. (2018) shows the importance of the Internet of Things (IoT) as a significant application in smart cities, which can be viewed as a key strategy to stimulate industrialization, communication, and urbanization issues. Alghais and Pullar (2018) outline the key issues of urban growth in Kuwait and explore whether new cities under consideration would be a better alternative to overcome those issues. Their interviews with government officials and private sector representatives concluded that traffic congestion and housing shortages are the two main officially recognized crucial urban issues in Kuwait. The two issues are projected to be resolved by initiating new cities given the government's urban strategic plan, rather than extending already prevailing urban districts. The new cities are supposed to use smart applications to support transport efficiency, i.e., the train transport initiative.

Their survey results confirmed the negative impact of urban growth, as indicated by the earlier two issues, namely, traffic congestion and housing shortages. We also confirmed the necessity of stimulating new smart city initiatives as a relevant solution for urban problems in Kuwait.

In this regard, the country deliberates urban development and undertakes planning alterations by initiating urban cities. The objective is to overcome most, if not all, urban population pressures. Kuwait strategy 2035 emphasizes this objective, where the country aims to be cogitated as a regional financial, commercial, and cultural midpoint, as reported earlier.

7.3 Smart City Initiatives

The *Saad Al-Abdullah* smart city initiative is one part of the new Kuwait 2035 plan. The plan is a group of measures aiming to boost the economy over the forthcoming years by reducing economic reliance on oil as the sole source of public revenues. Additionally, the plan aims to signify Kuwait's promising economic future by 2035 by building the country as a hub for being a regional financial, trade, tourism, and commercial midpoint. This would motivate more local and regional investments (see Smart City Hub, 2017; Urban Gateway, 2018).

In 2019, the *Saad Al-Abdullah City* was introduced as a smart city initiative that cost $4 billion. The city is the first smart and environmentally friendly city in Kuwait (green and smart city). The city project covers 50 square kilometers and is expected to accommodate 400,000 people and apply the internet network and information and communication technologies (ICTs) as connecting mechanisms for its services and roads (Jazzar, 2019). The city design format is shown in Fig. 1.

Fig. 1 Saad Al-Abdullah city

The city structure merged between smart technologies and classic urbanization. The city is designed to avoid visual pollution and to be connected to electrical energy through solar cells. The use of specific building supplies is to be within the financial ability of the citizens to afford (METenders, 2022). For implementation purposes, the city project is being initiated in partnership with South Korea, where the city is planned to be designed as the South Korean city of Bundang (see AlEnezi et al., 2018; News.kuwaittimes.net, 2016; Alghais & Pullar, 2018). Emphasis is placed on intellectual mobility, which focuses on traffic management and infrastructure monitoring. Installing smart electricity and water meters that transmit data on time is an additional example of e-government services for 2018. Apart from *Saad Al-Abdullah*, eight more "smart cities" would be launched in due time.

8   Limitation Confronting Kuwait's Smart Cities Initiatives

As a crucial part of Kuwait Vision 2035, smart city initiation and transformation of Kuwait's cities into secure and digitalized sustainable cities in the future would require the following. First, significant coordination between all stakeholders, including the government, private sector, smart city solution providers, and engaging the community population. Second, provide stakeholders with more clarifications around the advantages of having smart cities and increase their awareness toward technology challenges and privacy concerns, digital security with coverage, and capacity. Third, introduce the relevant legislation, policies, funding sources, and business models to develop the existing infrastructure for water, energy, and transportation systems. Fourth, we show how the current smart city and the proposed eight future smart city initiatives would contribute positively to overcoming the challenges related to traffic and congestion, boosting economic development opportunities, enhancing public safety initiatives, achieving economic and environmental sustainability and transparency, and improving peoples' quality of life and standards of living. The abovementioned challenges, aside from others, should be addressed to benefit from smart cities' introduction to the community.

9   Conclusions

The growing urbanization rate across the world has initiated the need to present the smart city format with the intention of overcoming urban growth challenges. According to the urban modeling perspective, the smart city design would be more resilient, economically and environmentally sustainable, and more efficient in providing services that would lead to better quality of life (QoL) for residents. For the GCC region, the issue of the growing urbanization rate is not less important in comparison to the rest of the worldwide countries. As urbanization is expanding together with its confronts, the GCC countries are deeply interested in initiating smart cities and adopting digitalized economies. These initiatives, and alike, are now becoming vision-oriented initiatives for urban management processing in the GCC region.

Testing for the objectives of the current study is motivated by many attributes. Smart city initiatives are still in their go-forward stages in GCC countries, and the impact of those initiatives toward overcoming urban population pressures is not evident thus far. Additionally, smart cities are supposed to achieve integration between people, technology, and information, hence boosting the quality of life (QoL) of residents in terms of education, health, transportation, public and private services, public safety, and the like; however, this is not approved thus far as per the current status of the GCC countries.

Overall, the current study identifies differences among smart city concepts, while smart city structure can be indicated by some smart core constituents, various attributes, core themes, and infrastructure. To judge how smart is a city, the study concludes that there is variation among regions regarding what proxies can be applied to indicate the level of city smartness concern. For the GCC experience, including Kuwait, smart city initiatives are already well documented in their national urban strategic plans. For Kuwait, transforming cities into secure and digitalized sustainable cities would require overcoming significant limitations that may include coordination between all stakeholders, increasing stakeholders' awareness toward technology challenges and privacy concerns, digital security with coverage and capacity, introducing the relevant legislation, policies, funding sources, and business models to develop the existing infrastructure for water, energy, and transportation systems, etc., and showing how the current smart city and the proposed eight future smart city initiatives would contribute positively to overcoming the challenges related to traffic and congestion, boosting economic development opportunities, enhancing public safety initiatives, achieving economic and environmental sustainability and transparency, and improving peoples' quality of life and standards of living.

References

Abellá-García, A., Ortiz-de-Urbina-Criado, M., & De-Pablos-Heredero, C. (2015). The ecosystem of services around smart cities: An exploratory analysis. *Procedia Computer Science, 64*, 1075–1080.

Albino, V., Berardi, U., & Dangelico, R. (2015). Smart cities: Definitions, dimensions, performance, and initiatives (pp. 3-21). | Published online: 04 Feb 2015. Retrieved from https://doi.org/10.1080/10630732.2014.942092

AlEnezi, A., AlMeraj, Z., & Manuel, P. (2018). Challenges of IoT based new generation smart-government. *Journal of Informatics and Mathematical Sciences, 10*(3), 53–544.

Alghais, N., & Pullar, D. (2017). Modeling future impacts of urban development in Kuwait with the use of ABM and GIS. *Transportation, GIS, 20*(20), 20–42.

Alghais, N., & Pullar, D. (2018). Projection for new city future scenarios: A case study for Kuwait. *Heliyon, 4*, e00590. https://doi.org/10.1016/j.heliyon.2018.e00590

Alkandari, A., & Alshaikhli, I. (2012). Designing a smart city in Kuwait: An initial study. *Journal of Advanced Computer Science and Technology Research, 2*(3), 116–126.

Al-Nakib, F. (2014). Revisiting Hadar and Badu in Kuwait: Citizenship, housing and the construction of a dichotomy. *International Journal of Middle East Studies, 46*, 5–30.

Alshalfan, S. (2013). *The right to housing in Kuwait: An urban injustice in a socially just system*. London School of Economics and Political Science. Available: https://eprints.lse.ac.uk/id/eprint/55012

Al-Thani, S., Skelhorn, C., Amato, A., Koc, M., & Al-Ghamdi, S. (2018). Smart technology impact on neighborhood form for a sustainable Doha. *Sustainability, 10*, 4764.

Asmyatullin, R., Tyrkba, K., & Ruzina, E. (2020). Smart cities in GCC: Comparative study of economic dimension. *IOP Conference Series: Earth and Environmental Science, 459*, 062045. https://doi.org/10.1088/1755-1315/459/6/062045. International Science and Technology Conference "EarthScience". IOP Publishing.

Balta-Ozkan, N., Boteler, B., & Amerighi, O. (2014). European smart home market development: Public views on technical and economic aspects across the United Kingdom, Germany and Italy. *Energy Research & Social Science, 3*, 65–77.

Barth, J., Fietkiewicz, K., Gremm, J., & Hartmann, S. (2017). Informational urbanism. A conceptual framework of smart cities. In *Proceedings of the 50th Hawaii international conference on system sciences*. http://hdl.handle.net/10125/41496. ISBN: 978-0-9981331-0-2 CC-BY-NC-ND.

Berg, J. (2017). Sharing cities: A case for truly smart and sustainable cities. *New Political Science, 39*, 417–419.

Bufetova, A. (2013). Structure of urbanization and trends in intraregional differentiation of standard of living. *World of Economics and Management, 13*, 57–66.

Carillo, J., Yigitcanlar, T., García, B., & Lönnqvist, A. (2014). Knowledge and the city. In *Concepts, applications and trends of knowledge-based urban development*. Routledge.

Charfeddine, L., Al-Malk, A., & Al Korbi, K. (2018). Is it possible to improve environmental quality without reducing economic growth: Evidence from the Qatar economy? *Renewable and Sustainable Energy Reviews, 82*, 25–39.

Chauhan, S., Agarwal, N., & Kar, A. (2016). Addressing big data challenges in smart cities: A systematic literature review. *Info, 18*(4), 73–90.

Cheu, R., Mondragon, O., Nazarian, S., Carrasco, C., Gates, A., Cabrera, S., Villanueva-Rosales, N., Ferrugut, C., Taboada Jimenez, H., & Balal, E. (2015). *Research challenges toward the implementation of smart cities in the United States*. Report No. CAIT-UTC-060 (pp. 1–66). Center for Advanced Infrastructure and Transportation, Rutgers, The State University of New Jersey.

Chourabi, H., Nam, T., Walker, S., & Ramon Gil-Garcia, I. (2012). Understanding smart cities: An integrative framework. *The 45th Hawaii International Conference on System Sciences - 2012*. https://doi.org/10.1109/HICSS.2012.615

Cohen, B. (2012). *What exactly is a smart city?* Available on: www.fastcoexist.com/1680538/whatexactly-is-a-smart-city. Accessed 15 Dec 2017.

Constro Facilitator. (2021). *Smart city-elements, features, technology and government approach*. Available: https://www.constrofacilitator.com

Dakkak, N. (2016). *Why you should explore Hawalli in Kuwait* [online]. The Culture Trip Ltd. Available: https://theculturetrip.com/middle-east/Kuwait

de Wijs, L., Witte, P., & Geertman, S. (2016). How smart is smart? Theoretical and empirical considerations on implementing smart city objectives-A case study of Dutch railway station areas, innovation. *The European Journal of Social Science Research, 29*(4), 424–441.

European Commission. (2021). *Smart cities*. Available online: https://ec.europa.eu/info/eu-regional-and-urban-development/topics/cities-and-urban-development/city-initiatives/smart-cities_en

Fernandez-Guell, J., Collado-Lara, M., Guzman-Arana, S., & Fernandez-Anez, V. (2016). Incorporating a systematic and foresight approach into smart city initiatives: The case of Spanish cities. *Journal of Urban Technology, 23*(3), 43–67.

Foth, M., Choi, J., & Satchell, C. (2011). Urban informatics. In *Proceedings of the ACM 2011 conference on computer supported work* (pp. 1–8). ACM.

Giffinger, R., Fertner, C., Kramar, H., Kalasek, R., Pichler-Milanovic, N., & Meijers, E. (2007). *Smart cities – Ranking of European medium-sized cities.* Centre of Regional Science.

Griffiths, S. (2017). A review and assessment of energy policy in the Middle East and North Africa region. *Energy Policy, 102,* 249–269.

Hall, R., Bowerman, B., Braverman, J. Taylor, J., Todosow, H., & Wimmersperg, U. (2000, September 28). The vision of a smart city. In *2nd international life extension technology workshop*, Paris, France.

Halleux, V. (2016). *Briefing: Energy Union: The regional and local dimension PE 568.356.* 2015. Available online: https://www.europarl.europa.eu/RegData/etudes/BRIE/2015/568356/EPRS_BRI(2015)568356_EN.pdf

Harrison, C., & Donnelly, I. (2011). A theory of smart cities. *Proceedings of the 55th Annual Meeting of the ISSS - 2011, Hull, UK, 55*(1). Retrieved from https://journals.isss.org/index.php/proceedings55th/article/view/1703

Jazzar, M. (2019). Smart Clean City – Kuwait. *International Research Journal of Innovations in Engineering and Technology (IRJIET), 3*(9), 1–7. ISSN (online): 2581-3048.

Jedlin'ski, M. (2014). The position of green logistics in sustainable development of a smart green city. *Procedia – Social and Behavioral Sciences, 151,* 102–111.

Letnik, T., Marksel, M., Luppino, G., Bardi, A., & Božicʼnik, S. (2018). Review of policies and measures for sustainable and energy efficient urban transport. *Energy, 163,* 245–257.

Madanipour, A. (2011). *Knowledge economy and the city. Spaces of knowledge.* Routledge.

Martin, C., Evans, J., & Karvonen, A. (2018). Smart and sustainable? Five tensions in the visions and practices of the smart-sustainable city in Europe and North America. *Technological Forecasting and Social Change, 133,* 269–278.

Maymir-Durcharme, F., & Angelelli, L. (2014). The smarter planet: Built on informatics and cybernetics. *Journal of Systemics, Cybernetics and Informatics, 12*(5), 49–54.

Meijer, A., & Bolívar, M. (2016). Governing the smart city: A review of the literature on smart urban governance. *International Review of Administrative Sciences (IRAS), 82,* 392–408.

Mercer LLC. (2014). *Quality of Living Worldwide City Rankings – Mercer Survey.* Montreal Canada. See http://www.mercer.com/content/mercer/global/all/en/newsroom/2014-quality-of-living-survey.html

METenders (2022). *Middle East Project Intelligence & Tenders.* METenders.com

Mohanty, S. (2016, July). Everything you wanted to know about smart cities. *IEEE Consumer Electronics Magazine.* https://doi.org/10.1109/MCE.2016.2556879

Monfaredzadeh, T., & Berardi, U. (2015). Beneath the smart city: Dichotomy between sustainability and competitiveness. *International Journal of Sustainable Building Technology and Urban Development, 6*(3), 14–156.

Monfaredzadeh, T., & Krueger, R. (2015). Investigating social factors of sustainability in a smart city. *Procedia Engineering, 118,* 1112–1118.

Muente-Kunigami, A., & Mulas, V. (2019). *Smart cities.* Available online: https://www.worldbank.org/en/topic/digitaldevelopment/brief/smart-cities

News.kuwaittimes.net. (2016). *South Saad AlAbdullah, Kuwait's first smart city* [online] at http://news.kuwaittimes.net/website/south-saad-alabdullah-kuwaits-first-smart-city

Přibyl, O., & Horák, T. (2015). Individual perception of smart city strategies. *Smart Cities Symposium Prague* (SCSP). https://doi.org/10.1109/SCSP.2015.7181550

Ringel, M. (2021). Smart city design differences: Insights from decision-makers in Germany and the Middle East/North-Africa region. *Sustainability, 13,* 2143. https://doi.org/10.3390/su13042143

Saxena, S., & Al-Tamimi, T. (2018). *Visioning "smart city" across the Gulf cooperation council (GCC) countries* (Vol. 20, No. 3, pp. 237–251). Emerald Publishing Limited. ISSN1463-6689 jFORESIGHTjPAGE237.

Sicilia, Á., Madrazo, L., Massetti, M., Plazas, L., & Ortet, E. (2017). An energy information system for retrofitting smart urban areas. *Energy Procedia, 136*, 85–90.

Smart City Hub. (2017). *South Korea exports smart city concept into Kuwait*. Available: https://Smartcityhub.com

Tok, E., Mohammad, F., & Merekhi, M. (2015). Crafting smart cities in the gulf region: A comparison of Masdar and Lusail. *European Science Journal, 2*, 130–140. Available: https://eujournal.org/index.php/esj/article/view/3702

Tranos, E., & Gertner, D. (2012). Smart networked cities? Innovation. *The European Journal of Social Science Research, 25*(2), 175–190.

United Nations DESA. (2018). *World urbanization prospects 2018*. UN.

Urban Gateway. (2018). *Transforming Kuwait's cities into secure & digitalized sustainable cities*. Available on: https://urbangateway.org/event/transforming-kuwait%E2%80%99s-cities-secure-digitalized-sustainable-cities

Open Access This chapter is licensed under the terms of the Creative Commons Attribution 4.0 International License (http://creativecommons.org/licenses/by/4.0/), which permits use, sharing, adaptation, distribution and reproduction in any medium or format, as long as you give appropriate credit to the original author(s) and the source, provide a link to the Creative Commons license and indicate if changes were made.

The images or other third party material in this chapter are included in the chapter's Creative Commons license, unless indicated otherwise in a credit line to the material. If material is not included in the chapter's Creative Commons license and your intended use is not permitted by statutory regulation or exceeds the permitted use, you will need to obtain permission directly from the copyright holder.

Smart Cities from an Indian Perspective: Evolving Ambitions

Yagyavalk Bhatt ⓘ and Jitendra Roychoudhury ⓘ

Abstract Indian urban infrastructure is in the middle of a massive build-up. For several decades, Indian cities were caught between the demands of an exploding population and the need to provide infrastructure in terms of health, education, transport, and services to meet the core needs of the citizenry. With increasing prosperity and leveraging the strengths of one of the world's largest and fast-growing economies, Indian policymakers seek to correct their previous underinvestment in city infrastructure. Smart City Mission, Swachh Bharat Mission, and Atal Mission for Rejuvenation and Urban Transformation are some of the policy vehicles planned for a broad transformation of India's urban agenda. This chapter highlights some of the policy initiatives focused on meeting the objectives of Smart Cities. These policy initiatives are expected to help address the current service delivery gap from an urban infrastructure standpoint. By incorporating technology, improving digital access, innovations in traffic management, investments in mobility solutions, and ensuring that the heritage of the Indian cities is maintained, India seeks to answer the challenge of urbanization of millions. These policy mechanisms and the lessons from their successes and failures constitute this chapter's core. Indian urban infrastructure developments are unique globally, primarily because of the varying range of urbanization across the country, the contextual rationale and the evolving aspirations of the policymakers, and the devolution of developmental powers to local bodies. The Indian experience of implementing the Smart Cities objectives would be unique in the world, given the scale of the massive investments and the millions of citizens whose lives are impacted.

Keywords India · Energy · Smart cities · Urbanization · Sustainability · Innovation · Transport · Climate adaptation

Y. Bhatt (✉) · J. Roychoudhury
King Abdullah Petroleum Studies and Research Center, Riyadh, Kingdom of Saudi Arabia
e-mail: yagyavalk.bhatt@kapsarc.org; jitendra.roychoudhury@kapsarc.org

© The Author(s) 2024
F. Belaïd, A. Arora (eds.), *Smart Cities*, Studies in Energy, Resource and Environmental Economics, https://doi.org/10.1007/978-3-031-35664-3_19

1 Introduction

Most Indian cities have evolved over hundreds of years and carry with them obsolete city planning practices, making it difficult for them to measure up to the development standards of modern cities of the twenty-first century. Hence, it is not surprising to see that Indian cities struggle with modern metrics of civic service delivery and overall ease of living. Most Indian cities struggle with delivering the basics that a major functioning city of any developed world takes for granted (The World Bank, 2011). There are many reasons for this, and this chapter tries to answer the malaise that afflicts Indian cities. The demand for urban spaces has also shifted as the economy has changed. Cities have struggled to answer this evolution of demand and hence have stagnated without political direction and accountability. Policy developments such as the Smart City Mission (SCM) in India are a belated reaction to this malaise. SCM is a developing mechanism used by policymakers to correct decades of neglect and apathy. Hence, from an Indian perspective, the SCM is more of a policy to absorb a civic approach in policymakers where they reach out to citizens and try to develop projects that meet residents' requirements.

For policymakers, the process of urbanization in itself has been an enormous challenge. As cities become the engines of economic growth, it becomes even more critical to answer the challenge posed by delivering the essential needs of a predominantly young, large, and growing population. Provision of drinking water, sanitation, functioning utilities and transport systems, waste collection and management, security, health, and education are the core services that make up civic delivery. These services have either faltered or been marked by their absence, as the forces of urbanization have overwhelmed the capacity and capability of civic authorities.

The SCM, launched by India in 2015, seeks to deliver these core infrastructure services in an urban setting and give citizens a decent quality of life. The mission does not strive to provide a single definition of a smart city and thus devolves it to the cities to define it as they would like to, given the constraint on resources that a developing economy with a burgeoning population faces. It also seeks to enable the private sector to be a part of urban growth and use the mission to build capacity and capability in local urban bodies. The aims and aspirations of the SCM are thus slightly different from the general understanding that a reader might have of a technologically advanced urban space. The SCM in India is more of a mechanism to ensure that modern digital infrastructure is used to help enable faster decision-making and provide transparency and services to its citizens.

This chapter examines the potential of smart cities from an Indian perspective, focusing on the drivers of their evolution in India. The chapter is split into five sections, with the first section elaborating on how India has approached the challenge of urbanization and the second section on the timeline of urbanization management initiatives ending with the SCM. The third section then focuses on the challenge of urban mobility solutions. The fourth section then progresses to specific cities as exemplars of the application of the SCM. It is then followed by the final section highlighting the challenges, insights, and opportunities faced with SCM implementation. The chapter concludes by discussing the ramifications of the SCM. We offer policy recommendations toward progressing the SCM, understanding that this is a challenging journey on which India has just started.

2 Smart City Initiative's Paradigm in India: Drivers of Its Evolution

> India needs to build a minimum of 500 new cities urgently. However, it has to be done from scratch to accommodate people who are on the move and to provide them better quality of life. – C. K. Prahalad (Press Trust of India, 2009).

With a population of almost 1.4 billion and a US\$ 3.2 trillion economy (2022), India is among the fastest-growing economies of its size globally. It aspires to grow to a US\$ 40 trillion economy by 2047, when India's population is expected to be almost 1.65 billion (Niti Aayog and Asian Development Bank, 2022). With a median age of nearly 29 years, India is also among the youngest countries globally (Nanda & Sharma, 2020). India's population demographics are primarily rural, with approximately 65% living in rural areas and 35% living in approximately 7900 towns (Jha, 2020). India's urbanization, however, has been increasing exponentially over the past decades. In 2001, approximately 290 million people lived in cities. By 2020, that figure had grown to more than 460 million, with expectations that it will reach nearly 814 million by 2050, the highest concentration of urban population globally (TERI, 2022). To cater to the increase in demand for urbanization and population growth, India must build new cities and simultaneously refurbish its current towns to be able to meet the aspirations and expectations of its population.

India currently has 42 cities with a population of more than one million, and the expectation is that by 2030, the number of such cities will grow to more than 68 cities (TERI, 2022). Several of India's current cities will turn into megacities, with New Delhi expected to become the most populous city globally (United Nations, 2022). This rapid rate of urbanization will have significant implications for policymakers, who will need to serve these citizens' requirements, including housing, transportation, energy, health, education, and municipal services, among many other needs.

2.1 Transformation of the Indian Economy

The Indian economy has also been transforming over the past decades. The share of the services sector's Gross Domestic Product (GDP) contribution has increased over the decades with a commensurate fall in the percentage share of agriculture, shadowing the trend of urbanization seen in India. Services contributed 44.3% of

GDP in the 1990s, which grew to 53.9% in 2021. In comparison, agriculture has seen a fall from 28.6% in the 1990s to approximately 20% (Reserve Bank of India, 2002) (Ministry of Finance, Government of India, 2022). This shift in economic growth is critical for India to pull its people out of poverty and increase its per capita GDP. It is interesting to note that Indian cities occupy only 3% of the land area while contributing almost 60% to the country's GDP, reflecting the increasing economic importance of the cities to India's financial future (Niti Aayog and Asian Development Bank, 2022). Large Indian cities have been able to leverage their participation in the global economy through increased connectivity, both physically (as seen in the increase in the number and size of international airports in India) and virtually due to the improvement in communication technology, thus becoming gateways of globalization.

The growth of the services sector, the increase in urbanization, and the change in the economic power of cities are also reflective of the critical importance of the Information Technology (IT) revolution for India. The increase in IT and IT-enabled services over the past two decades in India has been mostly an urban phenomenon and has enabled India to become a powerhouse globally, with annual revenues of almost US\$ 200 billion in 2021 (Ministry of Finance, Government of India, 2022). This massive growth in the services economy has fueled multiple secondary and tertiary sectors, with rising demand for commercial offices, employee housing, entertainment venues, and hospitality services. This demand, in turn, has acted as a magnet for a rural population seeking to earn an income and, as a result, has grown into a virtuous cycle where one sector feeds others, leading to an increase in economic activity, which attracts even more rural migrants to the urban centers. Indian cities have, thus, increasingly become a center of demographic and economic transformation.

2.2 Economic Migration from Rural India

However, the negative impact of this rapid urbanization has been increasingly felt in cities, where it has emerged as a critical policy and governance challenge. As cities increasingly provide a pathway out of poverty and into prosperity, mass urbanization has become a curse for them. Indian cities have generally been poorly administered and even poorly governed. The increasing pressure of economic migration has resulted in the growth of informal housing, unplanned expansions, traffic congestion, and the deterioration of law and order. The increased migration has also increased pressure on the city's civic infrastructure, leading to a severe decline in service quality. The increased migration has also increased inequality as the rural poor have opted to migrate to urban areas in search of economic opportunities.

The sudden and dramatic growth in urbanization has meant that it has often been with limited or no urban planning, with cities growing organically and haphazardly and with little oversight in terms of civic resources and services, leading to a demand and supply gap in terms of urban services. With the future growth of the urban population in mind, policymakers and city administrators have come to a belated understanding of the complex factors driving urbanization. To understand the uniquely Indian urbanization phenomenon, it is vital to significantly place it in a contextual setting from its political leadership.

2.3   Slow Neglect of Civic Institutional Capacities

> I regard the growth of cities as an evil thing, unfortunate for mankind and the world, unfortunate for England and certainly unfortunate for India. The British have exploited India through its cities. The latter have exploited the villages. The blood of the villages is the cement with which the edifice of the cities is built. – M.K. Gandhi (Jha, 2020)

Post-independence, India built several planned cities, including New Delhi, Chandigarh, Navi Mumbai, Kalyani, Noida, Gandhinagar, Jamshedpur, and Bhubaneshwar. These postindependence built cities were identified by austere architectural conformity reflecting the socialistic political philosophy that prevailed at the time of independence (Jha, 2020). Mostly, the city planning approach was divorced from the contextual setting of the city's location and reflected the top-down thought process that focused on conformity, diluting the local cultural, ethnic, and diverse ethos, which, for example, primarily defines most of the centuries-old cities in India, such as Kashi (Varanasi), Ujjain, Orchha, and Mysore (Pandey, 2020). These new cities were planned for modest populations, ranging from tens of thousands (Chandigarh) to hundreds of thousands (New Delhi). Rapid urbanization has meant that these cities today play host to millions, straining capacities and generating incentives for rent-seeking from the political class (Bhatia, 2018). At the same time, cities have continued to function under archaic laws and administrative structures, which are ill-equipped to deal with the emerging issues that haphazard urbanization has brought forth.

Policymakers and stakeholders have long struggled with the challenges of urban infrastructure development, given the substantial political economy involved and the lack of interest in city development. Indeed, most policymakers and stakeholders have only considered the real estate aspect of urban development, as it tends to help feed the political economy. India has had cities that have evolved over the centuries. Haphazard modernization and urbanization have impacted these cities as they expanded organically to cater to increased housing demand from urbanization pressures (Brookings India and Brookings Institution, 2016). Increased energy demand and pollution have exacerbated the urbanization pressures on these cities.

2.4   Rise of an Aspirational Youth Demographic

With a substantial share of the young population (65% of India's population is below the age of 35) and increasing prosperity, Indian energy consumption is rising consistently, with massive cities feeding that demand (Our World in Data, 2021). Energy use in India has almost doubled since 2000 (Fig. 1), with 80% of energy demand coming from coal, oil, and biomass; however, India's energy use and emissions per capita are currently less than half the world's average (International Energy Agency, 2021). To manage the ever-growing urban energy usage and improve the quality of life, the Indian government's key priorities have recently switched to urban planning through policy intervention to cater to sustainable infrastructure and intelligent information technology.

As the urban population increases, energy-efficient technologies will be needed in all sectors, more importantly in the building, transport, water, and waste management sectors (Brookings India and Brookings Institution, 2016). Furthermore, India's commitment to the Paris Agreement and its net-zero targets made the task even more challenging. Now, urban development should cater to the sustainability aspects of infrastructure development, considering emissions.

Cities in India need urgent incentives to revitalize their administrative structures and capital to invest in solely needed civic infrastructure. As the Indian economy transforms into one where the share of the services sector is expected to continue to increase, with a falling share from agriculture, migration toward cities would continue to grow exponentially. This is where policymakers have belatedly tried to focus their attention on policy initiatives focused on understanding how urbanization is happening from a developmental viewpoint and ensuring that the development connects to the existing civic infrastructure. The intention behind these policy developments is to involve the local city administration much more granularly by

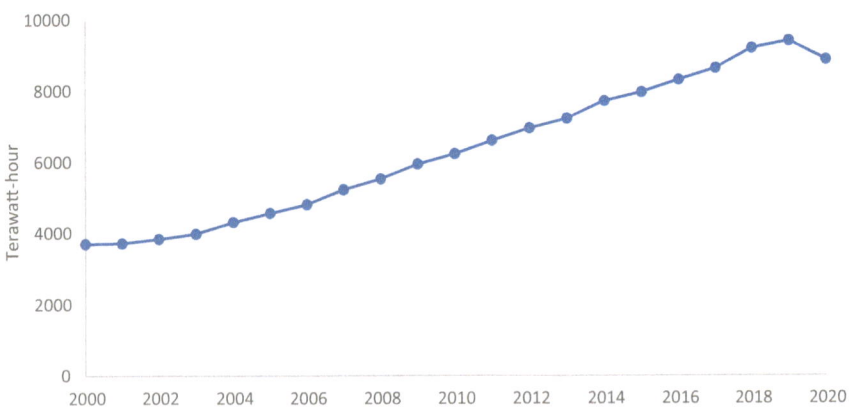

Source: Our World in Data based on BP and Shift Data Portal

Fig. 1 India's primary energy consumption. (Source: Our World in Data based on BP and Shift Data Portal)

focusing on bottom-up, citizen-based initiatives while simultaneously providing an opportunity for the private sector to be involved in these developments. The involvement of the private sector is critical given the quantum of investments needed.

2.5 Cities as Engines of Economic Growth

Urbanization in India has exploded over the past few decades, resulting in a massive infrastructure build ramping up. For several decades, postindependence in 1947, Indian administrators and policymakers were caught between the overall demands of a growing, predominantly rural population and the need to provide civic amenities and infrastructure in terms of health, education, transport, and services to meet the core needs of the city residents (NITI Aayog, 2021). Decision-making during that period was very hierarchical, reflecting the socialistic political milieu that existed at the time. India is predominantly rural, and policymakers have focused on the needs of the vast rural majority, with cities suffering from benign neglect. Recently, with increasing prosperity and leveraging the strengths of one of the world's largest and fast-growing economies, Indian policymakers now seek to correct their previous underinvestment in city infrastructure. SCM, Swachh Bharat Mission, Atal Mission for Rejuvenation and Urban Transformation (AMRUT), and Housing for All are some of the policy vehicles planned for a broad transformation of India's urban agenda. This chapter is intended to provide an overview and highlight some of the policy initiatives focused on meeting the objectives of smart cities. These policy initiatives are expected to help address the current service delivery gap from an urban infrastructure standpoint. By incorporating technology, improving digital access, innovations in traffic management, investments in mobility solutions, and ensuring that the cultural heritage of Indian cities is maintained, India seeks to answer the challenge of urbanization. These policy mechanisms and the lessons from their successes and failures constitute this chapter's core. Indian urban infrastructure developments are globally unique, primarily because of the varying degree of urbanization across the country, the contextual rationale and the evolving aspirations of policymakers, and the devolution of developmental powers to local bodies. Urbanization is an increasingly emerging market phenomenon. The developed world has urbanized and is now facing decaying and stagnant urbanization. The Indian experience of implementing the Smart Cities objectives would be unique in the world, given the scale of massive investments and the millions of citizens whose lives are impacted. The lessons learned from this uniquely Indian journey are an essential topic of inquiry for researchers worldwide.

3 Smart Cities Concept in India

Indian policymakers have generally failed to understand the urbanization process and the economic forces that drive the development of urban bodies and aid in forming cities. This is not surprising, as the administrators of cities are often political appointees, who seldom are from the cities themselves and often lack an understanding of the drivers shaping their city's development. Adding to this lack of experience has been a top-down focus of policymaking, leading to a mismatch between service delivery and acceptability levels. This section focuses on the policy development pathway leading to the SCM.

3.1 Government Policies and Objectives

India developed the Integrated Development of Small and Medium Towns (IDSMT) program in 1979, which continued until 2005, when it was subsumed into the Urban Infrastructure Development Scheme for Small and Medium Towns (UIDSSMT). The IDSMT program was meant to improve the infrastructure in towns and small urban centers, reduce migration pressure on larger cities and increase economic opportunities by creating public assets in small and medium towns (Town and Country Planning Organisation, 2022).

India passed the 74th Constitutional Amendment Act in 1992, which enabled the formal recognition of urban local bodies (Ministry of Housing and Urban Affairs, 1992). This act was meant to empower citizenry at the local urban body level, transform how the government engaged with citizens, and help shape the way public service was delivered (National Institute of Urban Affairs, 2016). However, since its enactment, the government has shown marginal progress in formulating supportive policies and initiatives to enhance city life.

The Scheme of Infrastructural Developments in Mega Cities was launched in 1993–94, aiming to ensure infrastructural developments in Mumbai, Kolkata, Chennai, Bangalore, and Hyderabad. These infrastructural investments focused primarily on water supply and sanitation, roads and bridges, city transport, and waste management (MoHUA, 1994). This program was subsumed into the Jawaharlal Nehru National Urban Renewal Mission (JNNURM) and formally closed in 2007.

The current smart city policies by policymakers are mainly to correct their original laggardness in urban policymaking space and ensure that the infrastructure invested can sustain future urbanization (Brookings India and Brookings Institution, 2016). India has been trying to correct its approach to urbanization through multiple approaches. These iterations continue to play out, with little effect on the ground, given the massive pressures urbanization creates through the vast numbers involved.

3.2 Jawaharlal Nehru National Urban Renewal Mission (JNNURM)

In 2005, India recognized the need for an urban development policy by launching JNNURM. It was India's most ambitious urban initiative to encourage reforms and fast-track the development of cities (Ministry of Housing and Urban Affairs, 2005). The JNNURM had two submissions, Urban Infrastructure and Governance (UIG) and Basic Services for the Urban Poor (BSUP) for large cities, and two subschemes, Urban Infrastructure Development Scheme for Small and Medium Towns (UIDSSMT) and Integrated Housing and Slum Development Program (IHSDP) for small and medium towns. To facilitate the mission and ensure improvement in urban governance and service delivery, India under the JNNURM mandated thirteen reforms, six for urban local bodies (ULBs) and seven for state governments. Furthermore, ten optional reforms were also created, which were expected to be undertaken by ULBs, state governments, and other relevant government authorities (Ministry of Housing and Urban Affairs, 2005).

JNNURM's primary objective was to create economically productive, efficient, equitable, and responsive cities. Furthermore, the program was designed to integrate the development of infrastructure services through the creation of sustainability projects by accelerating the flow of investment (Ministry of Housing and Urban Affairs, 2005). In its plan, JNNURM covered more than 65 cities across India for 7 years (2005–2012), with 367 and 213 projects approved under UIG and BSUP, respectively, with total financial assistance of approximately US$ 14.52 billion.[1] Furthermore, under the two subschemes, a total of 62 projects were approved with comprehensive financial assistance of nearly US$ 0.58 billion. However, most of the projects were not completed during the initial proposed 7 years of the mission due to a lack of competent bidders for the projects, project clearance delays, land acquisition, and delays in shifting utilities and pipelines (Grant Thornton, 2011). Therefore, the scheme was extended by 2 years to facilitate project completion (up to 2014).

Under the National Action Plan on Climate Change (NAPCC) that was launched in 2008, India established eight national missions identifying and focusing on multiple priorities as it tried to advance action on its climate ambitions (Ministry of Finance, Government of India, 2022). The National Mission on Sustainable Habitat (NMSH) was one of those eight missions. The objective of the NMSH was to develop sustainable habitat standards, promote energy efficiency as a part of the urban planning standards, strengthen the automotive fuel economy standards, and encourage and incentivize both the purchase of fuel-efficient vehicles and the use of public transportation. The NMSH, in turn, is being implemented through the Atal Mission on Rejuvenation and Urban Transformation (AMRUT), Swachh Bharat Mission (SBM), and Smart Cities Mission (SCM).

---

[1] Average US$ to INR rates were taken for the year 2005 as it was launch month of the initiative$.

3.3   Atal Mission for Rejuvenation and Urban Transformation (AMRUT)

In 2014, the newly elected government discontinued JNNURM and launched the Atal Mission for Rejuvenation and Urban Transformation (AMRUT) in 2015. AMRUT was comparable to JNNURM in its objective; however, the scale of ambition and scope of AMRUT was far more expansive. The AMRUT mission covered more than 500 cities with a total financial outlay of approximately US$ 7.6 billion.[2] Furthermore, the central government also limited its grant allocation to 33% of total project costs for cities with a million-plus population and 50% for cities with less than a million population. Additionally, to accelerate implementation and correct the policy issues that hindered JNNURM, the AMRUT mission enabled state governments to sanction projects under their jurisdiction. One of the significant moves under AMRUT was to task ULBs with overseeing individual projects with the involvement of third-party agencies and authorities to identify a private investor for civic services and infrastructure development (Ministry of Housing and Urban Affairs, 2015a). In addition, in 2021, India launched AMRUT 2.0, "Making Cities Water Secure," with a financial outlay of approximately US$ 40 billion[3] over 5 years. The AMRUT 2.0 guidelines have been formulated to assist Indian states and Union Territories (UT) in making Indian cities secure by adopting the principles of circular economy, promoting conservation, and rejuvenation of surface and groundwater bodies (Minister of Housing and Urban Affairs, 2021). As of June 2022, AMRUT has completed 4335 projects with a total financial outlay of approximately US$ 4.5 billion and awarded 1522 new projects (Minister of Housing and Urban Affairs, Government of India, 2021).

3.4   National Heritage City Development and Augmentation Yojana (HRIDAY)

In 2015, along with AMRUT, India also launched National Heritage City Development and Augmentation Yojana (HRIDAY). This scheme aims to undertake strategic and planned development of heritage cities to improve the overall quality of life, covering 12 heritage cities: Ajmer (Rajasthan), Amaravati (Andhra Pradesh), Amritsar (Punjab), Badami (Karnataka), Dwarka (Gujarat), Gaya (Bihar), Kancheepuram and Velankanni (Tamil Nadu), Mathura and Varanasi (Uttar Pradesh), Puri (Odisha), and Warangal (Telangana). Under this scheme, the total financial outlay was covered by the Central Government (Ministry of Housing and Urban Affairs, 2015a, b, c, d). As of 2021, HRIDAY has completed 77 projects with

---

[2] Average US$ to INR rates were taken for June 2015 as it was launch month of the initiative.

[3] Average US$ to INR rates were taken for October 2021 as it was launch month of the initiative.

a total financial outlay of approximately US$ 60 million (National Heritage City Development and Augmentation Yojana, 2022).

3.5 Smart Cities Mission (SCM)

In 2015, the Ministry of Housing and Urban Affairs (MoHUA),[4]India, launched India's SCM and its guidelines, along with other initiatives such as AMRUT and HRIDAY (MoHUA, 2015). With the mission launch, India joined the trend in developing smart cities with other countries. The primary objective of the SCM is to promote cities that can provide their citizens with the basic civic infrastructure (water supply, electricity, sanitation, urban mobility, transport solutions, housing, digital connectivity, education, health, governance, safety, and security) along with a decent quality of life, a sustainable environment and enabling application of intelligent solutions. In the mission, India proposed building and developing at least 100 smart cities across the country over 5 years. India adopted a three-way strategy of city improvement through retrofitting, redevelopment, and city extension through greenfield expansion by investing in innovative, smart technologies to develop sustainable urban hubs.

The primary objective that the SCM seeks to achieve is to build capacity at a local level and ensure that the planning and management of the city are devolved to the city itself. This would ensure that while there is coordination between the federal and state institutions in terms of project management and investment mechanisms, the city's development planning is bottom-up with the engagement of citizens and their specific development needs (Fig. 2).

The plan was to redevelop and restore the existing cities, which included the slum area, and develop a liveable space by improving the overall quality of life. Furthermore, it enables municipalities to improve the framework and services by incorporating technology and data solutions to boost the socioeconomic growth of the city, state, and country. To provide much-needed thrust to the mission, India will provide approximately US$ 7.6 billion[5] in financial assistance over the 5 years of the mission announcement (Ministry of Housing and Urban Affairs, 2015a). Furthermore, matched financial assistance will also be provided by state governments under which smart cities were allocated. The main objectives of the smart city plan were to promote mixed land use, expand housing opportunities, promote transit-oriented development, and create sustainable and green urban cores (Ministry of Housing and Urban Affairs, 2015b).

India has also launched other initiatives to fast-track and support SCM development. The India Urban Data Exchange (IUDX), an open-source platform, is one of

---

[4]The government of India merged the urban development and housing and urban poverty alleviation ministries to a single ministry to Ministry of Housing and Urban Affairs in 2017.

[5]Average US$ to INR rates were taken for June 2015 as it was launch month of the initiative.

Smart Cities Mission Strategy

Redevelopment Project	Retrofitting project	Greenfield project
• City renewal • Old slum areas • Core city areas • Area-50 acres (min)	• City improvement • Local area planning • 24x7 water supply • Area - 500 acres (min)	• City expansion • satellite towns • integrated township • Area-250 acres (min)

City wide smart solutions
- Electronic service delivery
- Intelligent traffic management systems
- Smart metering and management
- Integrated multimodal transport systems
- Video crime monitoring

Source: Ministry of Housing and Urban Affairs, India

Fig. 2 Smart city mission strategies. (Source: Ministry of Housing and Urban Affairs, India)

the transformative initiatives that enable the exchange of data between city departments, government agencies, citizens, and the private sector to address complex urban challenges (Ministry of Housing and Urban Affairs, 2021). Another major initiative introduced to support smart cities is "Digital India," a flagship program of the Government of India with a vision to transform India into a digitally empowered society and knowledge economy (Government of India, 2015) (Fig. 3).

Under the SCM, India has identified approximately 5151 projects with an investment of approximately US$ 32.8 billion (Ministry of Housing and Urban Affairs, 2015c). This investment will be generated through private sector participation in the project's tendering process with financial incentives from central and state governments.

4 Urban Mobility: Transport Focus Under the Smart Cities Mission

SCM is one of the largest and most ambitious urban development plans laid out by the Government of India. To enable SCM, the government of India is implementing the approach through planned financial outlay. Under SCM, India provided financial assistance of US$ 7.6 billion[6] over the 5 years of the mission announcement. Furthermore, on a matching basis, an equal amount will have to be contributed by the state or ULB under which the smart city was allocated. However, SCM targets

---

[6]Average US$ to INR rates were taken for June 2015 as it was launch month of the initiative.

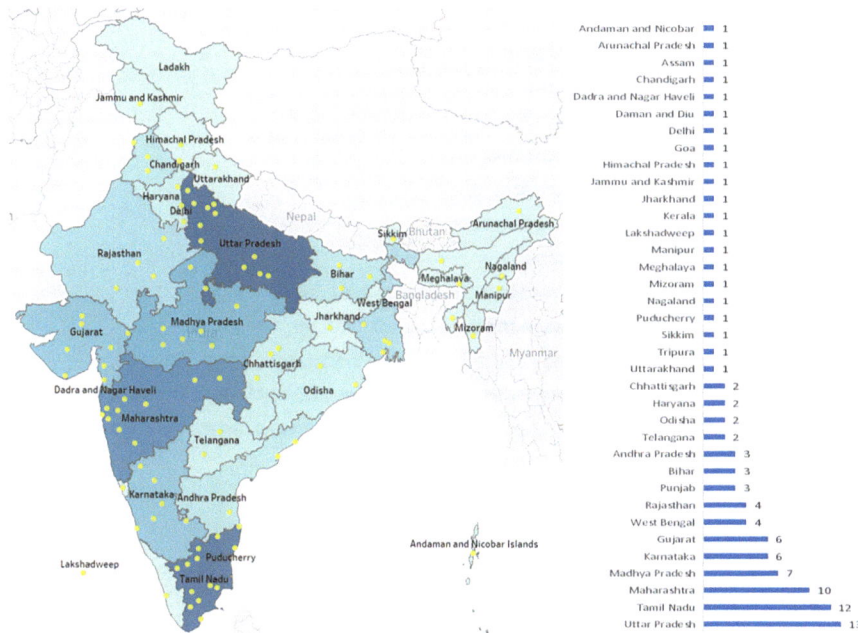

Source: Author's visualization of data from Ministry of Housing and Urban Affairs, Government of India

Fig. 3 Planned Smart Cities in India under the Smart Cities Mission. (Source: Author's visualization of data from Ministry of Housing and Urban Affairs, Government of India)

Table 1 Phasewise investment plan for 100 smart cities under SCM

	Phase 1	Phase 2	Phase 3	Phase 4	Total
No. of selected cities	20	40	30	10	100
Total no. of projects	829	1959	1891	472	5151
Investment (in INR crores)	48,064	83,698	57,393	15,863	205,018
Investment (in US$ billion[a])	7.7	13.4	9.2	2.5	32.8

Source: Ministry of Urban Development, Government of India
[a]Average US$ to INR rates were taken for June 2015 as it was launch month of the initiative

100 cities in total. For that, India has an estimated investment of approximately US$ 32.8 billion, which will be divided into four phases (Table 1) of the SCM to cater to 100 cities in total (Ministry of Housing and Urban Affairs, 2015d). Furthermore, the investment plan by the central government includes 5151 projects with a time frame of 2023. (Ministry of Housing and Urban Affairs, 2015c).

The critical sectors under which most of the funds were allocated are area development, urban mobility, economic development, IT, and energy (Fig. 4).

Even though the SCM plan looks daunting and ambitious, especially in India's context, India and state governments completed approximately 587 projects in 2019

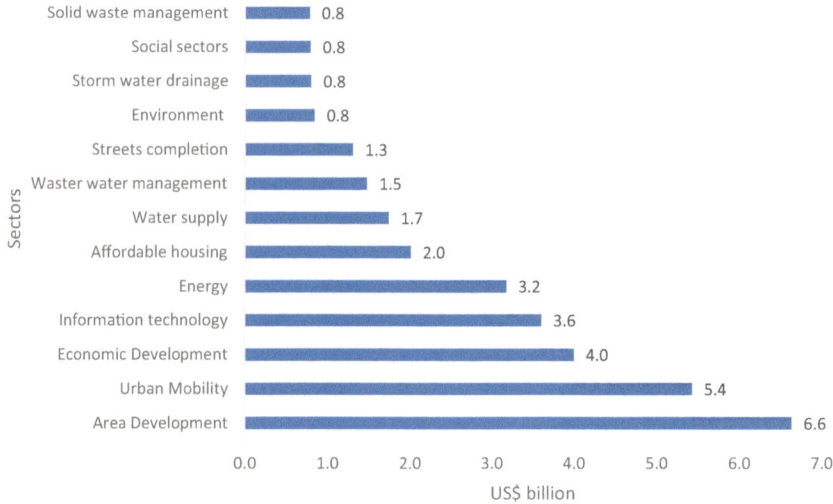

Source: Ministry of Urban Development, Government of India

Fig. 4 Investment by sectors in US$ billions under the Smart Cities Mission. (Source: Ministry of Urban Development, Government of India)

with an investment of approximately US$ 17.3 billion. Furthermore, 2005 projects are ongoing, and 2725 are under tendering.

Meeting Smart Mobility Needs: Upgrading Transport Infrastructure for the Digital Age

In Indian cities, the transportation sector, especially road transport, is a significant contributor to air pollution due to the combustion of diesel and petrol (Dubash, 2017). As urban air quality declines, the risk of its adverse impacts on health affects people who live in cities, consequently contributing to low economic productivity (United Nations, 2016). To address these issues and support the plan under SCM, central and state governments have started developing the policy framework around smart urban mobility. Through supporting policies, the government authorities in India have begun encouraging the uptake of electric mobility and public bike sharing, public transport, and multimodal transit hubs.

Electric Mobility

The Indian transport ministry announced a goal to transition from new sales of gasoline and diesel vehicles to 100% plug-in electric vehicles (PEVs) by 2030. However, India adjusted the aspirational target of 100% to achieve a 30% market share for PEVs by 2030. To achieve these targets, the central government has introduced various initiatives. In 2019, the central government allocated a budget of US$ 1.4 billion[7] over 3 years to promote electric vehicle (EV) deployment and charging

---

[7] Average US$ to INR rates were taken for June 2019 as it was launch month of the initiatives.

infrastructure. This budget is part of the recent Faster Adoption and Manufacturing of Hybrid and Electric Vehicles (FAME) II scheme. Approximately 86% of this funding is set aside to encourage the consumer to adopt, and approximately 10% is allocated to fund charging infrastructure. Before the FAME II scheme, in the first phase of FAME, the government supported the adoption of 278,000 EVs in different forms with a total incentive of approximately US$ 47 million (Dua et al., 2021).

Furthermore, to support the SCM mission, the Ministry of Housing and Urban Affairs and SCM launched web-based information and guidance documents to promote EVs in cities under the SCM mission. This document aims to help smart cities understand and encourage EVs by providing cities with information on the benchmark, measures, and strategies through which the city will be able to promote electrification in the transportation sector. This document also guides developing capacities to better understand the EV sector in terms of policy perspective, design, financing, technology, and overall development (Ministry of Housing and Urban Affairs, 2016a). Currently, there are 20 Indian states that have proposed or adopted EV policies for their respective states. As of 2021, the total registered EVs by volume in all segments stood at approximately 3.13 lakh units (0.313 million) (Business Today, 2022).

Public Bicycle-Sharing Program

Several cities have adopted public bicycle-sharing (PBS) programs to meet the requirements of short urban commutes and solve last-mile connectivity issues for public transport. However, PBS programs suffer from myriad problems, ranging from vandalism of the bikes, ill maintenance of assets, and lack of policy and infrastructure support (e.g., lack of dedicated cycling lanes) to disinterest from public authorities after the initial launch (Manish, 2019). Despite these challenges, PBS schemes have been active in several cities, such as Chandigarh, Mysuru, Pune, and Bengaluru. The stakeholders can use the PBS scheme as a part of service deployments toward citizens (Bhubaneshwar has such a scheme) or to increase ridership in public transport solutions (e.g., metros or buses) by acting as a feeder service (George, 2021). PBS schemes being trialed across India have various operating models and devised different revenue streams to manage the challenges involved. Some SCM cities have incorporated PBS programs with low-impact transport solutions to ensure diversity in public transport offerings. The challenges that PBS programs share across the country highlight the need for substantive government support and intervention to ensure the creation of dedicated infrastructure, coherent policy development, and financial support and incentives where applicable. Increasing adaptation and trialing of PBS programs across India, even in smaller cities, point toward policymakers' continued interest in cycling as a potential transport solution.

Metrorail-Urban Rail as a part of Transit

Energy-efficient and sustainable forms of public transport are essential for cities' resiliency and sustainable growth. Metro rails, as a form of efficient urban public transport, have been a part of the Indian mass public transport system for several decades. The first implementation was in the eastern city of Kolkata in 1974 (EPW,

2019). However, while earlier, they were limited to the main cities, these systems are increasingly being implemented in tier 2 and tier 3 cities, with intermodal connectivity solutions being implemented in tandem. India plans to build over 1000 km of metro rail lines and develop metro rail systems in 50 cities (Construction World, 2022). The Metro Rail Policy of 2017 is the enabling framework under which the focus has been on the expansion of metro rail networks while at the same time focusing on the indigenization of the rolling stock infrastructure to reduce costs. To ensure that the metro rails being implemented are financially sustainable, the Value Capture Finance Policy Framework of 2017 was implemented to enable stakeholders to pursue financial sustainability (Dudeja, 2021). The National Transit Oriented Development Policy also focuses on promoting public transport as a means of reducing the carbon footprint from transport and helping develop inclusive and integrated habitats (MoHUA, 2022). Mass rapid urban public transport systems such as metro rail and further enhancements of regional rapid transport systems will be a crucial feature and an enabling mechanism to ensure that Indian cities of the future are not gridlocked by traffic and are more sustainable and environmentally friendly from an emissions viewpoint.

5   Transformation of Indian Cities: Select Case Studies

While the scale of the SCM is massive, this section has selected eight cities based on their geographical distribution across India (Fig. 5) to highlight their approaches to SCM.

Chandigarh, Union Territory
Chandigarh is a union territory that serves as the joint capital of the states of Punjab and Haryana. Chandigarh is one of the first planned cities of independent India. It serves as a connecting hub to the major states of North India, namely, the National Capital Territory (NCR) of Delhi, Haryana, Punjab, and Himachal Pradesh. It is also one of the greenest cities in India (Singh, 2015). Chandigarh Smart City Limited, a nodal agency, has identified 63 projects with an investment of INR 1910.52 crore (US\$ 271.35 million) and INR 282.29 crore (US\$ 41 million) for pancity projects and area-based development projects, respectively. The Chandigarh SCM targets drinking water and wastewater projects, solid waste management, urban retrofit and redevelopment, mobility, energy reforms, and pancity information and communications technology projects (Chandigarh Smart City Limited, 2022a).

Chandigarh's smart city development under SCM is an excellent example for other cities in India. Chandigarh's e-governance project created a knowledge-based society wherein every citizen could access the benefits of digitization to benefit from the exchange of information and access to government departments, leading to a better quality of life. The e-governance project costs approximately INR 11 crore (US\$ 1.56 million), and all services were live in 2020 (Chandigarh Smart City Limited, 2022a). The Integrated Command and Control Center (ICCC) project was

Source: Author's analysis

Fig. 5 Selected Indian cities for case studies under the SCM. (Source: Author's analysis)

initiated in 2020 to enhance safety and security, improve the efficiency of municipal services and promote a better quality of life for society. The project includes adaptive traffic control, geographic information system implementation, smart parking, public bike sharing, optic fiber cable networks, and automatic number plate recognition. The ICCC project costs approximately INR 294.94 crore (US$ 41.9 million) and is under execution (Chandigarh Smart City Limited, 2022b).

Gwalior, Madhya Pradesh
Gwalior, renowned for its rich cultural heritage and related tourism, is a city in Madhya Pradesh. Gwalior faces the three most important issues common across most Indian cities. These issues are urban mobility, waste management, and economic development. Gwalior's smart city plan includes initiatives to boost connectivity and accessibility, a resilient urban ecosystem through renewable energy, and efficient urban infrastructure without diluting the city's identity and culture to overcome these issues. Because of its location, Gwalior is also viewed as a city that

could help reduce the urbanization pressure on New Delhi by acting as a counter magnet. This would help to reduce the migration pressure on New Delhi and enable the development of Gwalior as a city on its own. Like other cities under SCM, Gwalior also faces challenges such as lack of information on SCM guidelines, institutional challenges, spatial disparity, mismatch in planning and execution, and finally, lack of private sector participation for investment (S. Gupta, 2019).

Shillong, Meghalaya

Shillong is a hill town and the capital city of the northeastern state of Meghalaya. It is one of the 100 cities identified by the Government of India to be developed as a smart city. The vision of the smart city program is to transform Shillong into a cultural and economic hub in Meghalaya with a focus on tourism and culture and to make it a liveable, clean, green, inclusive, modern, safe and citizen-friendly, and well-governed city (Shillong Smart City Limited, 2022). Under the SCM, the government of India released approximately INR 160 crore (US$ 22.7 million), out of which INR 55 crore (US$ 7.8 million) and a matching amount by the government of Meghalaya has been transferred to special purpose vehicles (SPVs), making it a total of US$ 15.6 million for project implementation. The projects include management of traffic, solid waste and parking, rainwater harvesting, area development, improved living conditions for the urban poor, retrofitting improvement of access roads to local streets, pedestrianization and bazaar streets, assured water supply, and IT connectivity and digitalization (Ministry of Housing and Urban Affairs, 2022).

Ahmedabad, Gujarat

The city of Ahmedabad has emerged as Gujarat's central technological and innovation development city. Gujarat is one of the most prosperous states in the country; Gujarat's gross state domestic product (GSDP) is estimated at INR 1,879,826 crore (US$ 259.25 billion) in 2022 (Gujarat, 2022). Ahmedabad was selected among the top 20 smart cities to receive funding from MoHUA for projects under its smart city proposal. The city proposal includes pancity and area-based development initiatives focusing on infrastructure and the objective to improve public safety and surveillance, traffic management, shared services quality, emergency response, and real-time tracking of services. In its proposal, Ahmedabad proposed 22 projects covering area-based development and a pancity project valued at approximately INR 2255 Crore (approximately US$ 300 million). The projects cover the redevelopment of slums through the development of utility networks, wastewater treatment, and residential development. Furthermore, Ahmedabad proposed implementing smart features through solar energy waste segregation, intelligent traffic management, and smart parking energy efficiency. Ahmedabad also presented projects such as a smart transit integrated transit management platform and command and control center for real-time tracking (Ministry of Housing and Urban Affairs, 2016b).

Indore, Madhya Pradesh

Indore represents the largest economy in Central India, with GDP standing at US$ 14 billion. It is a commercial and trading capital of the state, with its history spanning over five centuries and significant footprints in every commercial sector

(Incredible India, 2022). However, Indore has poor urban spatial planning, with residential and industrial areas developed without adequate supporting infrastructures such as public open spaces, education, healthcare, and a good road network. Furthermore, in Indore, every third resident in the city is a slum dweller. Indore also faces ever-growing traffic congestion and air quality issues with inadequate public transport. With its strengths and weaknesses, Indore was one of the 20 cities to be adopted under the first phase of SCM. To facilitate and implement these projects, in 2016, the State government established a company, "Indore Smart City Development Limited (ISCDL)." The company's objective is to plan, design, develop, implement, manage, maintain, operate, and monitor the Smart City Development projects for the city of Indore.

Mumbai, Maharashtra

Mumbai is India's commercial capital and economic powerhouse, with the city's GDP expected to be almost US$ 230 billion by 2030 (Das, 2022). It is also one of the most vulnerable cities to climate change hazards such as sea-level rise. Mumbai is subjected to storm surges in almost every rainy season, and flooding during these periods is quite common. To increase resiliency and tackle the challenges of the future, Mumbai has developed the Mumbai Climate Action Plan (MCAP), a 2050 roadmap for the city to tackle the challenges of climate change through adaptation and mitigation strategies (MACP, 2022). The MCAP aligns well with India's net-zero pledge, seeking to achieve net-zero emissions by 2050. Mumbai is a part of the C40 Cities Network, one of the five cities in India that is a member of this association of 97 cities globally. The other cities are Delhi, Kolkata, Chennai, and Bengaluru. Mumbai's municipal corporation, the Brihanmumbai Municipal Corporation (BMC), prepared the plan with inputs from the C40 Cities Network and the World Resources Institute. The MCAP plan focuses on key action areas: sustainable waste management, urban greening and biodiversity, urban flooding and water resource management, energy and buildings, air quality, and sustainable mobility. For Mumbai to deliver on its 2050 roadmap, it must upgrade the capacities of its municipal corporation and seek to work with stakeholders across the political and business spectrum to match the ambitions spelled out in the document. Mumbai's Climate Action Plan is a bold initiative that will need to be reviewed periodically to understand the overall impact of policy actions.

Vishakhapatnam, Andhra Pradesh

Vishakhapatnam is an important port city on the east coast of India. Located in Andhra Pradesh, the city acts as a gateway to the hinterland of central India. It is also an important industrial city with major industrial sectors. Vishakhapatnam leveraged grants from the U.S. Trade and Development Agency (USTDA) in February 2016 to develop its urban renewal ambitions (AECOM, 2016). The Greater Visakhapatnam Smart City Corporation Ltd. It is the executing body for the development of urban projects, which include street lighting, refurbishment of classrooms, schools, and parks, development of civic open spaces, and upgrading and reinvigorating the city's heritage areas. The city is investing in digital infrastructure to help support its ambitions of evolving into a more sustainable and business-friendly city.

Mangalore, Karnataka

Mangaluru, also known as Mangalore, is the largest city in the south Kannada district of Karnataka. Mangaluru is a major port city and commercial center. Mangaluru was selected under the 2nd round of the SCM in 2016. Mangaluru smart city projects focus on developing waterfronts, roads, and fisheries to promote trade and thus help improve the economy (Mangaluru Smart City Limited, 2022a). Mangaluru Smart City Limited, a nodal agency, has identified a total of 57 projects, of which 18 projects have been completed and 39 are ongoing. The total cost of all 57 projects was approximately INR 943.4 crore (US$ 134 million). As a waterfront city, for Mangaluru, it is essential to work on waterfront development projects. Mangaluru undertook one such project under SCM. In 2022, the waterfront development project was awarded the contract for the INR 185 crore project (US$ 26.3 million) under ten packages. The project's critical features include promenade development, a biodiversity park, and creating a bird-watching area with a cycling pathway of 2.1 km (Hindu, 2022). Mangaluru, like other smart cities, is developing an ICCC platform. The platform will facilitate the seamless flow of information through a centralized command and control center and enable the transformational needs of smart cities across areas of energy consumption, transportation, healthcare, connectivity, sustainability, and environment management. The CCC platform estimated cost is approximately INR 74 crore (US$ 10.5 million), and it is under execution (Mangaluru Smart City Limited, 2022b).

6   Implementation of Smart Cities in India: Challenges, Insights, and Optimism

Given the legacy and heritage of Indian cities coupled with the system of governance, the smart city initiative is expected to be a long process that will evolve as policy makers increasingly engage with citizens. The most significant learning and takeaway from the entire process of the SCM have been its engagement methodology with citizens and generating a ground-up understanding of local civic issues and requirements, which enable a much more focused approach toward financing and implementation (Sharma, 2017).

However, the challenges that exist and persist are formidable. Civic planning systems and institutions and capabilities are archaic. These were developed during the socialist era of a postindependence period when centralized planning was the norm, and the development policies designed were top-down. Now, with more private sector participation and involvement in the delivery of core infrastructure requirements, the old structure is incapable of evolving. Incorporating a new urban governance structure while ensuring that the current level of services continues to be maintained will be a massive challenge. It is imperative to have institutional capacity building in municipal councils to ensure continuity. Such councils are also crucial for developing financial instruments in the form of municipal bonds. Such mechanisms can also lead to civic institutions reforming due to market require-ments, mainly financially self-sufficient and financially accountable, with credit rat-ings impacting the civic service levels directly and indirectly (Samal, 2021). However, for such services to be improved, the civic bodies need to be improved. This can be through a focus on capacity building and capability building—improv-ing municipal staffing and local governance, strengthening local urban bodies, and incorporating political accountability at the city level.

SCM, however, has shown that policymakers are willing to learn and absorb the lesson and engage with citizens to develop projects with local context and usage models in mind. This bodes well for the civic landscape in India, as the infrastruc-ture that is going to be built is one that will be of use and value to citizens and hence will improve civic participation in the urban planning process. The structure of the SCM, which has focused on developing small areas with local consultation and a phasewise approach, with a retro fitment, refurbishment, and redevelopment approach, would make it much more inclusive of regional civic and urban aspira-tions than previous policy mechanisms. This is satisfying policy development and is quite a change from the earlier top-down policy and development model that was in vogue. The increasing usage of digital infrastructure and supporting frameworks such as satellite imagery is helping municipal bodies target better revenue collec-tions, thus helping boost their financial conditions. In addition, in a way, also mak-ing them politically responsible as a result (Hindustan Times, 2021). SCM, with its focus on investments in digital infrastructure and leveraging such strengths, would help to make the entire process transparent and thus be more inclusive by a partici-patory approach.

SCM has also enabled civic bodies to reach out to cities globally to partner with, learn from, and better absorb urban planning lessons, rather than seeking to reinvent the wheel and often commit the same mistakes (Sharma, 2016). Having a foreign partner in the planning approach enables civic bodies, often entangled in current civic issues, to peer over the horizon in terms of being prepared for future growth challenges and opportunities. As Indian cities are increasingly becoming engines of growth for the economy, this is of critical importance (IBEF, 2021).

7   Conclusion and Policy Recommendations

A challenge observed to be shared across the cities in India is the diversity of pro-grams and schemes, with often overlapping authorities and administrative powers. SCM, AMRUT, Housing for All, SBM, HRIDAY, and Digital India have organiza-tional structures and funding avenues. There is no perceptible convergence of these programs on the ground. As a result, they often tend to be inefficient or work at cross purposes, leading to duplication, bureaucratic delays, and mission creep. Coupled with the lack of capacity and capability at the municipal level, this often results in a lack of collaboration. ULBs and city and state governments have seldom focused on building such skill sets, and the lack of expertise has led to project delays.

The civic authorities lack updated information regarding their areas. This includes updated and digitized property boundaries, roads, underground utilities, land use records, etc. Even tier 2 and tier 3 towns in India cannot access such updated, real-time databases. Developing such databases would help civic authorities manage the urban sprawl that is continuously expanding in their cities and be able to deliver civic services to their residents.

SCMs are primarily focused on being able to attract private investments so that authorities can initiate public-private partnerships to bridge investment gaps. However, the lack of skill and capabilities that stymie civic authorities also extends to private investors, who are themselves struggling with such issues. This results in private investors not being able to identify and support projects. Second, private investors tend to focus on short-term profit maximization, which results in them prioritizing short-term projects. However, long-term projects that focus on the core delivery of civic services are severely underfunded and need both capital and human resource infusion. This mismatch in expectations often leads to the incorrect choice of private investors for specific projects, often leading to rework and project delays. Often, civic authorities are not able to conduct a suitable due diligence on private sector investors' capabilities, as they lack such skills, and without a suitable cost-benefit analysis and prudent risk analysis and mapping, private investors are saddled with poorly drafted projects that cannot be implemented or executed, again leading to delays and cost overruns.

At its heart, the SCM and the associated initiatives are an ambitious urban renewable and refurbishment agenda. The intention behind the retrofitting of existing cities is to meet the aspirations of an increasingly demanding urban population and build the capacity and capability to meet the challenges of a future where sustainability, equity, and resiliency become increasingly important. Indian policy action in support of the Smart Cities objectives provides a unique opportunity to learn from, especially for developing and least developed countries. Many Indian cities involved in SCMs and their often-unique urban challenges provide researchers seeking to create mechanisms that help manage urban development and growth with substantial study material and cases. The Indian urban development plan, its growth trajectory, and the subsequent impact on cities will have an essential bearing as India evolves toward a more resilient, inclusive, and sustainable future.

References

AECOM. (2016). *Vizag smart city.* https://aecom.com/projects/vizag-smart-city/
Bhatia, G. (2018, January 29). *Rebuilding our cities.* Accessed 5 Aug 2022. https://www.thehindu.com/opinion/op-ed/rebuilding-our-cities/article22544965.ece
Brookings India, & Brookings Institution. (2016). *Building smart cities in India.* Brookings.
Business Today. (2022, March 8). Registered EV sales cross 1 lakh mark in 2022; Electric 2W sales grow 433%. *Business Today.* Accessed 8 July 2022. https://www.businesstoday.in/auto/

story/registered-ev-sales-cross-1-lakh-mark-in-2022-electric-2w-sales-grow-433-325131-2022-03-08

Chandigarh Smart City Limited. (2022a, September 12). *Chandigarh Smart City Limited.* Chandigarh Smart City. Accessed 12 Sept 2022. https://www.chandigarhsmartcity.in/web/guest/documents

Chandigarh Smart City Limited. (2022b, September 12). *Integrated command and control center.* Chandigarh Smart City Limited. Accessed 12 Sept 2022. https://www.chandigarhsmartcity.in/implementation-of-iccc

Construction World. (2022, September 5). *The construction of 1,000km rail work is in progress: PM Modi.* https://www.constructionworld.in/transport-infrastructure/metro-rail-and-railways-infrastructure/the-construction-of-1-000km-rail-work-is-in-progress%2D%2Dpm-modi/36212

Das, S. (2022, April 4). *Developing a successful climate action plan for Mumbai.* Accessed 15 July 2022. https://www.thenatureofcities.com/2022/04/04/developing-a-successful-climate-action-plan-for-mumbai/

Dua, R., Hardman, S., Bhatt, Y., & Suneja, D. (2021). Enablers and disablers to plug-in electric vehicle adoption in India: Insights from a survey of experts. *Energy Reports, 7,* 3171–3188.

Dubash, N. K. (2017). *Air pollution in Indian cities: Understanding the causes and the knowledge gaps.* Centre for Policy Research.

Dudeja, V. P. (2021, December 20). *Why urban India needs transit-oriented development?* https://www.businessworld.in/article/Why-Urban-India-Needs-Transit-Oriented-Development-/20-12-2021-415337/

EPW. (2019, November 30). *Creating a '21st Century World': Will Metro Systems Create 'Smart Cities'?* https://www.epw.in/engage/article/creating-21st-century-world-metro-systems-smart-city

George, A. (2021, Janaury 14). *How can public bike-sharing initiatives thrive in Indian cities?* Accessed 15 July 2022. https://wri-india.org/blog/how-can-public-bike-sharing-initiatives-thrive-indian-cities

Government of India. (2015, July 1). *Digital India.* Digital India. Accessed 7 July 2022. https://www.digitalindia.gov.in/

Grant Thornton. (2011). *Appraisal of Jawaharlal Nehru National Urban Renewal Mission.* Grant Thornton.

Gujarat. (2022, July 07). *Indian Brand Equity Foundation.* Accessed 7 July 2022. https://www.ibef.org/states/gujarat-presentation

Gupta, S. (2019). Smart City Paradigm in India: Gwalior A Case Study. *Humanities and Social Sciences Reviews,* 341–347. https://core.ac.uk/download/pdf/268004618.pdf

Hindu, The. (2022, April 29). Waterfront development project under Mangaluru Smart City Mission takes a leap. Accessed 13 Sept 2022. https://www.thehindu.com/news/cities/Mangalore/waterfront-development-project-under-mangaluru-smart-city-mission-takes-a-leap/article65366836.ece

Hindustan Times. (2021, October 14). *PMC expects to collect more property tax from 15 wards via digitisation.* https://www.hindustantimes.com/cities/pune-news/pmc-expects-to-collect-more-property-tax-from-15-wards-via-digitisation-101634221188179.html

IBEF. (2021). *Smart cities mission.* https://www.ibef.org/government-schemes/smart-cities-mission

Incredible India. (2022, July 7). *Indore.* Incredible India. Accessed 7 July 2022. https://www.incredibleindia.org/content/incredibleindia/en/destinations/indore.html

International Energy Agency. (2021). *India Energy Outlook 2021.* International Energy Agency. https://www.iea.org/reports/india-energy-outlook-2021

Jha, R. (2020, December 5). *The impact of Gandhi on post-Independence treatment of cities.* Accessed 5 Aug 2022. https://www.orfonline.org/expert-speak/the-impact-of-gandhi-on-post-independence-treatment-of-cities/

MACP. (2022, March). *Mumbai Climate Action Plan.* Accessed 15 July 2022. https://mcap.mcgm.gov.in/

Mangaluru Smart City Limited. (2022a, September 12). *Mangaluru Smart City Limited.* Accessed 12 Sept 2022. https://www.mangalurusmartcity.net/about-mscl

Mangaluru Smart City Limited. (2022b). *Mangaluru-The coastal confluence where nature and opportunity meet.* Mangaluru Smart City Limited.

Manish, K. (2019, December 24). *Story of public bicycle sharing in India from beginning.* Accessed 14 July 2022. https://urbanvoices.in/story-of-public-bicycle-sharing-in-india/

Minister of Housing and Urban Affairs. (2021). *Atal mission for rejuvenation and urban transformation 2.0.* Minister of Housing and Urban Affairs.

Minister of Housing and Urban Affairs, Government of India. (2021, October 1). *Atal Mission for Rejuvenation and Urban Transformation (AMRUT).* AMRUT. Accessed 7 July 2022. http://amrut.gov.in

Ministry of Finance, Government of India. (2022, January). *Economic Survey 2021–2022.* Accessed 29 July 2022. https://www.indiabudget.gov.in/economicsurvey/

Ministry of Housing and Urban Affairs. (1992). *The Constitution (74th Amendment) Act.* Ministry of Housing and Urban Affairs.

Ministry of Housing and Urban Affairs. (2005). *JNNURM. Mission.* Ministry of Housing and Urban Affairs.

Ministry of Housing and Urban Affairs. (2015a). *Atal mission for rejuvenation and urban transformation.* Ministry of Housing and Urban Affairs.

Ministry of Housing and Urban Affairs. (2015b). *Heritage city development and augmentation yojana.* Ministry of Housing and Urban Affairs.

Ministry of Housing and Urban Affairs. (2015c). *Smart cities – Mission statement and guidelines. Mission and guidelines.* Ministry of Urban Development.

Ministry of Housing and Urban Affairs. (2015d). *Smart cities mission – An overview of implementation.* Ministry of Housing and Urban Affairs.

Ministry of Housing and Urban Affairs. (2016a). *Electric mobility policy workbook.* Ministry of Housing and Urban Affairs.

Ministry of Housing and Urban Affairs. (2016b). *List of projects as per Smart City Proposal: Ahmedabad.* Ministry of Housing and Urban Affairs.

Ministry of Housing and Urban Affairs. (2021, February 23). *India Urban Data Exchange.* Accessed 6 July 2022. https://nudm.mohua.gov.in/

Ministry of Housing and Urban Affairs. (2022, September 12). *Shillong.* Accessed 12 Sept 2022. https://smartcities.gov.in/node/176

MoHUA. (1994). *Infrastructure development.* Accessed 9 Aug 2022. https://mohua.gov.in/upload/uploadfiles/files/infr_develop.pdf

MoHUA. (2015, June 25). *About smart cities mission.* Accessed 12 July 2022. https://smartcities.gov.in/about-the-mission

MoHUA. (2022). *Metro Rail.* https://mohua.gov.in/upload/uploadfiles/files/Metro%20Rail%20_%20MoHUA.pdf

Nanda, P. K., & Sharma, N. C. (2020, September 1). *India's young demography won't be that young by 2036.* Accessed 5 Aug 2022. https://www.livemint.com/news/india/india-s-young-demography-won-t-be-that-young-by-2036-11598959442533.html

National Heritage City Development and Augmentation Yojana. (2022, February 3). *Press Information Bureau.* Accessed 7 July 2022. https://pib.gov.in/PressReleaseIframePage.aspx?PRID=1795154

National Institute of Urban Affairs. (2016, April 9). *Comparison of Smart Cities Programs in India and the US.* Accessed 15 July 2022. https://niua.org/cidco/comparison-of-smart-cities-programs-in-india-and-the-us/

NITI Aayog. (2021). *Reforms in urban planning capacity in India.* NITI Aayog.

Niti Aayog and Asian Development Bank. (2022). *Cities as engines of growth.* Niti Aayog and Asian Development Bank. Accessed 5 Aug 2022. https://www.niti.gov.in/sites/default/files/2022-05/Mod_CEOG_Executive_Summary_18052022.pdf

Our World in Data. (2021, July 23). *India: Energy Country Profile*. Accessed 15 July 2022. https://ourworldindata.org/energy/country/india

Pandey, S. (2020, July 8). *Our post-Independence cities are so ugly: Didi Contractor*. Accessed 5 Aug 2022. https://www.thehindu.com/society/our-post-independence-cities-are-so-ugly-didi-contractor/article31980225.ece

Press Trust of India. (2009, November 16). *India needs to build 500 new cities: Prof Prahlad*. Accessed 1 Aug 2022. https://business.rediff.com/slide-show/2009/nov/06/slide-show-1-india-needs-to-build-500-new-cities.htm

Reserve Bank of India. (2002, January 14). *Services in the Indian growth process*. Accessed 5 Aug 2022. https://www.rbi.org.in/scripts/PublicationsView.aspx?id=10895

Samal, M. (2021, June 13). *Rethinking the concept of municipal bonds to improve urban governance in India*. https://journals.sagepub.com/doi/full/10.1177/09749306211023623

Sharma, N.. (2016, May 13). *Ambassadors of 150 countries invited to Investment Summit*. https://economictimes.indiatimes.com/news/economy/infrastructure/smart-city-programme-ambassadors-of-150-countries-invited-to-investment-summit/articleshow/52247146.cms

Sharma, N. (2017, June 2). *Indian smart cities can become lighthouses for world, says London School of Economics*. https://economictimes.indiatimes.com/news/economy/infrastructure/indian-smart-cities-can-become-lighthouses-for-world-says-london-school-of-economics/articleshow/58952093.cms?from=mdr

Shillong Smart City Limited. (2022, September 12). *Shillong Smart City*. Accessed 12 Sept 2022. http://sscl.meghalaya.gov.in/

Singh, J. P. (2015). Chandigarh: Transformation of a modern city to a smart city. In *Sustainable built environment*. Sustainable Built Environment.

TERI. (2022). *Urbanisation*. Accessed 5 Aug 2022. https://www.teriin.org/resilient-cities/urban-isation.php

The World Bank. (2011, July 14). India's urban challenges. *India's Urban Challenges*.

Town and Country Planning Organisation. (2022, March 24). *Integrated development of small & medium towns*. Accessed 9 Aug 2022. http://tcpo.gov.in/integrated-development-small-medium-towns

United Nations. (2016). *UN health agency warns of rise in urban air pollution*. United Nations.

United Nations. (2022). *Around 2.5 billion more people will be living in cities by 2050, projects new UN report*. Accessed 5 Aug 2022. https://www.un.org/es/desa/around-25-billion-more-people-will-be-living-cities-2050-projects-new-un-report

Open Access   This chapter is licensed under the terms of the Creative Commons Attribution 4.0 International License (http://creativecommons.org/licenses/by/4.0/), which permits use, sharing, adaptation, distribution and reproduction in any medium or format, as long as you give appropriate credit to the original author(s) and the source, provide a link to the Creative Commons license and indicate if changes were made.

The images or other third party material in this chapter are included in the chapter's Creative Commons license, unless indicated otherwise in a credit line to the material. If material is not included in the chapter's Creative Commons license and your intended use is not permitted by statutory regulation or exceeds the permitted use, you will need to obtain permission directly from the copyright holder.

Correction to: Smart Cities

Fateh Belaïd and Anvita Arora

Correction to:
F. Belaïd, A. Arora (eds.), *Smart Cities*, Studies in Energy,
Resource and Environmental Economics,
https://doi.org/10.1007/978-3-031-35664-3

The book was published without removing the title "General Secretariat of Dubai Executive Council" of author Jamila El Mir in Chapter 5. The title has now been removed.

The book was published by reversing the author's name in Chapter 12. The author name has now been corrected as **Moinse Dylan**.

The updated original versions of these chapters can be found at
https://doi.org/10.1007/978-3-031-35664-3_5
https://doi.org/10.1007/978-3-031-35664-3_12

© The Author(s) 2024
F. Belaïd, A. Arora (eds.), *Smart Cities*, Studies in Energy, Resource and
Environmental Economics, https://doi.org/10.1007/978-3-031-35664-3_20

Index

© King Abdullah Petroleum Studies and Research Center 2024

F. Belaïd, A. Arora (eds.), *Smart Cities*, Studies in Energy, Resource and Environmental Economics, https://doi.org/10.1007/978-3-031-35664-3

385